EXS 82

Evolutionary Ecology of Freshwater Animals

Concepts and Case Studies

Edited by B. Streit
T. Städler
C.M. Lively

Springer Basel AG

Editors:

Prof. Dr. Bruno Streit
Dr. Thomas Städler
Abteilung Ökologie und Evolution
Fachbereich Biologie
J.W. Goethe-Universität
Siesmayerstrasse 70
D-60054 Frankfurt
Germany

Dr. Curtis M. Lively
Department of Biology
Indiana University
Bloomington, Indiana 47405
USA

Front cover:

Top: The threespine stickleback, *Gasterosteus aculeatus*, from western North America. The phenotype is one of many different phenotypes found in fresh water, believed to be descended from marine and anadromous populations.

Center: Two individuals of *Potamopyrgus antipodarum*, a New Zealand prosobranch snail. This phenotypically variable species (e.g., with smooth or keeled shells) harbors both sexual and clonal lineages.

Bottom: Two mature zooids from a colony of *Cristatella mucedo*, a freshwater bryozoan. In the transverse section of the colony a young retracted zooid (right) and two floatoblasts are depicted.

Library of Congress Cataloging-in-Publication Data
Evolutionary ecology of freshwater animals : concepts and case studies
  / edited by B. Streit, T. Städler, C.M. Lively.
      p.   cm. -- (Experientia. Supplementum ; 82)
      Includes bibliographical references and indexes.
      ISBN 978-3-0348-9812-6   ISBN 978-3-0348-8880-6 (eBooK)
      DOI 10.1007/978-3-0348-8880-6

      1. Frewhwater animals--Evolution.   2. Freshwater animals--Ecology.
  3. Freshwater animals--Evolution--Case studies.   4. Freshwater animals--Ecology--Case studies.
  I. Streit, Bruno.   II. Stadler, Thomas.   III. Lively, C. M. (Curtis M.)   IV. Series: Experientia.
  Supplementum ; v. 82.
  WL 141.E96   1997
  591.76--dc21

Deutsche Bibliothek Cataloging-in-Publication Data
Evolutionary ecology of freshwater animals : concepts and case
studies / ed. by B. Streit ... - Basel ; Boston ; Berlin : Birkhäuser,
1997
  (EXS: 82)
  ISBN 978-3-0348-9812-6

EXS. - Basel ; Boston ; Berlin : Birkhäuser
  Früher Schriftenreihe
  Fortlaufende Beil. zu: Experientia

© 1997 Springer Basel AG
Originally published by Birkhäuser Verlag in 1997
Softcover reprint of the hardcover 1st edition 1997
Printed on acid-free paper produced from chlorine-free pulp

ISBN 978-3-0348-9812-6

9 8 7 6 5 4 3 2 1

Contents

Evolutionary processes following colonizations

List of Contributors

Christine A. Andrews, Department of Ecology and Evolution, State University of New York, Stony Brook, NY 11794-5245, USA

Michael A. Bell, Department of Ecology and Evolution, State University of New York, Stony Brook, NY 11794-5245, USA

Christer Brönmark, Department of Animal Ecology, Ecology Building, Lund University, S-22362 Lund, Sweden

Jonas Dahl, Department of Ecology, Ecology Building, Lund University, S-22362 Lund, Sweden

Dieter Ebert, Institut für Zoologie, Universität Basel, Rheinsprung 9, CH-4051 Basel, Switzerland

Göran Englund, Center for Ecology, Evolution and Behavior, T.H. Morgan School of Biological Sciences, University of Kentucky, Lexington, KY 40506-0225, USA

Larry Greenberg, Department of Ecology, Ecology Building, Lund University, S-22362 Lund, Sweden

Philippe Jarne, Lab. Génétique et Environnement, Institut des Sciences de l'Evolution, Université Montpellier II, Place E. Bataillon, F-34095 Montpellier Cedex 5, France

Steven G. Johnson, Department of Biological Sciences, University of New Orleans, New Orleans, LA 70148, USA

Jukka Jokela, ETH-Zürich, Experimental Ecology, ETH-Zentrum NW, CH-8092 Zürich, Switzerland

Mathew A. Leibold, Department of Ecology and Evolution, University of Chicago, 1101 E 57th Street, Chicago, IL 60637, USA

Curtis M. Lively, Department of Biology, Indiana University, Bloomington, IN 47405, USA

Beth Okamura, School of Animal and Microbial Sciences, The University of Reading, Whiteknights, PO Box 228, Reading RG6 6AJ, UK

Stephanie J. Schrag, Department of Biology, Emory University, Atlanta, GA 30322, USA

Klaus Schwenk, Netherlands Institute of Ecology, Centre for Limnology, Rijksstraatweg 6, NL-3631 AC Nieuwersluis, The Netherlands

Andrew Sih, Center for Ecology, Evolution and Behavior, T.H. Morgan School of Biological Sciences, University of Kentucky, Lexington, KY 40506-0225, USA

Piet Spaak, EAWAG/ETH, Department of Limnology, Überlandstr. 133, CH-8600 Dübendorf, Switzerland

Thomas Städler, Abteilung Ökologie und Evolution, Fachbereich Biologie, J.W. Goethe-Universität, Siesmayerstraße 70, 60054 Frankfurt, Germany

Alan J. Tessier, Kellogg Biological Station, Michigan State University, Hickory Corners, MI 49060, USA

Norbert Walz, Institut für Gewässerökologie und Binnenfischerei, Abteilung Limnologie von Flußseen, Müggelseedamm 310, D-12587 Berlin (Friedrichshagen), Germany

Michael J. Winterbourn, Department of Zoology, University of Canterbury, Private Bag 4800, Christchurch, New Zealand

David E. Wooster, Center for Ecology, Evolution and Behavior, T.H. Morgan School of Biological Sciences, University of Kentucky, Lexington, KY 40506-0225, USA

Evolutionary Ecology of Freshwater Animals
ed. by B. Streit, T. Städler and C. M. Lively
© 1997 Birkhäuser Verlag Basel/Switzerland

Introduction

This volume provides an overview of several aspects of the evolutionary ecology of freshwater animals. Evolutionary ecology views ecological problems in the light of evolutionary processes, emphasizing natural selection and adaptation, species interactions, and life-history features, including behavioral interactions and dispersal. Central to this approach is the importance of heritable variability for phenotypic traits within and among natural populations. Both the physical and the biotic environment with their respective complexities compose the ecological arena in which populations and species evolve.

As an integrating field, evolutionary ecology draws on several subdisciplines of biology, among them general ecology, population biology, population genetics, life-history theory, and biogeography and paleontology. Owing to its conceptual richness and emphasis, evolutionary ecology is thus a rapidly growing, diversifying field at the nexus of various disciplines. The aim of this series of articles is to highlight some of the current central hypotheses and approaches in the field, using freshwater model systems as a unifying theme.

What is the motivation to study problems of evolutionary ecology in freshwater animals? Actually, **freshwater biota** and their respective lotic or lentic environments represent appealing model systems for a variety of reasons:

- The general ecology of freshwater systems (limnology), and thus the general ecological background of the respective species or assemblage of species under investigation, has been studied extensively for a long time. The rigorous study of these systems has a tradition dating back more than a century, stressing the ecosystem approach earlier than studies in other major habitat types. Ecological limnologists have pioneered quantitative studies of energy flow, nutrient cycling, and related aspects of the autecology of some model component species.
- Experimental work on various freshwater organisms is frequently easier, especially with planktonic species, than comparable work on terrestrial organisms. The latter typically live in heterogeneous environments with stronger temporal and spatial fluctuations of environmental variables, such as small-scale temperature changes. Many lake-inhabiting planktonic species with small body size are characterized by rapid generation turnover, facilitating a range of observational and experimental studies.

- Freshwater ecosystems are often transient in nature, at both ecological and evolutionary timescales. Freshwater environments, especially in the tropics, may be subject to annual droughts and/or flooding, and most freshwater systems have suffered extensive rearrangements due to glacial (or pluvial) cycles during the Pleistocene. This physical instability should have consequences for the genetic and phenotypic differentiation of populations and for selection on various life-history traits, promoting variants that can cope with unpredictable or seasonally changing environmental variables.
- Freshwater animals display a variety of reproductive modes that deviate from biparental sexuality, such as ameiotic parthenogenesis in cladocerans, rotifers and some prosobranch snails, self-fertilization in pulmonate snails, and vegetative cloning in bryozoans and other phyla. This striking diversity in often dominant constituents of freshwater communities offers exceptional opportunities to uncover the ecological and evolutionary causes and consequences of these alternative reproductive modes, and is accessible to both comparative and experimental approaches. Bryozoans and various other phyla are further characterized by their modular, rather than unitary, growth form and life history.
- Theoretical concepts in evolutionary ecology were originally based on studies of terrestrial rather than aquatic taxa, largely neglecting the wealth of information gathered on freshwater communities. One of the goals of this series of 12 chapters is to facilitate communication between ecologists and evolutionary biologists experienced in, respectively, freshwater, marine, and terrestrial systems.

The **first section**, focusing on "Ecosystem structure and trophic interactions", deals with lakes and rivers, plankton and benthon, and comprises four contributions. Leibold and Tessier examine the role of habitat partitioning by species in the genus *Daphnia*, which represent commonly important grazers in lake ecosystems. Within the framework of trophic interactions and "cascades", the authors focus on the often neglected interactions among resource competitors within a given trophic level (algal grazers). By developing a graphical model of habitat partitioning under resource competition mediated by habitat-specific predators, they argue that the observed patterns of habitat partitioning in *Daphnia* can be interpreted as reflecting the conflicting demands of avoiding predators and using environments that maximize growth.

Winterbourn synthesizes studies on the persistence and resilience of invertebrate faunas in New Zealand mountain streams of contrasting physical stability and flow variability, as well as other studies on faunal responses to catchment-scale disturbances and the temporal stability of stream faunas following afforestation with exotic conifers. He presents evidence suggesting that the faunas of many New Zealand mountain streams are dominated by the same widely distributed species characterized by life-history flexibility, lack of habitat specificity, and strong colonizing abilities. Winterbourn hypothesizes that the resilience of New Zealand mountain stream faunas, and the lack of strong habitat specialization of many species, can be explained in terms of their evolutionary history in changeable environments.

Brönmark, Dahl and Greenberg review work on benthic freshwater food chains in both lakes and streams, thus complementing earlier studies in pelagic freshwater habitats. The authors identify a variety of complex trophic interactions, i.e., sequences of biotic interactions that functionally link species via an intermediate species. Among them are trophic cascades, interaction modifications, and indirect commensalism. The strength of these trophic interactions varies greatly among studies, depending on habitat complexity, the type and intensity of environmental stresses, the ways in which prey defend themselves from predators, and the degree of trophic omnivory in the food chain. While it is clear from this synopsis that complex trophic interactions play an important role in structuring benthic communities in lakes and streams, the authors call for whole-system, long-term manipulations in order to achieve a deeper understanding of these processes.

Wooster, Sih and Englund concentrate on the influence of predators on invertebrate prey density in streams, focusing on behavioral responses of prey to the presence of predators. The authors review evidence revealing that prey dispersal responses to predators are important in driving patterns of predator impact on local prey density. In principle, predators may either increase of decrease prey dispersal rates. Importantly, the review shows that the spatial scale of empirical studies has a strong influence on the degree to which prey dispersal affects the "apparent" predator impact on prey density; only large-scale studies are expected to reflect mostly direct (consumptive) effects of predators on prey density.

The **second section** focuses on "Aspects of life-history evolution", exemplified by rotifers, daphniids, and bivalves. Each of the three contributions in this section approaches the general theme from a different perspective.

Walz reviews the ecological role of rotifers in freshwater plankton communities and their interactions with other species, both food particles and predators. He stresses the importance of competition for algal resources (e.g., with daphniids), of food particle size, and of food concentrations. At high food concentrations, rotifers may exhibit high maximum growth rates and thus rapid buildup of large populations; this ability increases with body size. On the other hand, many species are food generalists and feed on the same algal size range as other zooplankton groups; because daphniids generally have lower threshold food concentrations than rotifers, the latter may be outcompeted at lower food concentrations.

Ebert addresses fundamental concepts in life-history evolution, such as size at birth, growth rates, and age and size at maturity. Drawing on studies using *Daphnia magna* as a model system, he highlights the proximate mechanisms initiating maturation in organisms growing via a series of instars. Ebert suggests that a certain "threshold size" must be reached before maturation is attained, a type of maturation mechanism that has important implications for the evolution of age and size at maturity. A threshold mechanism would serve to canalize size at maturity, which in turn would imply higher variances for instar number and age at maturity. A decoupling of maturation phenotype from variation in size at birth and juvenile growth rate means that variation in size at birth can evolve independently from the evolution of size at maturity.

Jokela reviews the theory of optimal energy allocation among maintenance, somatic growth, and reproduction, focusing on long-lived, iteroparous, indeterminately growing bivalves. He develops the concept of "priority rank" of allocation targets and discusses the relevant timescales of allocation decisions (among-season vs within-season). The priority rank of allocation targets may depend on a multitude of factors, both external and internal to the organism. Selection favors individuals whose allocation pattern is closer to the theoretical optimum than the mean allocation pattern of the population. Empirical and experimental studies suggest that priority ranks among allocation targets are an important component of the life-history strategy in long-lived bivalves.

The **third section** highlights "Population biology and reproductive modes". The empirical examples are drawn from cyclically clonal daphniids, partially self-fertilizing pulmonate snails, parthenogenetic prosobranch snails, and the colonial bryozoans, which combine biparental sexual reproduction with vegetative cloning.

Schwenk and Spaak devote themselves to the important problem of hybridization between related species. Their model system, lake-inhabiting *Daphnia*, is different from the traditional examples of interspecific hybridization in that ameiotic parthenogenesis may allow hybrids to persist as clonal lineages, independent of possible sexual infertility (hybrid breakdown). Likewise, the patchy distribution of freshwater habitats results in a patchy or "mosaic" distribution of hybrids and parental taxa over a large geographic area. Hybrids in the *D. galeata* complex tend to exhibit a combination of parental characters, enabling them to persist in ecological niches that arise seasonally due to changes in predation and food regimes. Hybrids seem to be of recent and multiple origin and occasionally backcross with the parental taxa. Evolutionary consequences may arise from repeated backcrossing, potentially resulting in patterns of reticulate evolution.

Städler and Jarne address the population biology and genetics of freshwater snails, concentrating on partial selfing in the hermaphrodite pulmonates. The authors first review theoretical models predicting genetic structure in subdivided populations, with an emphasis on the effects of inbreeding. This section is of general relevance for empirical studies on the distribution of genetic variability in patchily distributed populations, regardless of the organism under study. Empirical population genetics data in freshwater pulmonates suggest a marked loss of within-population genetic variability under selfing, compared with predominantly outcrossing populations. Städler and Jarne also consider the genetic and demographic factors thought to influence mating system evolution, with a deliberate emphasis on inbreeding depression. They also evaluate the approaches available for estimating mating system parameters and inbreeding depression, and highlight recent work uncovering within-population heterogeneity in selfing rates.

Johnson, Lively and Schrag also focus on repoductive modes in freshwater snails, but from a more ecological perspective. The authors review their empirical studies on three species of freshwater snails that are concerned with the relationship between reproductive mode and the prevalence of trematode infection. These studies were designed to test the ecological hypotheses for the maintenance of biparental sex. The biogeographic evidence from two species (*Potamopyrgus antipodarum* and *Bulinus truncatus*) supports the Red Queen hypothesis, while the pattern in a third species (*Campeloma decisum*) suggests that parthenogenesis has been selected as a mechanism for reproductive assurance. The authors also discuss the taxonomic distribution of parthenogenesis in aquatic invertebrates, and suggest that brooding may be an exaptation for the evolution of parthenogenetic reproduction.

Finally, Okamura addresses the life history and population biology of freshwater bryozoans, with a particular emphasis on her own studies on populations of *Cristatella mucedo* in southern England. This work suggests that sexual reproduction is restricted to a relatively brief period each year, and may be foregone entirely in certain years and/or populations. Genetic markers have revealed that sexual reproduction is equivalent to inbreeding among genetically similar clonal stock within sites, generating little genetic variation. Okamura also discusses the impact of myxozoan parasites on host fitness, but concludes that the seemingly indiscriminate parasite attack of very similar bryozoan clones makes traditional Red Queen scenarios inapplicable. Rather, she argues that a metapopulation structure with frequent dispersal of asexual propagules might provide a short-term escape from parasites or other adverse conditions.

The **fourth section** on "Evolutionary processes following colonizations" draws attention to processes of adaptation, phenotypic change, and speciation in an explicitly historical framework. This section contains an extensive account of postglacial colonizations of freshwater habitats by anadromous fishes.

Bell and Andrews review the evolutionary consequences of such colonizations in various groups of anadromous fishes, with particular emphasis on the three-spined stickleback, *Gasterosteus aculeatus*. Deglaciation of the boreal Holarctic has created numerous opportunities to colonize fresh water, where strongly contrasting environmental conditions have favored rapid endemic radiations. The authors show that repeated colonizations of fresh water and rapid phenotypic evolution are general features of these radiations. Of the components of phenotypic evolution, trophic diversification has arguably been the most important. Complexes of anadromous fishes and their freshwater isolates probably form phylogenetic racemes in which anadromous populations represent phenotypically stable ancestors from which predictable sets of divergent, freshwater phenotypes have evolved repeatedly.

As editors, we hope that this series of case studies and critical conceptual syntheses will serve as a source of reference and ideas. It should be suitable for teachers and students in ecology, population biology, and the evolutionary biology of freshwater organisms. All authors have endeavored to compose critical evaluations of their own work as well as that of others, focusing on the conceptual issues of their respective topics. We hope that their efforts will stimulate further research and progress in the fields of aquatic ecology and evolutionary biology.

We would like to extend our gratitude to the referees who critically read and commented on the manuscripts, thus contributing significantly to the quality of the final version:

Theo Bakker, Maarten Boersma, Gary Carvalho, Rob Dillon, Tom Klepaker, Laurence Hurst, Philippe Jarne, Steve Johnson, Jukka Jokela, Jan Kozlowski, Mathew Leibold, Angus McIntosh, Vincent Resh, Karl-Otto Rothhaupt, Juha Tuomi, Robert Wallace, Mike Winterbourn.

In addition, we thank the four further referees who preferred to remain anonymous to the respective authors.

As editors we assume responsibility for the final version, including its shortcomings in content and organization. We hope that the reader will find the volume a helpful source of ideas and data concerned with the evolutionary ecology of freshwater animals.

Bruno Streit, Thomas Städler, Curtis M. Lively

University of Frankfurt, Germany,
and Indiana University, Bloomington, USA

Ecosystem structure and trophic interactions

Evolutionary Ecology of Freshwater Animals
ed. by B. Streit, T. Städler and C. M. Lively
© 1997 Birkhäuser Verlag Basel/Switzerland

Habitat partitioning by zooplankton and the structure of lake ecosystems

M. A. Leibold[1] and A. J. Tessier[2]

[1] Department of Ecology and Evolution, University of Chicago, 1101 E. 57th St., Chicago, IL 60637, USA and .
[2] W.K. Kellogg Biological Station and Department of Zoology, Michigan State University, Hickory Corners, MI 49060, USA

Summary. Models of trophic chains commonly simplify interactions among species in a given trophic level. However, resource competitors within a trophic level can often interact strongly with one another, and the mechanisms of interaction may modify how such competitors participate in trophic interactions. Here we examine the role of habitat partitioning by *Daphnia*, commonly important grazers in lake ecosystems. We review the history of research on habitat partitioning in *Daphnia*, and suggest that habitat partitioning commonly reflects the conflicting constraints of avoiding predators and using environments that maximize growth. Based on such conflicting demands, we develop a graphical model of habitat partitioning under resource competition mediated by habitat-specific predators. Previous work has supported many of the predictions of the model except for some effects on the participation of *Daphnia* in trophic interactions. Using a standard *D. pulex* clone bioassay of food quality, we present evidence that these deviations can be accounted for by variable responses of algal communities that can alter their edibility of *Daphnia*.

Introduction

The concept of the trophic cascade is rapidly becoming a dominant paradigm in the study of limnetic ecosystems. Trophic cascades occur when increases in the density of top predators alter the densities of organism in lower trophic levels such that organism that are an odd number of trophic levels lower are negatively affected but those that are an even number of levels lower are positively affected. In most theoretical models of trophic cascades, however, the way that species participate in trophic cascades has been greatly simplified and complications that might arise from behavioral plasticity of component species are rarely addressed (DeMelo et al., 1992; Neill, 1994). Critics of the concept of the trophic cascade have rightly pointed out that one of the more likely complications occurs when habitat structure modifies the way species interact with their resources and their predators (Arditi and Ginsberg, 1989). For example, in stream systems, Power (1992) has shown how the availability of a physical refuge can decouple predator-prey linkages and thereby alter the expression of a trophic cascade. While in principle the concept of habitat heterogeneity can be incorporated into models of trophic interactions, in practice what constitutes different habitats is defined by the organisms and how they perceive the relative costs and benefits of utilizing different environments.

In limnetic ecosystems, the most obvious and best studied type of habitat structure occurs in stratified lakes, wherein almost the entire suite of environmental variables changes as one proceeds from the shallow, warm epilimnion to the deep, cold hypolimnion. Plankton ecologists have long recognized the significance of this stratification to zooplankton communities because it could explain patterns of coexistence among closely related species. Early studies on habitat use in zooplankton communities focused on the potential for habitat partitioning to reduce interspecific competition (Tappa, 1965). However, more recent studies have led to the frequent view that patterns of habitat use are more closely related to predation risk and have consequently discounted the role of interspecific competition (Lampert, 1987). Here we wish to argue that competitive release and predation avoidance are not mutually exclusive and that they may interact in interesting and enlightening ways.

There has been a large body of work on the joint effects of resource availability and predation on zooplankton population abundance and community structure, especially for species of the genus *Daphnia* (e.g., DeMott, 1989; Gliwicz and Pijanowska, 1989). In parallel, studies of habitat use and of diel vertical migration by zooplankton have also focused increasingly on the joint effects of resource availability and of predation risk (Gliwicz and Pijanowska, 1988; Leibold, 1990; Dini and Carpenter, 1992). It seems reasonable to hypothesize that these two aspects of zooplankton biology might be closely linked. Though implicit in many studies, this linkage has not been explicitly developed in most previous work (Leibold, 1991).

In this paper we examine how models of density-dependent habitat behavior under predation risk are related to models of trophic cascades to make predictions about the role that habitat specialization by zooplankton can play in mediating their contribution to limnetic ecosystem structure. The enormous diversity in species foraging and habitat use, and in foodweb structure among lakes, has resulted in a literature burdened by specific cases. Our goal in starting with a very general model of habitat selection is to provide a framework within which we can explore the consequences of hypothesized trade-offs in habitat use. Hence, our approach is to adapt previously developed general models and unpack them (*sensu* Rosenzweig, 1991) for exploring the role of habitat partitioning by zooplankton in modifying their participation in trophic cascades. We also review how ideas on the link between habitat partitioning and trophic cascades have evolved since early observations revealed the importance of habitat use in zooplankton, and relate these observations to our work on the evolutionary ecology of competing *Daphnia* grazers in temperate lakes of the USA.

A review of habitat partitioning in zooplankton

Probably no other aspect of plankton biology has generated more research interest than vertical habitat use behavior (beginning with Weismann, 1877). Contrary to what is implied by the word plankton, most zooplankton species are quite capable of efficient, directed swimming, and exhibit distinct vertical habitat preferences at the level of individuals and populations. Furthermore, there is ample evidence that individual habitat choice is strongly modified by environmental conditions; i.e., there is much behavioral flexibility. Consequently, more than 100 years of research has sought to identify the proximate and ultimate environmental factors responsible for individual behavioral variation. A logical starting place in this work has been the search for general (among lake) patterns of population habitat use behavior.

Evident in this extensive literature on zooplankton depth distribution are three general features. First, there is often a relationship between the seasonal population dynamics of species and their vertical habitat use during certain times of the year (Wesenberg-Lund, 1926; Tappa, 1965). For example, *D. pulicaria* (a common species in North America) is typically restricted to the deep, cold hypolimnion during the summer but is more widely distributed in the spring and fall when it is more abundant. In contrast, a commonly coexisting species, *D. galeata mendotae*, is typically found in shallower, warmer waters but is most abundant only in summer. Second, the vertical habitat choice of most species changes with light level, resulting in diel vertical migrations (DVM), (Paffenhöfer and Price, 1988; Ringelberg, 1993). Although there is much variety to DVM (Hutchinson, 1967), the most common pattern is to descend into deeper water at dawn and ascend into shallower water at dusk. Third, both biotic (e.g., competition and predation) and abiotic factors (e.g., temperature and light) are important in determining habitat use by particular zooplankton species (Kerfoot, 1985; Lampert, 1989). The manner in which these factors combine in any given lake can determine the extent of habitat heterogeneity and the degree of habitat specialization and DVM (De Stasio et al., 1993). Some of these factors have received considerable individual attention, particularly with respect to variation in DVM, but few studies have attempted to synthesize their relative importance to the vertical structuring of multiple species assemblages.

Explanations for how biotic and abiotic factors differentially affect species habitat choices have considered aspects of species morphology, physiology, and life history (demography). In the first third of the 20th century, aspects of species body shape were considered of primary importance in understanding vertical habitat use (Wesenberg-Lund, 1926; Woltereck, 1932). Many populations of cladocerans and rotifers display impressive seasonal changes in adult body shape (i.e., cyclomorphosis) which appear to relate to their habitat depth. Forms occupying the surface waters of the epilimnion often develop extensive elongations of the body compared to forms that remain in the deeper hypolimnion (Juday, 1904; Woltereck, 1930). Arguments over the adaptive significance of cyclomorphosis focused on differences in the selective environment of warm, shallow waters compared to the deep, cold waters. While most of that debate considered abiotic factors such as temperature, light, surface film, and turbulence, the significance of vertical heterogeneity in resource levels was clearly recognized (Schädel, 1916; Woltereck, 1920).

The second half of this century brought first a focus on interspecific competition and niche diversification (Hutchinson, 1951) as an important explanation for vertical segregation and DVM of zooplankton species. The fact that species inhabiting the same lake are found to have their peak abundances segregated in time and/or vertical space (Tappa, 1965; Lane, 1975; Makarewicz and Likens, 1975) does not indicate competition as the sole driving force. It is, however, strong evidence that the species experience the different seasonal periods and vertical depth zones as distinct habitats. That is, the species must differ in how they perceive the relative costs and benefits of utilizing these different habitats. Likely reasons for such species differences should include aspects of physiology and life history (Allan, 1977), in addition to the morphological traits emphasized in the first part of the century. Unfortunately, little research has explicitly tested even the general concept of "local" adaptation of zooplankton species to particular seasons and depth habitats, let alone developed much understanding in terms of traits and trade-offs.

Before much experimental work had been conducted on species differences in response to food and abiotic factors that vary seasonally and with depth, a powerful biotic factor (predation) attracted the attention of zooplankton biologists almost to the exclusion of these other facets of habitat heterogeneity. Initially the focus was on fish planktivory, which proved to be a strong force influencing the size structure, and hence, species composition of plankton (Hrbácek et al., 1961; Brooks and Dodson, 1965; Galbraith, 1967). Invertebrate predators were also discovered to be effective agents of selective mortality in the plankton (Hall, 1964; Dodson, 1974; Lane, 1979; Zaret, 1980). A rich experimental literature rapidly developed around predation not only as an ecological force but also as an agent of natural selection in the evolution of zooplankton morphology and behavior.

While most planktivores are known to exhibit restricted depth distributions and fish in particular are effective planktivores only in well-lit, epilimnetic or metalimnetic waters, the significance of this to seasonal dynamics and vertical structuring of zooplankton communities as a whole has attracted somewhat limited research (Fast, 1971; Kitchell and Kitchell, 1980; Threlkeld, 1980; Wright and Shapiro, 1990; Tessier and Welser, 1991). In contrast, the significance of predation to DVM of zooplankton populations is well studied (e.g., Zaret and Suffern, 1976; Gliwicz, 1986;

Neill, 1990; Lampert, 1993). The difference in emphasis is notable; the vast majority of research on zooplankton habitat choice is focused on the short-term behavior of DVM *per se*, and not on mean depth distributions, seasonal changes or consequences to population dynamics, competition, and community structure. It is not surprising then that the linkage among habitat heterogeneity in lakes, trophic interactions and ecosystem functioning is not well developed.

In summary, the past 100 years of research on zooplankton habitat use illustrate a shift in emphasis form abiotic factors to biotic factors. Unfortunately, rather than building upon previous work, the focus jumps from temperature, to light, to competition to predation. Although there has been a sustained focus on fitness trade-offs involving body morphology, less work addresses aspects of physiology or life-history adaptation. Not only has the emphasis shifted from a general concern with vertical depth distributions and seasonal dynamics to a focus on short-term DVM behavior, but most recent work describes proximate rather than ultimate mechanisms. For example, predation is considered an important ultimate explanation for DVM, but few studies have explored the ecological and evolutionary trade-offs involved in adapting to this force compared with the detail with which proximate mechanisms of light and kairomones have been explored. A single hypothesis concerning body size and foraging efficiency (size-efficiency hypothesis, Brooks and Dodson, 1965) has drawn sporadic study over the past 20 years, largely without resolution (see reviews by Hall et al., 1976; Bengtsson, 1987; DeMott, 1989; Rothhaupt, 1990). A potentially richer understanding would undoubtedly involve more detailed study of metabolic (Burns, 1985), demographic (e.g., Vanni, 1987) and life cycle (e.g., Hairston, 1987) trade-offs.

It is also unclear that all ultimate explanations for trade-offs in habitat use by zooplankton need be biotic. Certainly, competitors and predators are of major importance to the seasonal and vertical structuring of plankton assemblages. However, substantial evidence exists that adaptation to temperature (Allan, 1977; Carvalho, 1987), light (Hairston, 1976) and perhaps turbulence (Jacobs, 1987) also involves trade-offs in morphology, physiology and life history that may be individually important, or interact with biotic factors to shape habitat structuring of zooplankton. Again, a focus on short-term behavior (i.e., DVM) rather than the general nature of habitat segregation has hampered efforts to incorporate abiotic and biotic factors (e.g., Geller, 1986).

We offer one final observation on research trends in zooplankton habitat partitioning. While early researchers were aware of and concerned with genetic variation in habitat use and DVM within species and populations, they actively sought patterns from broad-scale comparisons of lakes. For example, Lozeron (1902, cited in Hutchinson, 1967) noted the importance of light transmission properties in different lakes to explaining variation in the magnitudes of DVM. Wesenberg-Lund (1926) emphasized the importance of lake size and basin shape in comparing species composition and habitat use behaviors. In an insightful paper, Woltereck (1932) compared the patterns of vertical habitat use and DVM in American and Canadian lakes to the situation in lakes of Europe. He illustrated the joint association of a phylogenetic contrast in assemblages of *Daphnia* on the two continents and a contrast of vertical habitat-use specialization.

Sixty years of further research has brought us no closer to understanding the broad patterns of variation described by these early researchers. In contrast, much current research has focused increasingly on short-term studies of single lake populations (see also comments by Neill, 1994). The benefit of small-scale research has been a detailed, experimental understanding of particular ecological and evolutionary mechanisms. For example, there are excellent laboratory studies of genetic variation in habitat use behavior within single *Daphnia* populations or of phenotypic plasticity within single clones (e.g., Dodson, 1988; De Meester, 1989; Haney, 1993; Loose, 1993), and field experiments with predators and resources explore short-term effects on zooplankton, including habitat use (e.g., Leibold, 1990; Dini and Carpenter, 1992). Though it is clear that there is substantial variation (both genetic and environmental) in habitat use and correlated ecological traits in *Daphnia* species (both within and among lake populations), there are virtually no studies that have examined how this variation affects interactions with other species. More importantly, the application of this improved understanding of genetic variation and ecological mechanisms to broad patterns of variation among lakes, regions, and continents is sparse.

Let us consider the pattern observed by Woltereck (1932) in more detail. He noted that in North American lakes, species of the *Daphnia pulex* complex were co-dominant perennial members of the plankton, whereas in Europe it was species of the *D. longispina* complex that solely dominated. He also noted that in North America the most common assemblage of *Daphnia* species included hypolimnetic specialists and DVM generalists, whereas in European lakes epilimnetic specialists and DVM generalists were more common. Consequently, in many North American lakes the epilimnion had few *Daphnia* grazers during the daylight hours, but routinely was occupied both day and night in European lakes. Woltereck felt that differences in morphology and physiology fundamental to the *pulex–longispina* separation were important in understanding habitat segregation. Unfortunately, a limited understanding of ecological mechanisms (e.g., predation) at that time made it difficult to propose explanations for either local or continental contrasts in zooplankton assemblages.

Numerous independent studies have re-enforced these early observations by Woltereck and greatly deepened our knowledge of the systematics and ecological mechanisms likely to underlie this continental contrast. For example, taxonomic diversity on both continents is now much better described (Lieder, 1986; Wolf, 1987; Colbourne and Hebert, 1996), and this finer level of resolution appears to strengthen the original comparison drawn by Woltereck. Our paleo-ecological understanding of the availability of glacial refuges for aquatic organisms (e.g., Stemberger, 1995) has grown to the extent that reasonable hypotheses can be proposed to explain the contrast in species dominance between North America and Europe. For example, there is a large contrast in fish planktivory on the two continents; European lakes contain a diversity of efficient, open-water fish planktivore species, which are largely missing from North American lakes. Despite our improved understanding of ecological details and historical influences, no macro-ecological studies (*sensu* Brown, 1995) have expanded upon Woltereck's (1932) initial ideas.

A simple model of habitat partitioning in zooplankton

Considering the likely complexity of interactions of biotic and abiotic factors affecting habitat use by zooplankton, it is surprising that modeling approaches have not been employed more widely. Iwasa (1982) and Gabriel and Thomas (1988) used evolutionary game theory (ESS) to explore DVM of zooplankton exposed to fish predation. In the latter, more sophisticated model, only a single pair of behavioral strategies is explored within the context of a single environmental contrast. Despite this simplicity, Gabriel and Thomas (1988) demonstrate that a mixture of two DVM strategies can be evolutionary stable under certain environmental conditions and ranges of parameter values. Clark and Levy (1988) and De Stasio et al. (1993) have made use of dynamic programming to derive optimal habitat use behaviors for mobile organisms in vertically stratified systems. These models focus on maximizing fitness of individuals given a detailed under-standing of how environmental factors influence growth, reproduction and mortality.

Both the optimality and ESS approaches can be considered specific models in the sense that they make use of numerous parameters to describe the consequences of individuals choosing particular habitat types. They link organismal traits (e.g., physiology, morphology, behavior) into fitness consequences by summing resource gains and subtracting mortality risks for particular environmental conditions. They do not, however, provide a very general context for the concept of species habitat segregation. Below we present a graphical model of habitat selection under the joint influence of resource competition and predation risk. Our goal is to provide a frame-work general enough for both broad scale (macroecological) studies and detailed investigations of trade-offs associated with zooplankton habitat specialization and DVM.

The model is closely related to that developed by Rosenzweig (1979, 1981, 1991; Pimm and Rosenzweig, 1981; Pimm et al., 1985; Rosenzweig and Abramsky, 1986; Brown and Rosenzweig, 1986) and especially by Brown (1990, 1992, 1997). However we focus more explicitly on the relative roles of predators and resource productivity and we link the model more directly to the participation of grazers in trophic cascades. We also use a graphical approach that is somewhat more general than the previous work and express the interactions in a graphical space defined by the abundances of resources in each habitat rather than the densities of competing consumers to facilitate links to mechanistic resource competition models (e.g., MacArthur, 1972; Tilman, 1982). The method we have used, however, reduces the possibility of examining resource partitioning by diet within habitats, whereas Rosenzweig's method does not.

To "unpack", the model for application to planktonic systems, we have also modified the model to include the possibility of short-term temporal use of alternate habitats by the consumers as occurs in zooplankton that cycle between habitats by undergoing diel vertical migration. Although this kind of cyclical use of habitats is a prominent feature of zooplankton behavior, it may also be important in other taxa (e.g., freshwater fish, Hall et al., 1979; long distance migrations by birds). The result is that the model more closely resembles a patch choice model with travel costs (see Brown, 1997) than a pure habitat choice model without travel costs (even though energetic costs of vertical migration in zooplankton appear to be very minor; Dawidowicz and Loose, 1992). The model can easily be modified to simpler situations that do not include this kind of cyclical habitat use.

The model

Consider zooplankton consumers that have three possible options in utilizing the water column (Fig. 1(A)). One option is to use only the epilimnion staying there both night and day, feeding exclusively on epilimnetic resources, but subject to habitat-specific predators (in freshwater system, this would be primarily thermophilic, visually-oriented planktivorous fish). Another option is to use only the hypolimnion, again staying there both night and day and subject to predation by a different set of habitat-specific predators (either other species of fish or, more commonly, invertebrate predators such as *Chaoborus*). The third option is to migrate between the two habitats on a diel basis. This third option can be advantageous if there is significant diel variation in predation risk in one or both of the habitats. For large species of zooplankton (i.e., *Daphnia*), the most likely source of such variation comes from the reduced feeding efficiency of visual predators under low nighttime light conditions. However, invertebrate predators are also known to induce such diel migrations in their zooplankton prey (Neill, 1990).

The choice of which strategy to use should depend on the potential contributions to growth and reproduction relative to the potential morality costs present in each habitat (see Brown, 1992, 1997 for a more general discussion). Assuming linear and additive potential contributions of epi-

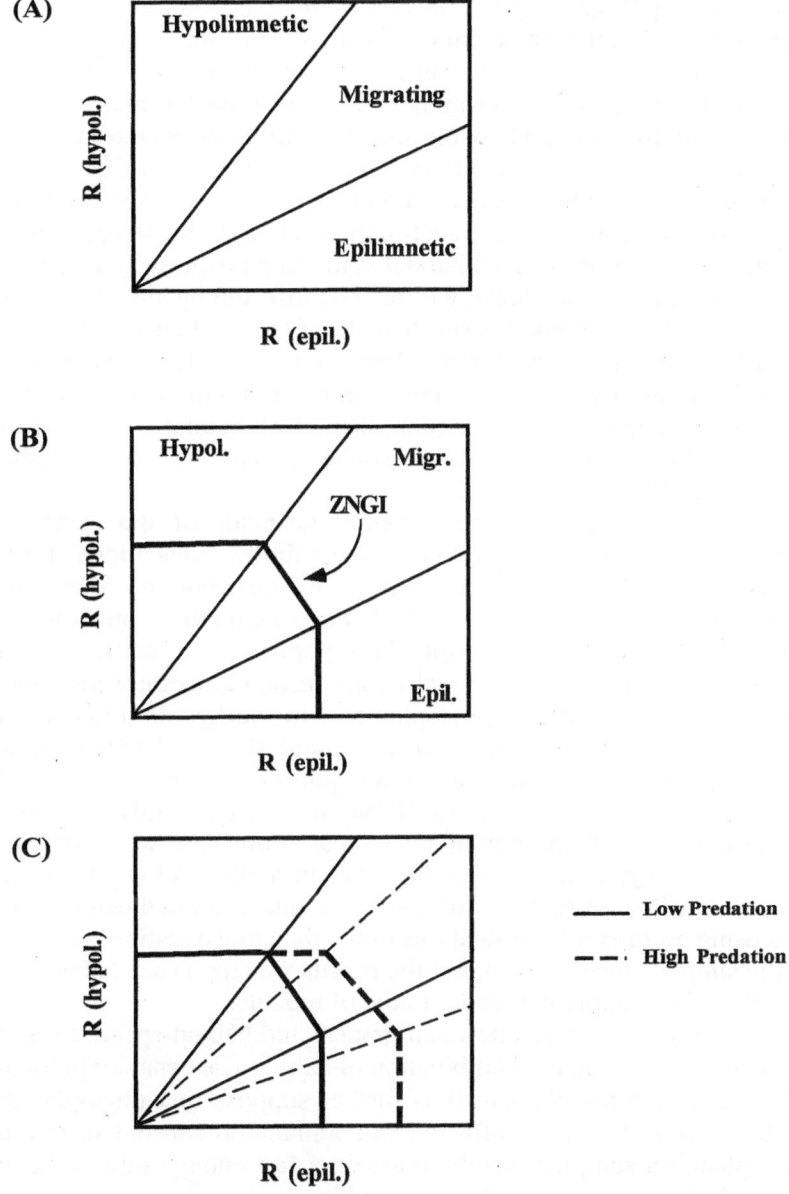

Figure 1. Resource-dependent habitat selection by an exploitative consumer feeding in two habitats (epilimnion and hypolimnion) subject to variation in predation by fish in the epilimnion. In all three figures the axes represent the density of resources (R) in reach of the two habitats. In 1(A), three regions in which the consumer should be either epilimnetic, hypolimnetic or should migrate between the two habitats are denoted for a constant (arbitrary) predation level. In 1(B), the zero net growth isocline (ZNGI) for the consumer is illustrated contingent on the behavior shown in 1(A). In 1(C), the effect of increased fish predation in the epilimnion on the behavior (habitat choice) and the zero net growth isocline is illustrated. See text for a more detailed explanation.

limnetic and hypolimnetic food resources to zooplankton growth and reproduction, and constant predation through time, the decision should depend on the ratio of phytoplankton resources in the two habitats (Fig. 1(A)). The very limited combinations of resource densities where the consumer is indifferent to choosing between different habitat strategies, broadly similar to Rosenzweig's (1979, 1981) "isolegs", should be constant proportions shown as thin lines in Figure 1 (A). If there were not differences in temperature or other factors, such conditions for indifference should occur when food density in each habitat is the same. However, the proportions that determine equivalence will also be affected by any other factor such as temperature or predation risk that alter fitness. Many features of the biology of the organisms can alter the shape of these "isolegs", but they will generally be monotonically increasing as shown in Figure 1 (Brown, 1992, 1997). For the simplest case we explore here, each habitat will be favored when resource ratios surpass some threshold value in that habitat (Gilliam and Frazer, 1987, 1988).

The model assumes that the dynamic behavior of the system of interacting zooplankton and phytoplankton follows conventional Lotka-Volterra-like predator-prey dynamics (e.g., Rosenzweig and MacArthur, 1963; MacArthur, 1972; Tilman, 1982) that depend on the habitat distributions of the zooplankton. If a zooplankton population is restricted to the epilimnion it will have conventional single-resource predator-prey interaction in the epilimnion and phytoplankton in the hypolimnion will be limited by other grazers or by competition. Similarly, zooplankton restricted to the hypolimnion only interact with hypolimnetic phytoplankton. If a zooplankton population undergoes DVM, it will be jointly limited by both hypolimnetic and epilimnetic resources: in this case we assume that epilimnetic and hypolimnetic resources act in a substitutive way (*sensu* Tilman, 1982). Precise formulation for the results described below can be derived using more exact formulations of the dynamic equations (especially for the simplest linear cases), but the results we emphasize below using graphical methods apply to a wider range of models.

Consequently, interactions between grazers and habitat-specific resources depend on the particular combination of habitat-use behavior predicted for any grazer species. We can thus further suppose that phytoplankton populations are significantly affected by zooplankton consumers and that both zooplankton and phytoplankton are on a fast enough time scale that their populations can approach near-equilibrium densities before being perturbed by environmental disturbances. The zooplankton should then be able to deplete phytoplankton levels in the habitats in which they are found so that the zooplankton's resource-dependent birth rates equal their death rates (see Gilliam and Frazer, 1987, for a somewhat different approach and a description of the behavioral dynamics involved). The conditions under which this will occur are shown in Figure 1 (B). The heavy line shown is the "zero net growth isocline" or "ZNGI" (*sensu* Tilman, 1982) of a zooplankton consumer. When the consumer is restricted to one habitat (the E and H regions in Fig. 1(B) the ZNGI is insensitive to resource densities in the alternate habitat. Only when the consumer uses both habitats by undergoing diel vertical migration (the M region in Fig. 1(B)) can the resources show substitutability (i.e., the consumer's ability to reduce the abundance of one resource depends on the abundance of the other resource). In this region (M in Fig. 1(B)) the slope will reflect the relative value of feeding and spending time in the two habitats and will be affected by factors such as temperature, mortality etc. as well as the duration of daylight versus nighttime.

Also, suppose that predation can vary in each habitat independently, and that predator dynamics are not tightly coupled to zooplankton dynamics, either because they are substantially regulated by other factors such as interference, predation, or other resources such as spawning grounds (see also Mittelbach et al., 1988), or because they are on much slower time scales than is the plankton and thus recover more slowly from perturbations. The effect of different predation levels within a habitat is to alter both the behavior of the consumer (position of lines of habitat selection indifference in Fig. 1(C)) and its ability to depress resources in the habitat where the predator is found (ZNGIs in Fig. 1(C)). The ZNGI in the habitat where the predator is absent will not change, except to reflect the different resource densities required to change the behavior of the consumer under different predation levels. The ZNGIs for the consumer when it utilizes the habitat with the predator will be affected to reflect the greater resource requirements which result in an equilibrium between birth and death rates under those conditions as shown in Figure 1(C).

Long-term coexistence of two such zooplankton consumers requires that their ZNGIs intersect and that each consumer consumes more of the resource which most limits its growth relative to the alternate consumer (Tilman, 1982). This latter condition is generally the case in this model because of the tight relationship between habitat use and resource consumption, i.e., consumers have to be in the same habitat as the resources that they consume. There are many possible configurations of ZNGIs of the two consumers that can result in stable coexistence (Fig. 2). In most of them the two consumers have different habitat distributions and one of the consumers excludes the other consumer from one habitat by depressing resources below the other consumer's requirements to match its death rate in that habitat. The one possibility where this does not occur and the species coexist in spite of having the same habitat distribution is when both species migrate and each derives a greater relative benefit from resources in different habitats (Fig. 2(C)). The variety of outcomes shown in Figure 2 will arise under different environmental conditions and under situations in which differences between the two competing species reflect different trade-offs. The "ghost of competition past," segregation between species with no overlap in habitat use only occurs when the species have very different levels of specialization in each habitat (Fig. 2(E)). There is also the

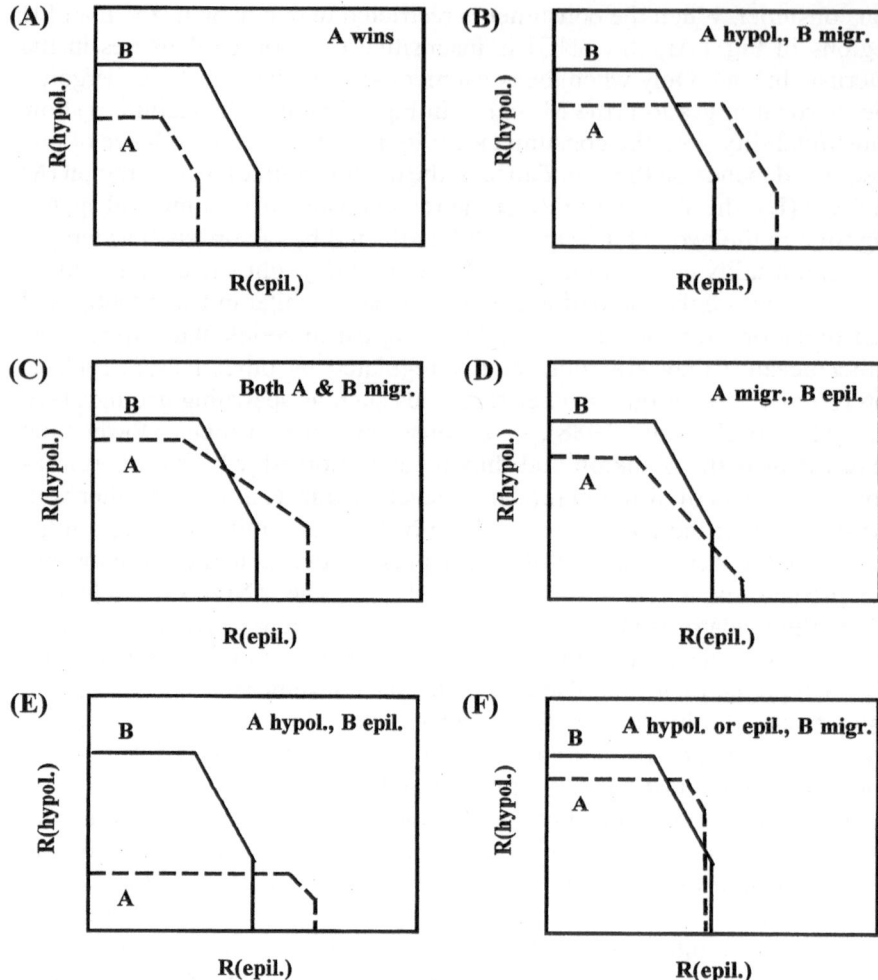

Figure 2. Various possible configurations of the zero net growth isoclines for two coexisting competitors. In 2(A), one species is the competitive dominant in all habitats and can drive the other species to extinction. In 2(B) through 2(F), the two species can coexist. The behavioral outcome for each case is shown. In 2(F), there are two possible stable points depending on initial conditions.

interesting possibility of alternate stable states shown in Figure 2(F) that occurs when a grazer whose susceptibility to predators differs little between night and day, interacts with a competitor whose susceptibility differs much more substantially between night and day. Such a situation may explain the occasional observation of species pairs that show reversals in relative patterns of habitat use from lake to lake (e.g., Woltereck, 1932; Kratz et al., 1987).

The outcome of competition among grazers in the model is strongly context-dependent. For example, different temperatures in the two habitats could alter the slope of the isolegs that regulate habitat selection behavior and the ZNGIs of each of the species. Since the effects of temperature (and many other similar abiotic factors) on *per capita* reproduction and growth are non-linear (usually humped with an intermediate optimum that can depend on resource density), the effects on the isolegs and ZNGIs are potentially complex. This is particularly so when the species involved have different temperature optima. Under such conditions, transitions among alternate configurations shown in Figure 2 may not be uni-directionally related to temperature gradients. However, for any given set of temperatures, the model can be used to interpret how habitat partitioning and impacts on resource levels relate to predation and resource productivity.

As was shown by MacArthur (1972, see also Rosenzweig, 1973; Tilman, 1982), resource productivity does not affect the standing crops of the resources in this model since it does not affect the point at which the ZNGIs intersect. Resource productivity also does not affect the habitat distributions of the consumers in the model since these are also completely determined by the ways habitat-specific mortality rates balance with resource densities and not by the productivity of the resources. Instead, resource productivity affects the consumer abundances. The effects on consumer densities can be seen by considering that in high-productivity environments the specific productivity of the resources will be higher than in low-productivity environments. Since standing crops are the same in both situations this will also be directly reflected in a difference in net productivity. Consequently, the birth rates of the consumers are also equal (and proportional to resource abundances) and the extra productivity can only be maintained if there are more consumers present in the high-productivity environment. This prediction depends on having only one resource type in each habitat. When the habitats contain more than one resource species and they differ in edibility, the ability of the consumer to respond to resource productivity may be compromised by the response of less-edible resource types (Phillips, 1974; Leibold, 1989).

Resource densities and consumer habitat distributions are much more strongly affected by variation in the mortality risk in the two habitats, and this kind of variation can be responsible for different possible configurations of ZNGIs of the two consumers (Fig. 2). One particularly interesting pattern of habitat segregation between two consumers occurs when the consumers have traits that reflect a trade-off between competitive ability and susceptibility to predation in one of the habitats (e.g., Jacobs, 1977; Tessier, 1986). One possible scenario which can result is shown in Figure 3. In this situation one species can competitively exclude the other in both habitats when predation is low, because of its superior ability to harvest and utilize resources in both habitats. However, as predation increases in one of the habitats, its ability to depress resources in that habitat will

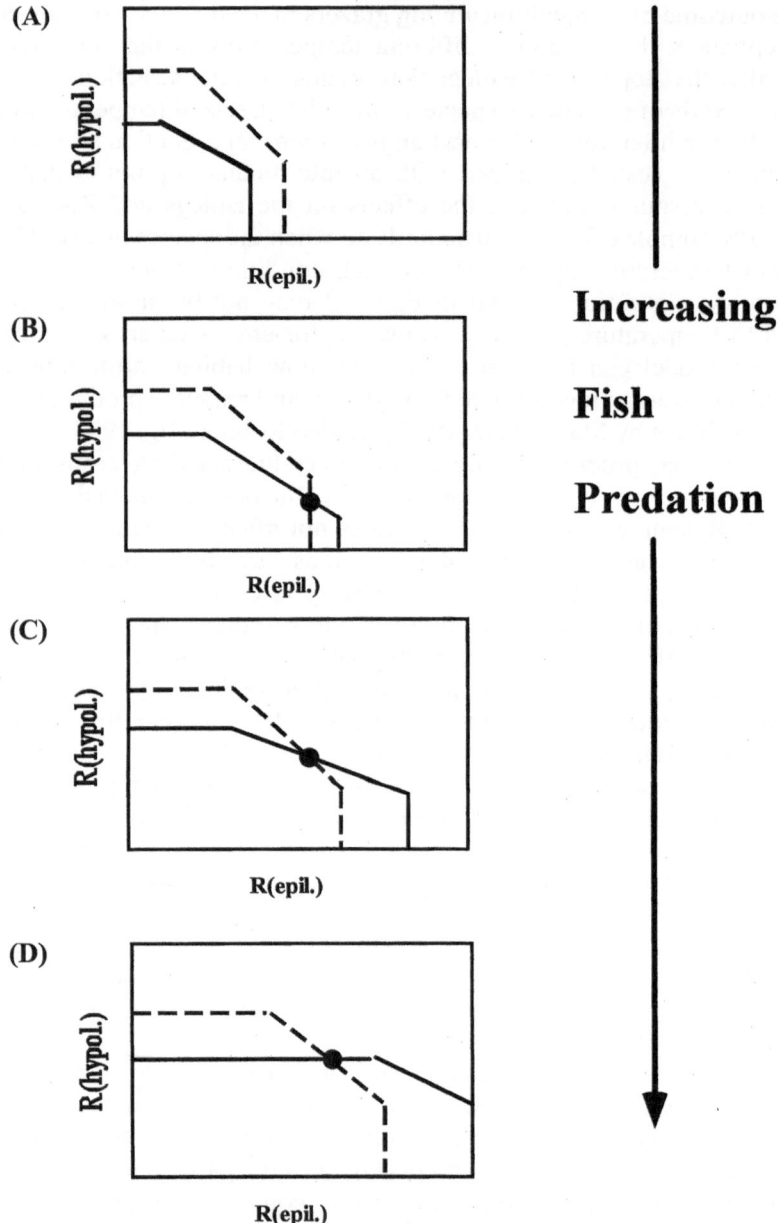

Figure 3. A scenario for variable outcome of competition in environments that differ in fish predation intensity. The scenario depends on a trade-off between competitive ability (A (solid line) is competitively superior to B (stippled line) in the absence of predation but is more sensitive to fish predation). In 3(A), A completely excludes B. In 3(B), B can exist by being completely epilimnetic whereas A shifts to migrating between the two habitats. In 3(C), both species can coexist and migrate together; stability is only possible when each species consumes relatively more in the habitat that most limits their growth. In 3(D), A is hypolimnetic, whereas B migrates.

be reduced by its greater sensitivity to predation. As such, its ZNGI would shift to the right at a faster rate than the other, less vulnerable consumer. If the difference in sensitivity to predation is large enough, this shift can result in coexistence of the two species by allowing the competitively inferior species to invade the high-predation habitat. As predation continues to increase in that habitat, the ability of the first species to utilize the habitat is decreased, and its distribution is more severely constrained to the predator-free habitat, as shown in Figure 3.

Thus, although habitat segregation between species is maintained by exploitative competition as modeled by Rosenzweig (1979, 1981) and Brew (1982), the particular patterns of habitat partitioning and their relation to consumer and resource abundances depend on the level of predation (mortality risk) and the productivity of the resource base. More specifically, in this trade-off scenario, there should be a loose relationship between habitat distributions and resource abundances with predation intensity. Conversely, habitat distributions and resource densities should not be strongly affected by resource productivity. However, consumer abundances should be affected by both predator abundances and resource productivity. The qualitative predictions of the model (Figs 1–3) are fairly robust to other possible assumptions. An important exception is the possible effects of heterogeneous resources that differ in edibility on the prediction that nutrients should not affect resource densities (Leibold, 1989). If resource heterogeneity within habitats is large enough, the effective resource densities (the weighted contributions that they have on growth and reproduction of the consumers) should not differ with resource productivity, but the actual densities of resources could increase with resource productivity as their average edibility decreases.

This model can be applied to a system of two competing *Daphnia* feeding on resources from two different habitats (the shallow, warmer epilimnion and the deeper, colder hypolimnion) in lakes with epilimnetic fish predators to make the following set of predictions (summarized in Tab. 1). Fish densities should correlate with epilimnetic phytoplankton densities because the death rate they impose on epilimnetic or migrating *Daphnia* compromises *Daphina's* ability to suppress algal resources (i.e., location of the equilibrium point shifts to the right with increasing fish predation in Fig. 3). Fish should not affect the densities of hypolimnetic phytoplankton unless there are habitat shifts by hypolimnetic *Daphnia* specialists (i.e., no shifts in the vertical position of the equilibrium point as long as species A is hypyolimnetic in Fig. 3). As a consequence, the growth and per-capita reproduction of epilimnetic or migrating *Daphnia* will be enhanced (as required to match their higher death rate near equilibrium), whereas the growth and reproduction of hypolimnetic *Daphnia* will not be affected. Epilimnetic or migrating *Daphnia* will be less abundant when fish are abundant, whereas hypolimnetic *Daphnia* might increase if they compete with migrating *Daphnia* that are thus reduced. Fish densities should also

Table 1. Model predictions for a system consisting of two competing species of *Daphnia* that segregate by depth in the water column exposed to variation in epilimnetic planktivorous fish density and overall water nutrient levels. Predictions assume a trade-off in fish predator vulnerability and competitive ability

---

Model Predictions:

Fish predation in epilimnion:

1) increases phytoplankton resource densities in epilimnion, no strong effect on hypolimnion unless substantial niche shifts by *Daphnia*
2) increases *Daphnia* growth and reproduction in epilimnion
3) decreases epilimnetic and migrating *Daphnia* abundances, may lead to increased hypo-limnetic *Daphnia*
4) changes patterns of habitat use by component *Daphnia* species, hypolimnetic species may invade the epilimnion when fish decrease

Higher nutrient levels:

5) increase population sizes of all *Daphnia* strongly
6) do not strongly affect algal standing crops or *Daphnia* growth and reproduction in either habitat
7) do not affect patterns of habitat use by either species of *Daphnia*

Removal of one competitor species:

8) increases the abundance of the remaining consumer
9) allows the remaining consumer to expand its habitat distribution

---

be associated with changes in the habitat distributions of *Daphnia* as described above (Fig. 3).

In the model, the eventual distributions of the two competing *Daphnia* are determined by competition between them. Thus, if one of the species is removed or suppressed, the abundance of the other species should increase. Since much of the competitive effect occurs by habitat preemption, the distribution of the unsuppressed species should expand into the habitat previously occupied by the suppressed species.

Evaluation of the model predictions

Previous work manipulating fish, nutrients, and daphniids in experimental enclosures has supported the predictions of this model (Leibold, 1989, 1991; Leibold and Tessier, 1997). These results, obtained in experiments lasting about 6 weeks are, however, not entirely consistent with data from comparative analyses among lakes that differ in predation or nutrients levels. These comparative studies show that variation in habitat partitioning and in the relative abundance of *Daphnia* is correlated with the intensity of fish predation and with variation in the size of the hypolimnetic refuge as predicted (Leibold and Tessier, 1991; Tessier and Welser, 1991). However, we do not find very good evidence that this variation in fish predation or hypolimnetic refuge size is linked to variation in epilimnetic algal density as predicted by the model. Conventionally, variation in phytoplankton is thought to be more closely related to nutrient levels in lakes. One possible explanation for the discrepancy is that heterogeneity of algal resources within habitats is also important. We here describe additional tests of the model that explore such complications.

Regardless of the particular hypothesized trade-offs in adaptation of competing species to the habitats, the above model makes two general predictions about responses to differences in primary productivity. Variation in productivity of the algal resource due to changes in external or internal nutrient loading should not affect equilibrium standing crop of edible resources for zooplankton, nor habitat distribution of those zooplankton. As shown by Leibold (1989, 1996), however, competition among algal species that differ in resistance to grazing (edibility) and ability to compete for limiting nutrients would likely result in a positive relationship between nutrient loading, grazer biomass, and algal biomass. The latter increases due to an increasing proportion of inedible forms in the algal community (potentially mediated by a number of mechanisms including taxonomic and genotypic shifts, changes in physiological and morphological status, or shifts mediated via the effects of higher trophic levels on nutrient regeneration; reviewed by Sterner, 1989). These shifts in phytoplankton composition can alter both absolute and relative performances of different zooplankton taxa (see DeMott, 1989). Hence, more refined tests of the model's prediction require a direct measure of the edible fraction (or qualitative value) of the total resource base, a difficult proposition considering the multi-faceted aspects of algal edibility (e.g., taxonomy, size, shape, biochemical constitution).

We have approached this issue by employing a single, representative clone of *D. pulex* as an assay organism to quantify resource richness in different lakes and depth habitats. By collecting the natural resource and feeding it to juveniles of this clone under controlled laboratory conditions (e.g., temperature, photoperiod, maternal acclimation), we can use specific weight gain (Tessier and Goulden, 1987) as a bioassay measure of resource conditions. Tsao and Tessier (unpublished data) have shown that growth rate of this daphniid clone is representative of the growth performance of different *Daphnia* species exposed to a wide range of natural resource conditions. Desmarais (1996) has further demonstrated that growth rate performance of this clone is a good predictor of full life-table performance of daphniids as different as *C. reticulata* and *D. rosea* feeding on a wide range of natural plankton resources.

We employed the specific growth rate of this *D. pulex* clone to quantify resource availability in the epilimnion of seven lakes during late summer (August and September) 1995. The lakes differed in trophic status as measured by total phosphorus concentration during spring turnover (April–May), ranging from oligotrophy to mesotrophy (Fig. 4). All seven lakes were similar in predation regime in that they contained bluegill sun-

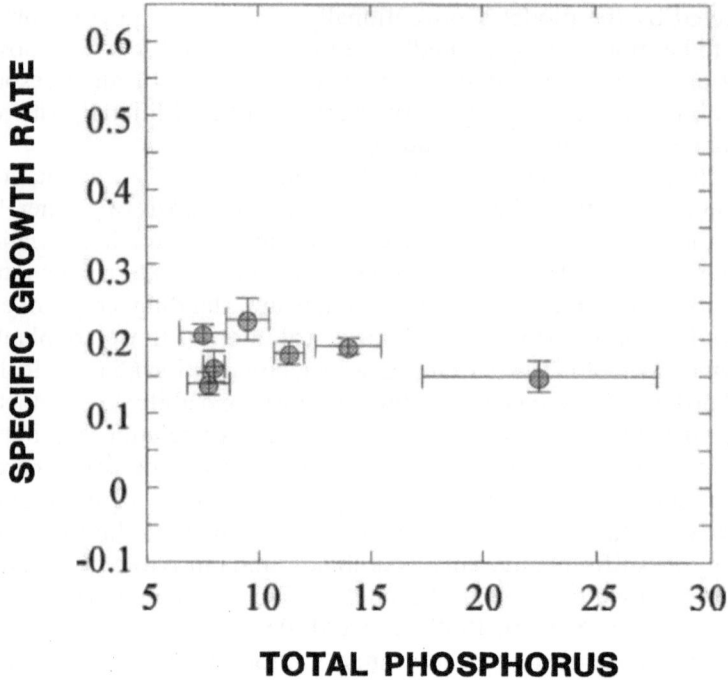

Figure 4. Resource levels in the epilimnion of seven dimictic lakes in Barry County, Michigan, during August and September as a function of total phosphorus (µg/L) during spring turnover (March–May). Resources measured as specific growth rate (µg/µg/day, mean±S.E.) of juveniles of a standard clone of *D. pulex* in lab bioassay. Scale range for growth rates indicates the observed range (−0.1 to 0.65) expressed by *Daphnia* in these lakes over the course of the entire year.

fish as the major fish planktivore, and were thermally stratified with a deep-water refuge to allow the *Daphnia* populations to escape this warm-water predator during the day. As predicted by the model, resource conditions in all lakes were essentially identical in late summer. Further, resource availability was very low compared to the natural range of growth-rate variation expressed by this *D. pulex* clone in a wider range of resource conditions.

Not only do resource conditions for grazers appear insensitive to resource productivity (evaluated as nutrient loading), but the model prediction of consistent habitat segregation despite variation in nutrient loading also holds. When we examine lakes that contain the same two *Daphnia* species, but differ dramatically in trophic status, there is remarkably little variation in vertical habitat segregation (Fig. 5). Interestingly, even the pattern of DVM is similar in these lakes, which although they differ in trophic status are all thermally stratified and contain bluegill sunfish as the major planktivore. Though there is some variation among lakes, this variation in habi-

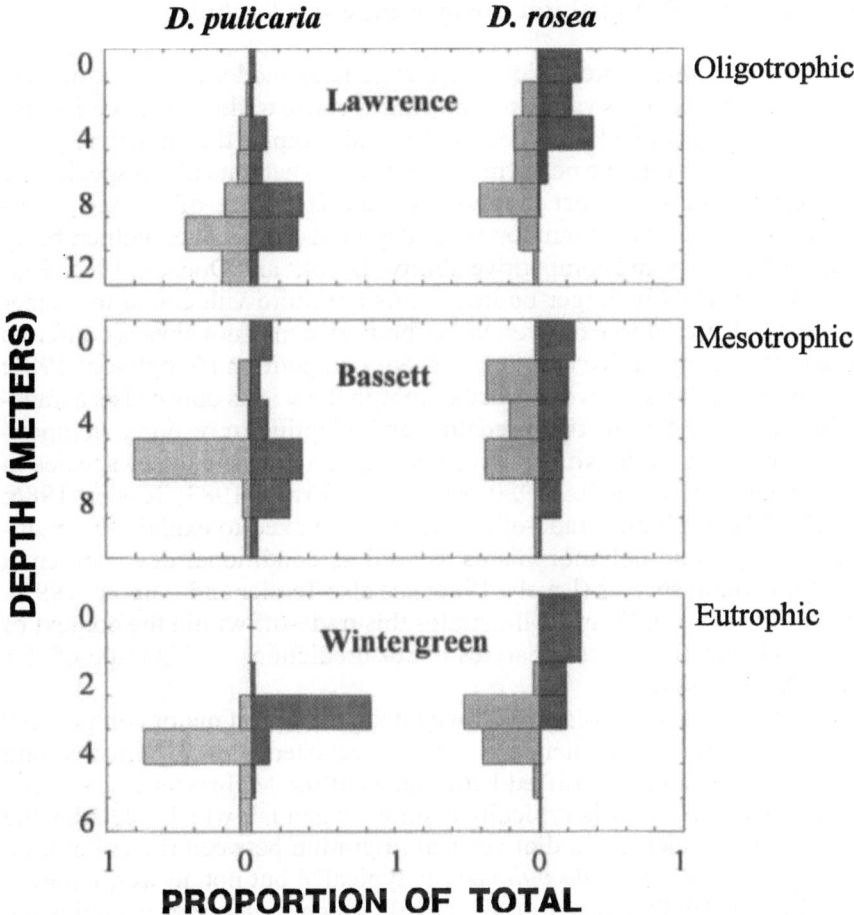

Figure 5. Comparison of diel depth distribution of *D. rosea* and *D. pulicaria* in three southwest Michigan lakes that differ in trophic status. Total phosphorus values for the lakes during spring turnover are: Lawrence −9.0 µg/L, Bassett −19.0 µg/L, Wintergreen −333.0 µg/L. Data for Lawrence and Bassett lakes collected during July 1990; analogous July data for Wintergreen taken from Threlkeld (1979). Loss of *D. pulicaria* from the bottom waters with increased trophy is a consequence of deep-water anoxia.

tat partitioning and DVM across a trophic gradient is much smaller than the substantial shift in such behaviors that occurs when one examines these species across lakes that differ in the intensity of predation risk (Leibold and Tessier, 1991). In comparisons involving lakes that differed in predation risk (evaluated as fish density), we found lakes in which *D. pulicaria* was hypolimnetic and *D. rosea* migrated when fish densities were high, and lakes in which *D. pulicaria* migrated and *D. rosea* was strictly epilimnetic when fish densities were low (Leibold and Tessier, 1991).

Specific trade-offs in habitat use by competing *Daphnia*

Several of the major predictions of our general model appear validated, giving confidence of its value in exploring specific trade-offs in habitat use by competing zooplankton species. Viewed simply, the model requires some trade-off in relative performance in the two habitats if two species are to coexist. Perhaps the most common hypothesized trade-off among generalized filter-feeding zooplankton (e.g., daphniids) is between vulnerability to fish planktivory and competitive ability (Brooks and Dodson, 1965; Hall et al., 1976). Clearly, larger-bodied forms are more vulnerable to a size-selective fish predator. However, larger body size may not always confer an advantage in exploitative (resource-based) competition (Bengtsson, 1987; DeMott, 1989). Irrespective of whether morphology is its only cause, a trade-off between adapting to fish predators and adapting to resource competition has been an attractive hypothesis to explain coexistence of species in lakes (Jacobs, 1977; Seitz, 1980; Mills and Forney, 1983; Tessier, 1986; Leibold, 1991). Similar trade-offs have been invoked to explain the maintenance of genetic polymorphisms as well as conditional developmental strategies within species (Lively, 1986; see also Tessier and Leibold, 1997). As discussed above, Figure 3 illustrates this trade-off within the context of our model, and Table 1 summarizes major predictions of this trade-off for a two-*Daphnia* system.

Over the past several years, we have attempted to test major components of this trade-off as an explanation for the coexistence of *D. pulicaria* and *D. rosea* in the small, stratified kettle lakes of the Midwestern U.S. (e.g., Fig. 5). *Daphnia rosea* is typically a summer annual, which occupies the epilimnion or undergoes a diel vertical migration between the epilimnion and hypolimnion. *Daphnia pulicaria* is typically, but not always, a perennial (Cipolla, 1980; Geedey et al., 1996), which in summer occupies the metalimnion or hypolimnion. Many aspects of the ecology of these two species suggest the general model and the specific trade-off of predator vulnerability – competitive ability hold (Leibold and Tessier, 1997). Population densities of these species are jointly influenced by predators and food resources (Leibold, 1991), and the species do compete for limiting resources (Hu and Tessier, 1995). Further, the species show asymmetric patterns of limitation by food and predators, which are closely related to differences in body size, pigmentation, and habitat use (Leibold and Tessier, 1991). *Daphnia pulicaria* is generally larger and more pigmented than *D. rosea*, making it more attractive to fish predators. However, by restricting its habitat use largely to the deep, cold water (refuge), *D. pulicaria* experiences low mortality from fish predation compared to *D. rosea* (Leibold and Tessier, 1991, 1997).

Interestingly, fish are not the only important predators in these lakes; an invertebrate predator (*Chaoborus*) is also commonly found and forages upon *Daphnia*. Unlike fish, *Chaoborus* forages at night and moves throughout the entire water column. However, it also functions as a size-selective predator, but in contrast to fish prefers smaller-sized prey. In experimental studies, Gonzalez and Tessier (unpublished data) have shown that the smaller *D. rosea* is more vulnerable to *Chaoborus* compared to *D. pulicaria*. Overall, there is a dichotomy between these daphniids in predation risk due to the combination of body size and habitat-use behaviors. Small size and migratory behavior result in high predation rate on *D. rosea* by both *Chaoborus* and fish, whereas large size and deep-water habitat choice result in low predation rate on *D. pulicaria*.

A difference in vulnerability to the epilimnetic fish predator and overall differences in predation mortality are only half the hypothesized trade-off. The species were also hypothesized to differ in competitive ability with the larger *D. pulicaria*, the superior form. Within the context of the habitat model, this part of the hypothesis implies that *D. pulicaria* is restricted to the hypolimnion, primarily because it is more vulnerable to the epilimnetic fish predator. More specifically, competition from *D. rosea* in the epilimnion reduces the profitability of the epilimnion habitat for *D. pulicaria*, but in the absence of fish planktivory, *D. pulicaria* should expand its habitat use and displace *D. rosea* from both habitats. Tests of this component of the trade-off have not consistently supported it. In laboratory studies using a single food resource, *D. pulicaria* appears to perform better than *D. rosea*, although the magnitude of the difference between species varies with food level and temperature (Leibold, unpublished data). Field life-table studies (Threlkeld, 1979, 1980; Leibold, 1991) also indicated that temperature and food quality may interact to influence the relative competitive abilities of these two species. In an explicit test of the habitat partitioning model, Leibold (1988, 1991) observed that fish removal did result in some movement of *D. pulicaria* into the epilimnion and apparent displacement of *D. rosea*. However, a similar fish manipulation in another lake at a different time of year revealed no such habitat shift (Gonzalez and Tessier, unpublished data).

We are now aware of at least two complicating factors indicating that a simple trade-off between predator vulnerability and competitive ability is too simple a framework for explaining habitat partitioning in *Daphnia* species. First, there is substantial variation within species and populations in traits like body size (Leibold and Tessier, 1991; Tessier et al., 1992; Spitze, 1995), habitat-use behavior (De Meester, 1993; Leibold et al., 1994) and demographic traits that influence exploitative ability (Tessier and Leibold, unpublished data). It is not yet understood whether the genetic trade-offs (correlations) among these traits are the same within as between species. More importantly, variation in these traits among clones within species implies that habitat partitioning among species may reflect a balance between intra- and interspecific competition. The difference between two lakes in the habitat response of *D. pulicaria* to fish removal may reflect differences in the range of genetic variation within each population.

Second, an additional trade-off in adaptation to epilimnetic and hypolimnetic habitats appears to also be involved. Hu (1994) observed that the competitive superiority of *D. pulicaria* over *D. rosea* varied seasonally. Under cold conditions of spring and early summer, *D. pulicaria* was the superior form, but at the peak of summer stratification, *D. rosea* was better able to exploit the resource conditions of the epilimnion. At least two factors may underlie this shift in competitive ability: resource quality and temperature. It is likely that both are involved (Hu and Tessier, 1995), but recently Desmarais (1996) has demonstrated that differences in the quality of resources from high- and low-predation habitats is important in explaining reversal in the relative performance of different daphniid species.

Conclusions

The applicability of trophic cascades has been questioned by a number of critics. Though there is often experimental support for the notion (reviewed by Leibold, 1989), it is also clear that a number of processes (reviewed by Power, 1992) can damp responses by lower trophic levels to top predators (McQueen et al., 1986). An important next step in developing a better understanding of trophic cascades is to model these individual processes and work to evaluate them in natural communities. Here we have focused on habitat structure and its possible effects on grazer assemblages that can coexist via habitat partitioning. Our goal has been to present ways to link overall patterns involving trophic cascades (indirect effects of carnivorous predators on plants) to shifts in the species composition and habitat use patterns of intermediate grazers. Though we do not believe that effects mediated via habitat shifts are the only mechanism at work in our lakes, we have tried to present evidence that there are strong links between trophic cascades and patterns of habitat use and resource competition in deep lakes of North America.

Our empirical work illustrates both the strength and the shortcomings of the simple model developed above. First, experimental data show that grazer-resource dynamics on a relatively short time scale can be predicted relatively well by the model in a stratified lake (Leibold, 1991), even though there were effects of nutrient or predation treatments on phytoplankton composition (Leibold, 1989). Second, comparative analyses of lakes that differ in fish density, or size of the hypolimnetic refuge from fish predation generally show patterns of habitat partitioning and relative abundance of the two *Daphnia* species that are consistent with the model predictions (Leibold and Tessier, 1991; Tessier and Welser, 1991). Third, comparative analyses also show that nutrient levels generally have little effect on patterns of habitat partitioning unless they are associated with changes in fish predation as predicted by the model. However, nutrient enrichment typically increases epilimnetic phytoplankton levels, which is not predicted by the model. Our bioassay experiments with a standard clone of *D. pulex*, however, show that the effects of nutrients on epilimnetic phytoplankton do not enhance the overall resource quality as perceived by the grazers. More phytoplankton biomass does not necessarily indicate any change in resource conditions for *Daphnia*. Thus, heterogeneity of algal resources may interact with grazer habitat partitioning under predation risk and alter resource quality in ways not included in our model.

Variation in resource quality is potentially important because it suggest that *D. pulicaria* and *D. rosea* may interact via mechanisms related to different potential trade-offs. There is some evidence that the trade-off might involve reduced predation risk associated with traits that reduce exploitation ability (Leibold, 1991), but there is also evidence (Hu, 1994; Desmarais, 1996) that the trade-off might involve relative abilities to compete for high- and low-quality resources, and at high and low temperatures. An important question, then, is whether these different factors (type and intensity of predation, temperature, food quality and quantity) covary (in space and time) in relatively simple ways. For example, there is good evidence that resource quantity and quality is typically highest in the relatively cool water conditions of spring and fall lake turnover, when fish planktivory is also lowest (Sommer, 1989). Hence, it is not surprising to note that the consistent differences in vertical habitat use by *D. rosea* and *D. pulicaria* we described are associated with these species also being consistently different in seasonal population dynamics. *Daphnia pulicaria* is the fall-winter-spring form and *D. rosea* is typically the summer form. These taxa undoubtedly differ in multiple ways in their adaptation to predators, temperature, food, etc. But if these ecological factors can be collapsed (due to covariation) into a smaller number of environmental gradients that capture most of the seasonal and vertical habitat differences, then general models of habitat partitioning such as we present can provide a robust framework for understanding species coexistence.

Variation among lakes in the ecological conditions important in defining habitat structure and the interactions among *Daphnia* species has important potential consequences to their overall role in trophic cascades. For example, evidence from the work of Hu (1994) and Desmarais (1996) indicates that *D. rosea* is better able to utilize the low-quality phytoplankton taxa that predominate in the epilimnion of stratified lakes than is *D. pulicaria*. Furthermore, *D. rosea's* greater use of surface waters makes it more vulnerable to predation by fish than is *D. pulicaria*, which largely takes refuge in the hypolimnion. Hence, we would expect *D. rosea* to play a more effective role in translating variance in planktivory into variance in phytoplankton abundance (i.e., trophic cascade). However, in a shallow unstratified lake, or under conditions of cooler temperatures and higher quality phytoplankton resources, *D. pulicaria* might prove the more effective participant mediating trophic cascades.

Our work shows that understanding the habitat structure of lakes in terms of trade-offs in adaptation by zooplankton is useful in making predictions about the strength of trophic interactions in stratified lakes. The general model we present suggests ways that shifts in habitat use, mediated via predator-dependent resource competition among *Daphnia*, could alter their participation in trophic cascades. It seems likely that this habitat-perspective can also be useful in elucidating other aspects of community dynamics in lakes. Seasonal succession, for example, seems to occur in quite distinct ways in eutrophic lakes (where there is a seasonal reduction in the size of the hypolimnetic refuge due to anoxia) versus oligotrophic stratified lakes (where there is generally always a refuge present). The description of patterns and processes involved in plankton seasonal succession (Sommer et al., 1986; Sommer, 1989) are intimately related to shifting roles of resources and predators on zooplankton, especially *Daphnia*. Therefore, a better understanding of how habitat use alters the participation of *Daphnia* in trophic cascades can potentially link the among-lake patterns, such as those we discuss here, with seasonal and interannual dynamics within lakes.

Acknowledgments
We thank the editors for the invitation to write this paper and J. Brown, E. Werner, D. Hall, C. Lively, J. Jokela and an anonymous reviewer for comments. This research was supported by grants BSR-9007597, DEB-9421539 and DEB-9509004. We also thank N. Consolatti and P. Woodruff for ongoing technical support. This is K.B.S. contribution number 821.

References

Allan, J.D. (1977) An analysis of seasonal dynamics of a mixed population of *Daphnia*, and the associated cladoceran community. *Freshw. Biol.* 7:505–512.

Arditi, R. and Ginsberg, L.R. (1989) Coupling in predator-prey dynamics: Ratio dependence. *J. Theor. Biol.* 139:311–326.

Bengtsson, J. (1987) Competitive dominance among Cladocera: Are single-factor explanations enough? *Hydrobiologia* 145:19–28.

Brew, J. (1982) Niche shifts and the minimization of competition. *Theor. Popul. Biol.* 22:367–381.

Brooks, J.L. and Dodson, S.I. (1965) Predation, body size and the composition of plankton. *Science* 150:28–35.

Brown, J.H. (1995) *Macroecology.* University of Chicago Press, Chicago.

Brown, J.S. (1990) Habitat selection as an evolutionary game. *Evolution* 44:732–746.

Brown, J.S. (1992) Patch use under predation risk: I. Models and predictions. *Ann. Zool. Fennici* 29:301–309.

Brown, J.S. (1997) Game theory and habitat selection. *In*: L. Dugatkin and H.K. Reeve (eds): *Game Theory; in press.*

Brown, J.S. and Rosenzweig, M.L. (1986) Habitat selection in slowly regenerating environments. *J. Theor. Biol.* 123:151–171.

Burns, C.W. (1985) The effects of starvation on naupliar development and survivorship of three species of *Boeckella* (Copepoa: Calanoida). *Arch. Hydrobiol. Ergebnisse Limnol.* 21:297–309.

Carvalho, G.R. (1987) The clonal ecology of *Daphnia magna* (Crustacea: Cladocera). II. Thermal differentiation among seasonal clones. *J. Anim. Ecol.* 56:469–478.

Cipolla, M.J. (1980) Sexual dynamics of *Daphnia* populations in six Michigan lakes. *Dissertation,* Northwestern University, Evanston, IL, USA.

Clark, C.W. and Levy, D.A. (1988) Diel vertical migrations by juvenile sockeye salmon and the antipredation window. *Am. Nat.* 131:271–290.

Colbourne, J.K. and Hebert, P.D.N. (1996) The systematics of North American *Daphnia* (Crustacea: Anomopoda): A molecular phylogenetic approach. *Phil. Trans. Roy. Soc. Lond* B351:349–360.

Dawidowicz, P. and Loose, C.J. (1992) Cost of swimming by *Daphnia* during diel vertical migration. *Limnol. Oceanogr.* 37:665–669.

De Meester, L. (1989) An estimation of the heritability of phototaxis in *Daphnia magna* Straus. *Oecologia* 78:142–144.

De Meester, L. (1993) The vertical distribution of *Daphnia magna* genotypes selected for different behaviour: Outdoor experiments. *Arch. Hydrobiol. Ergebnisse Limnol.* 39: 137–155.

DeMelo, R., Francis, R. and McQueen, D.J. (1992) Biomass predation: Hit or myth? *Limnol. Oceanogr.* 37:192–207.

DeMott, W.R. (1989) The role of competition in zooplankton succession. *In*: U. Sommer (ed.): *Plankton Ecology: Succession in Plankton Communities.* Springer-Verlag, Berlin, pp 195– 252.

Desmarais, K. (1996) Performance trade-offs across a natural resource gradient. *M.Sc. Thesis*, Michigan State University, East Lansing, MI, USA.

De Stasio, B.T., Jr., Nibbelink, N. and Olsen, P. (1993) Diel vertical migration and global climate change: A dynamic modeling approach to zooplankton behavior. *Verh. Internat. Verein Limnol.*

Dini, M.L. and Carpenter, S.R. (1992) Fish predators, food availability and diel vertical migration in *Daphnia. J. Plankton Res.* 14:359–377.

Dodson, S.I. (1974) Zooplankton competition and predation: An experimental test of the size-efficiency hypothesis. *Ecology* 55:605–613.

Dodson, S.I. (1988) The ecological role of chemical stimuli for the zooplankton: Predator-avoidance behavior in *Daphnia. Limnol. Oceanogr.* 33:1431–1439.

Fast, A.W. (1971) Effects of artificial destratification on zooplankton depth distribution. *Trans. Amer. Fish. Soc.* 100:355–359.

Gabriel, W. and Thomas, B. (1988) Vertical migration of zooplankton as an evolutionary stable strategy. *Am. Nat.* 132:199–216.

Galbraith, M.G., Jr. (1967) Size-selective predation on *Daphnia* by rainbow trout and yellow perch. *Trans. Amer. Fish. Soc.* 96:1–10.

Geedey, C.K., Tessier, A.J. and Machledt, K. (1996) Habitat heterogeneity, environmental change, and the clonal structure of *Daphnia* populations. *Funct. Ecol.* 10:613–621.

Geller, W. (1986) Diurnal vertical migration of zooplankton in a temperate great lake (L. Constance): A starvation avoidance mechanism? *Arch. Hydrobiol., Suppl.* 74:1–60.

Gilliam, J.F. and Frazer, D.F. (1987) Habitat selection under predation hazard: A test of a model with stream-dwelling minnows. *Ecology* 68:1856–1862.

Gilliam, J.F. and Frazer, D.F. (1988) Resource depletion and habitat segregation by competitors under predation hazard. *In*: L. Persson and B. Ebenman (eds): *Size-Structured Populations: Ecology and Evolution.* Springer-Verlag, Berlin, pp 173–184.

Gliwicz, Z.M. (1986) Predation and the evolution of vertical migration in zooplankton. *Nature* 320:746–748.

Gliwicz, Z.M. and Pijanowska, J. (1988) Effect of predation and resource depth distribution on vertical migration of zooplankton. *Bull. Mar. Sci.* 43:695–709.

Gliwicz, Z.M. and Pijanowska, J. (1989) The role of predation in zooplankton succession. *In*: U. Sommer (ed.): *Plankton Ecology: Succession in Plankton Communities.* Springer-Verlag, Berlin, pp 253–298.

Hairston, N.G., Jr. (1976) Photoprotection by carotenoid pigments in the copepod *Diaptomus nevadensis. Proc. Natl. Acad. Sci. USA* 73:971–974.

Hairston, N.G., Jr. (1987) Diapause as a predator avoidance adaptation. *In*: W.C. Kerfoot and A. Sih (eds): *Predation: Direct and Indirect Impacts on Aquatic Communities.* University Press of New England, Hanover, NH, pp 281–290.

Hall, D.J. (1964) An experimental approach to the dynamics of a natural population of *Daphnia galeata mendotae. Ecology* 45:94–112.

Hall, D.J., Threlkeld, S.T., Burns, C.W. and Crowley, P.H. (1976) The size-efficiency hypothesis and the size structure of zooplankton communities. *Annu. Rev. Ecol. Syst.* 7:177–208.

Hall, D.J., Werner, E.E., Gilliam, J.F., Mittelbach, G.G., Howard, D., Doner, C.G., Dickerman, J.A. and Stewart, A.J. (1979) Diel foraging behavior and prey selection in the golden shiner (*Notomigonus crysoleucas*). *Can. J. Fish. Aquat. Sci.* 36:1029–1039.

Haney, J.F. (1993) Environmental control of diel vertical migration behaviour. *Arch. Hydrobiol. Ergebnisse Limnol.* 39:1–17.

Hrbácek, J., Dvoraková, M., Korínek, V. and Procházková, L. (1961) Demonstration of the effect of fish stock on the species composition of zooplankton and the intensity of metabolism of the whole plankton association. *Verh. Intern. Verein. theor. angew. Limnol.* 14:192–195.

Hu, S.S. (1994) Competition in a seasonal environment: *Daphnia* population dynamics and coexistence. *Dissertation*, Michigan State University, East Lansing, USA.

Hu, S.S. and Tessier, A.J. (1995) Seasonal succession and the strength of intra- and interspecific competition in a *Daphnia* assemblage. *Ecology* 76:2278–2294.

Hutchinson, G.E. (1951) Copepodology for the ornithologist. *Ecology* 32:571–577.

Hutchinson, G.E. (1967) *A Treatise on Limnology,* vol. II, *Introduction to Lake Biology and Limnoplankton.* Wiley and Sons, New York.

Iwasa, Y. (1982) Vertical migration of zooplankton: A game between predator and prey. *Am. Nat.* 120:171–180.

Jacobs, J. (1977) Coexistence of similar zooplankton species by differential adaptation to reproduction and escape in an environment with fluctuating food and enemy densities. II. Field analysis of *Daphnia. Oecologia* 30:313–329.

Jacobs, J. (1987) Cyclomorphosis in *Daphnia. Mem. Its. Ital. Idrobiol.* 45:325–352.

Juday, C. (1904) The diurnal movement of plankton Crustacea. *Trans. Wis. Acad. Sci., Arts and Let.* 14:534–568.

Kerfoot, W.C. (1985) Adaptive value of vertical migration: Comments on the predation hypothesis and some alternatives. *Contrib. Mar. Sci. Suppl.* 27:91–113.

Kitchell, J.A. and Kitchell, J.F. (1980) Size-selective predation, light transmission and oxygen stratification: Evidence from the recent sediments of manipulated lakes. *Limnol. Oceanogr.* 25:1137–1140.

Kratz, T., Frost, T. and Magnuson, J. (1987) Inferences from special and temporal variability in ecosystems: Long-term zooplankton data from lakes. *Am. Nat.* 129:830–846.

Lampert, W. (1987) Vertical migration of freshwater zooplankton: Indirect effects of vertebrate predators on algal communities. *In:* W.C. Kerfoot and A. Sih (eds): *Predation: Direct and Indirect Impacts on Aquatic Communities.* University Press of New England, Hanover, NH, pp 291–299.

Lampert, W. (1989) The adaptive significance of diel vertical migration by zooplankton. *Funct. Ecol.* 3:21–27.

Lampert, W. (1993) Ultimate causes of diel vertical migration of zooplankton: New evidence for the predator-avoidance hypothesis. *Arch. Hydrobiol. Ergebnisse Limnol.* 39:79–88.

Lane, P.A. (1975) The dynamics of aquatic systems: A comparative study of the structure of four zooplankton communities. *Ecol. Monogr.* 45:307–336.

Lane, P.A. (1979) Vertebrate and invertebrate predation intensity on freshwater zooplankton communities. *Nature* 280:391–393.

Leibold, M.A. (1988) Habitat structure and species interactions in plankton communities of stratified lakes. *Dissertation*, Michigan State University, East Lansing, MI, USA.

Leibold, M.A. (1989) Resource edibility and the effects of predators and productivity on the outcome of trophic interactions. *Am. Nat.* 134:922–949.

Leibold, M.A. (1990) Resources and predation can affect the vertical distribution of zooplankton. *Limnol. Oceanogr.* 35:938–944.

Leibold, M.A. (1991) Trophic interactions and habitat segregation between competing *Daphnia* species. *Oecologia* 86:510–520.

Leibold, M.A. (1996) A graphical model of keystone predators in food webs: Trophic regulation of abundance, incidence, and diversity patterns in communities. *Am. Nat.* 147:784–812.

Leibold, M.A. and Tessier, A.J. (1991) Contrasting patterns of body size for *Daphnia* species that segregate by habitat. *Oecologia* 86:342–348.

Leibold, M.A. and Tessier, A.J. (1997) Experimental compromise and mechanistic approaches to the evolutionary ecology of interacting *Daphnia* species. *In:* W.J. Resetarits and J. Bernardo (eds): *Issues and Perspectives in Experimental Ecology.* Oxford University Press, Oxford; *in press.*

Leibold, M.A., Tessier, A.J. and West, C.T. (1994) Genetic acclimation and ontogenetic effects on habitat selection behavior in *Daphnia pulicaria*. *Evolution* 48:1324−1332.

Lieder, U. (1986) *Bosmina* und *Daphnia*: Taxonomisch kritische Gruppen unter den Cladoceren (Crustacea, Phyllopoda). *Limnologia* 17:53−66.

Lively, C.M. (1986) Canalization versus developmental conversion in a spatially variable environment. *Am. Nat.* 128:561−572.

Loose, C.J. (1993) *Daphnia* diel vertical migration behavior: Response to vertebrate predator abundance. *Arch. Hydrobiol. Ergebnisse Limnol.* 39:29−36.

Lozeron, H. (1902) Sur la répartition verticale du plankton dans le lac de Zürich, de décembre 1900 à décembre 1901. *Vjschr. Naturf. Ges. Zürich* 47:115−198.

MacArthur, R.H. (1972) *Geographical Ecology: Patterns in the Distribution of Species*. Harper and Row, New York.

Makarewicz, J.C. and Likens, G.E. (1975) Niche analysis of a zooplankton community. *Science* 190:1000−1002.

McQueen, D.J., Post, J.R. and Mills, E.L. (1986) Trophic relationships in freshwater pelagic ecosystems. *Can. J. Fish. Aquat. Sci.* 43:1571−1581.

Mills, E.L. and Forney, J.L. (1983) Impact on *Daphnia pulex* of predation by young yellow perch in Oneida Lake, New York. *Trans. Amer. Fish. Soc.* 112:154−161.

Mittelbach, G.G., Osenberg, C. and Leibold, M.A. (1988) Trophic relations and ontogenetic niche shifts in aquatic ecosystems. *In:* L. Persson and B. Ebenman (eds): *Size-Structured Populations: Ecology and Evolution*. Springer-Verlag, Berlin, pp 219−235.

Neill, W.E. (1990) Induced vertical migration in copepods as a defence against invertebrate predation. *Nature* 345:524−526.

Neill, W.E. (1994) Spatial and temporal scaling and the organization of limnetic communities. *In.:* P.S. Giller, A.G. Hildrew and D.G. Raffaelli (eds): *Aquatic Ecology: Scale, Pattern and Process*. Blackwell Scientific, Oxford, pp 189−232.

Paffenhöfer, G.-A. and Price, H.J. (eds) (1988) Zooplankton behavior symposium. *Bull. Mar. Sci.* 43:325−872.

Phillips, O. (1974) The equilibrium and stability of simple marine systems. II. Herbivores. *Arch. Hydrobiol.* 73:301−333.

Pimm, S. and Rosenzweig, M.L. (1981) Competitors and habitat use. *Oikos* 37:1−6.

Pimm, S., Rosenzweig, M.L. and Mitchell, W. (1985) Competition for food selection: Field tests of a theory. *Ecology* 66:798−807.

Power, M. (1992) Habitat heterogeneity and the functional significance of fish in river food webs. *Ecology* 73:733−746.

Ringelberg, J. (ed.) (1993) Diel vertical migration of zooplankton. *Arch. Hydrobiol. Ergebnisse Limnol.* 39:1−222.

Rosenzweig, M.L. (1973) Exploitation in three trophic levels. *Am. Nat.* 107:275−294.

Rosenzweig, M.L. (1979) Optimal habitat use in two species competitive systems. *Fortschr. Zool.* 25:283−293.

Rosenzweig, M.L. (1981) A theory of habitat selection. *Ecology* 62:327−335.

Rosenzweig, M.L. (1991) Habitat selection and population interactions: The search for mechanism. *Am. Nat.* 137:S5−S28.

Rosenzweig, M.L. and Abramsky, Z. (1986) Centrifugal community organization. *Oikos* 46:339−348.

Rosenzweig, M.L. and MacArthur, R.H. (1963) Graphical representation and stability conditions of predator-prey interactions. *Am. Nat.* 97:209−223.

Rothhaupt, K.O. (1990) Resource competition of herbivorous zooplankton: A review of approaches and perspectives. *Arch. Hydrobiol.* 118:1−29.

Schädel, A (1916) Produzenten und Konsumenten im Teichplankton. *Arch. für Hydrobiol. und Planktonkunde* 11:404.

Seitz, A. (1980) The coexistence of three species of *Daphnia* in the Klostersee. I. Field studies on the dynamics of reproduction. *Oecologia* 45:117−130.

Sommer, U. (ed.) (1989) *Plankton Ecology: Succession in Plankton Communities*. Springer-Verlag, Berlin.

Sommer, U., Gliwicz, Z.M., Lampert, W. and Duncan, A. (1986) The PEG-model of seasonal succession of planktonic events in fresh waters. *Arch. Hydrobiol.* 106:433−471.

Spitze, K. (1995) Quantitative genetics of zooplankton life histories. *Experiential* 51:454−464.

Stemberger, R. (1995) Pleistocene refuge areas and postglacial dispersal of copepods of the northeastern United States. *Can. J. Fish. Aquat. Sci.* 52:2197—2210.

Sterner, R.W. (1989) The role of grazers in phytoplankton succession. *In:* U. Sommer (ed.): *Plankton Ecology: Succession in Plankton Communities.* Springer-Verlag, Berlin, pp 107—170.

Tappa, D.W. (1965) The dynamics of the association of six limnetic species of *Daphnia* in Aziscoos Lake, Maine. *Ecol. Monogr.* 35:395—423.

Tessier, A.J. (1986) Comparative population regulation of two planktonic cladocera *(Holopedium gibberum* and *Daphnia catawba). Ecology* 67:285—302.

Tessier, A.J. and Goulden, C.E. (1987) Cladoceran juvenile growth. *Limnol. Oceanogr.* 32:680—686.

Tessier, A.J. and Leibold, M.A. (1997) Habitat use and ecological specialization within lake *Daphnia* populations. *Oecologia* 104:561—570.

Tessier, A.J. and Welser, J. (1991) Cladoceran assemblages, seasonal succession and the importance of a hypolimnetic refuge. *Freshw. Biol.* 25:85—93.

Tessier, A.J., Young, A. and Leibold, M.A. (1992) Population dynamics and body size selection in *Daphnia. Limnol. Oceanogr.* 37:1—13.

Threlkeld, S.T. (1979) The midsummer dynamics of two *Daphnia* species in Wintergreen Lake, Michigan. *Ecology* 60:165—179.

Threlkeld, S.T. (1980) Habitat selection and population growth of two cladocerans in seasonal environments. *In:* W.C. Kerfoot (ed.): *Evolution and Ecology of Zooplankton Communities.* University Press of New England, Hanover, NH, pp 346—357.

Tilman, D. (1982) *Resource Competition and Community Structure.* Princeton University Press, Princeton, NJ, USA.

Vanni, M.J. (1987) Food availability, fish predation and the dynamics of a zooplankton community coexisting with planktivorous fish. *Ecol. Monogr.* 57:61—88.

Weismann, A. (1877) Das Tierleben im Bodensee. *Schr. Gesch. Bodensees Umgebung* 7:1—31.

Wesenberg-Lund, C. (1926) Contributions to the biology and morphology of the genus *Daphnia. K. danske Vidensk. Selsk. Skr., Naturw. Math. Afd. (ser. 8)* 11:9—250.

Wolf, H.G. (1987) Interspecific hybridization between *Daphnia hyalina, D. galeata,* and *D. cucullata* and seasonal abundances of these species and their hybrids. *Hydrobiologia* 145:213—217.

Woltereck, R. (1920) Beiträge zur Frage nach dem Einfluß von Temperatur und Ernährung auf die quantitative Entwicklung von Süßwasserorganismen. *Zoolog. Jahrb.* 38: p. 1 Abt. f. allg. Zool. u. Physiol.

Woltereck, R. (1930) Alte und neue Beobachtungen über die geographische und die zonare Verteilung der helmlosen und helmtragenden Biotypen von *Daphnia. Internat. Rev. ges. Hydrobiol. u. Hydrog.* 24:358—380.

Woltereck, R. (1932) Races, associations and stratification of pelagic daphniids in some lakes of Wisconsin and other regions of the United States and Canada. *Trans. Wis. Acad. Sci., Arts and Let.* 27:487—522.

Wright, D. and Shapiro, J. (1990) Refuge availability: A key to understanding the summer disappearance of *Daphnia. Freshw. Biol.* 24:43—62.

Zaret, T.M. (1980) *Predation in Freshwater Communities.* Yale University Press, New Haven.

Zaret, T.M. and Suffern, J.S. (1976) Vertical migration in zooplankton as a predator avoidance mechanism. *Limnol. Oceanogr.* 21:804—813.

Evolutionary Ecology of Freshwater Animals
ed. by B. Streit, T. Städler and C. M. Lively
© 1997 Birkhäuser Verlag Basel/Switzerland

New Zealand mountain stream communities: Stable yet disturbed?

M. J. Winterbourn

Department of Zoology, University of Canterbury, Private Bag 4800, Christchurch, New Zealand

Summary. In this chapter, the nature of stream invertebrate communities in the South Island, New Zealand is considered in relation to concepts of disturbance and stability. Disturbance is widely regarded as a primary organizing factor in physically "harsh" stream environments, but the definition of what constitutes a disturbance is not always clear. Both environmental and faunal stability can also be difficult to define and measure, and are influenced by the spatial and temporal scales of interest, and historical factors. To illustrate the relationships between stability of stream faunas and the role of disturbance as a regulatory force, four case studies are presented. Two studies consider the persistence and resilience of invertebrate faunas in streams of contrasting and similar physical stability and flow variability. The other two studies examine faunal responses to catchment-scale disturbances (logging and the development of land for forestry) and the temporal stability of stream faunas following afforestation with exotic conifers. These and other studies indicate that the faunas of many South Island mountain streams are dominated by the same widely distributed species characterized by life history flexibility, lack of habitat specificity, and strong colonizing abilities. They also tend to exhibit a high degree of stability in species composition over time despite many streams having physically unstable beds, variable and unpredictable discharge patterns, and changing vegetational settings. It is hypothesized that the resilience of New Zealand mountain stream faunas, and the lack of strong habitat specialization of many species, can be explained in terms of their evolutionary history in changeable environments.

Introduction

Many rivers and streams appear to be physically harsh environments. Animals risk being swept away by the current, floods scour and redeposit bed materials, and when discharge is low or intermittent aquatic organisms can either be under severe physiological stress or risk desiccation (Allan, 1995). Running waters are also changeable environments, often strongly influenced by external climatic factors. Flow patterns can vary markedly over periods of hours, days, or seasons, and within a single stream flow variability can also be great on both small and large spatial scales.

Stream communities are structured by a combination of biotic and abiotic factors. Some will be strongly influenced by competitive interactions or predation, whereas others will be more strongly affected by physico-chemical aspects of the environment (Allan, 1995). Because of their variability and physical instability, abiotic factors are likely to have a major influence on community composition in many mountain streams. Poff and Ward (1989) for example, predicted that communities of streams

characterized as perennially flashy (spate-prone) with low flow predictability would be regulated primarily in this way. Similar views have been
expressed by New Zealand stream ecologists (e.g., Winterbourn et al.,
1981; Quinn and Hickey, 1990; Scarsbrook and Townsend, 1993; Biggs,
1995) who have studied streams of this kind.

A significant contribution to the debate concerning the relative importance of abiotic and biotic factors in stream ecology was made by
Peckarsky (1983). She presented a model of "lotic community structure"
based on the marine intertidal model of Menge (1976) and perceived that
running water environments could be placed on a gradient ranging from
physically benign to harsh conditions. At the outset, Peckarsky acknowl-
edged that one cannot readily identify conditions that are unequivocally
harsh or benign to stream invertebrates, but for comparative purposes she
proposed that harsh means "physical-chemical conditions that impose
physiological problems for many stream invertebrates". Thus, natural
harshness may be imposed by seasonal or diel fluctuations that are
unpredictable, whereas a benign stream would be one that could "potentially support a highly productive consumer community". This does not mean
that benign streams lack disturbances or other environmental fluctuations,
but that they are not of such intensity, frequency or predictability as to
be the primary regulators of invertebrate distribution, abundance and
diversity. In contrast, Peckarsky (1983) hypothesised that predation should
be the most important regulator of community structure in benign stream
environments (as proposed by Menge for intertidal habitats), whereas at
intermediate levels of environmental harshness or disturbance, competition
should have a dominant role.

Streams in general have been described as disturbed and patchy environments (e.g., Townsend, 1989) since stream beds are rarely immobile and
discharge typically varies. However, only in recent years has disturbance
been explicitly and widely regarded as having a significant role in the
regulation of stream community composition, structure and production
(e.g., Resh et al., 1988; Allan, 1995).

But what exactly is disturbance? One definition that is widely quoted,
and seems to be widely accepted, is that of White and Pickett (1985), "a
disturbance is any relatively discrete event in time that disrupts ecosystem,
community, or population structure, and changes resources, substrate availability or the physical environment". This definition incorporates environmental fluctuations and destructive events whether or not they are seen as
"normal" for a particular system, and as such it differs from some other
definitions proposed by stream ecologists. For instance, Resh et al. (1988)
considered that predictability needed to be considered in a definition of
disturbance, and therefore proposed that only those events with a frequency
and predictability outside an ill-defined but predictable range should be
included. Subsequently, Poff (1992) argued that disturbances are by definition ecological events, but if their temporal distribution is "predictable

enough" then ecological responses may be small because organisms will have had the opportunity to adjust to them through natural selection. This is an important idea emphasized also by Cobb et al. (1992a, b), and one to which I will return in discussing the nature of New Zealand stream faunas.

From an ecological standpoint the White and Pickett (1985) definition is an appropriate one since disturbances result in habitat disruption, loss of resources, and a reduction in the sizes of populations. However, the idea that a disturbance should fall outside a predictable range of variation also has merit, and "unpredictable" disturbances typically have a much greater impact on running water ecosystems than predictable ones of smaller magnitude (Poff, 1992). Gore et al. (1990) considered both views of disturbance were valid and could be seen as applying at different spatial or temporal scales. They considered that the definition of disturbance as an unusual or unpredictable event was the more general one, as it implied recovery on the part of the community or ecosystem of interest. In contrast, although small (and predictable) disturbances may result in the removal of individuals and changes in resource distribution, they can be viewed as being an aspect of a system's normal behaviour. To my mind this does not mean that recovery is not needed, but simply that recovery is operating on a much smaller (spatial or temporal) scale, and for the invertebrate fauna may largely involve redistribution of individuals.

In practice most authors fail to define disturbance, although it is usually clear what scale of event is meant. For example, the eruption of Mount St. Helens was by any standards a megadisturbance (Anderson, 1992) on a grand scale, and ecosystem recovery clearly is required if hundreds of metres of stream bed are denuded by a debris torrent. On the other hand, variations in discharge culminating in floods and droughts are commonly described as disturbances since they change bed conditions, result in the loss or removal of resources (e.g., food materials), and "demand" a faunal response.

In addition to natural climatic, hydrological, and geological events, disturbances imposed on streams and their biotas are increasingly consequences of human activities. For example, agriculture, forestry, mining and urbanization all result in changes to the physical and biological environment, including resource and substrate availability consistent with the White and Pickett definition of disturbance, and are evident in New Zealand at a wide range of spatial and temporal scales (e.g., Fahey and Rowe, 1992; Winterbourn and Ryan, 1994). Two case studies concerning catchment-level disturbances induced by logging and afforestation are incorporated in this chapter.

Some comment is also pertinent here about the term stability, which like disturbance has been used in variable and vague ways to describe changeability within both biota and the environment. In fact, the concepts of stability and disturbance are inextricably interwoven, and Meffe and

Minckley (1987) stated that stability refers to the propensity for the physical environment to be disturbed or altered. Death and Winterbourn (1994) considered that stability should be defined and measured so that it reflects the disturbance regimes of as many variables as possible, and noted that they may or may not be intercorrelated. To this end they developed a multivariate procedure to provide a stability score for stream sites based on six independently measured physical criteria. The channel stability index of Pfankuch (1975) is also a composite measure that requires the assessment of 15 visual features of the stream environment. Death and Winterbourn (1994) found that the Pfankuch index was strongly correlated with their multivariate stability score, and in two New Zealand studies the stream bed component of the index was correlated with species richness of the benthos (Winterbourn and Collier, 1987; Death and Winterbourn, 1995).

What criteria must be met if a biological assemblage is to be considered stable? Holling (1973) noted that assemblage stability has been interpreted either as persistence of assemblage structure, or as degree of constancy in numbers of organisms over time. Subsequently, Connell and Sousa (1983) used the term persistence to mean continued presence, particularly in the face of potentially destructive forces, and stability to denote relative constancy of numbers over time despite the recurrence of disturbances. Perceptions of stability in community structure are influenced by the temporal scale involved and definitions of relative constancy. Depending on the time scale of an investigation, assemblage stability may be a consequence of its persistence or its resilience, that is the ability of the assemblage of interest to recover rather than remain unchanged. Many stream communities are highly resilient (e.g., in recovering from flood events; Scrimgeour et al., 1988; Grimm and Fisher, 1989) and therefore can be described as stable on a time scale of months or years.

New Zealand streams have been described as physically dominated systems in which biological interactions take a secondary role to climatic and other physico-chemical factors (Winterbourn et al., 1981; Cowie 1985; Winterbourn, 1995). High but temporally variable rainfall, especially in the South Island mountains, results in fluctuating stream flows, high sediment loads and often physically unstable channels. Not surprisingly, perhaps, many mountain and forest steams are unproductive environments (Graesser, 1988; Winterbourn and Ryan, 1995) whose benthic faunas are dominated by food-generalists dependent largely on fine detritus and stone surface biofilms for food. Large particle detritivores (or shredders) are rare or absent from most upland South Island streams, a condition that has been linked with the poor retention characteristics of the streams (Linklater, 1995; Winterbourn, 1995). The faunas of these physically harsh and changeable lotic environments consist mainly of aquatic insects (predominantly Ephemeroptera, Diptera and Plecoptera) and include a group of widely distributed habitat generalists, many with flexible, poorly synchronized life histories (Winterbourn et al., 1981).

In this chapter I outline the results of several investigations of South Island stream invertebrate communities and discuss them in relation to disturbance and its role as a regulating force in running waters. The individual case studies address related issues at contrasting temporal and spatial scales and involve both natural and anthropogenically imposed disturbances.

Two "case-studies" consider stream communities in the Cass Basin and environs on the eastern side of the Southern Alps where numerous investigations have been undertaken in the last two decades (see Winterbourn, 1987; 1995). The first of these studies considers the effects of environmental stability on persistence and resilience of benthic invertebrate assemblages at 11 physically contrasting sites. Three Cass Basin streams are the focus of the second study which examines assemblage stability over time (5 years) in a climatically variable environment.

The other two case studies deal with catchment-scale disturbances on the western side of the Southern Alps, and consider the impact of changes consequent upon the felling of native forest and its replacement with exotic plantation forest. Both studies take a comparative approach. However, one considers a large number (26) of streams differing in time since the planting of exotic trees commenced, whereas the other discusses faunal recovery and stability in four differently managed catchments over a period of 14 years.

South Island mountain stream environments

New Zealand lies in the southwest Pacific between latitudes 35 °S and 47 °S. Most of its land area is concentrated in the North and South Islands, and the much smaller and most southerly Stewart Island. Nearly half of New Zealand consists of mountains or hills above 1200–1500 m, and about 50% of the land is steep and dissected by an extensive network of streams and rivers. Pleistocene glaciations and regular seismic events have had significant effects on the South Island landscape, and the roughly north-south orientation of the Southern Alps, which reach a maximum height of 3753 m, has a major impact on the climate. The region west of the Alps (Westland) where two of the studies to be discussed were undertaken is fully exposed to the prevailing westerly winds and is the wettest region in the country. Over 150 raindays occur annually and annual precipitation can exceed 10000 mm in places (Griffiths and McSaveney, 1983). More importantly perhaps from the perspective of stream ecology, the pattern of rainfall is unpredictable and both storm events and dry spells can (and do) occur at any time of year. The most significant feature of the rainfall is its variability. A consequence of this, the steepness of the mountains, and the rapid runoff from many catchments is that flood flows can be high, frequent and flashy. Furthermore, sediment yields in Westland rivers are amongst the highest in the world (Griffiths, 1979). On the eastern side of the Southern Alps where the other study sites were located, annual rainfall is lower, and flow variation tends to be less extreme. However, as in the west rainfall can be heavy in any month and unpredictable from year to year.

A further characteristic feature of stream waters in the South Island mountains is their low ionic content and overall high water quality (Mosley and Rowe, 1981; Winterbourn and Ryan, 1995). They are unaffected by acid precipitation and most can be described as calcium-sodium-bicarbonate waters (Close and Davies-Colley, 1990). A distinctive sub-group of streams prevalent on the West Coast have brown waters and low pH. They mostly arise on poorly drained terraces with highly leached acid soils, and their colour and acidity reflects high concentrations of dissolved organic acids derived from slowly decomposing vegetation (Collier, 1987). Many of the catchments of brown water streams are being developed for plantation forestry, an anthropogenic disturbance whose consequences for stream invertebrate faunas are discussed and evaluated below.

The streams I will consider east of the Southern Alps are mainly within the Cass Basin at altitudes of 580–790 m a. s. l., and form part of the large Waimakariri River system. They drain a glaciated landscape vegetated by tussock grassland and scrub, some improved pasture, and fragments of formerly extensive beech forest (*Nothofagus solandri* var. *cliffortioides*). Rainfall declines steeply from west to east across this region (average about 1300 mm y^{-1} at Cass) and its temporal distribution varies considerably from year to year (Greenland, 1977). The study streams west of the Main Divide are all close to the town of Reefton and form part of the extensive Grey and Buller River drainages of North Westland. Rainfall in the Maimai Experimental Area where one group of streams is located averages about 2400 mm y^{-1} (Moore, 1989). The West Coast streams respond rapidly to rainfall events and drain either native forest catchments or valleys that have been developed to support exotic plantation forestry at various times during the last 20 years.

The physical and climatic settings of South Island mountain streams mean that many of them have unstable channels and variable discharge regimes incorporating periods of both regular and rapidly changing flow. If physical disturbances such as spates and floods have overriding influences on community structure in streams and rivers as some authors suggest (e.g., Lake and Barmuta, 1986; Resh et al., 1988), then the South Island of New Zealand should be an outstanding place to examine this contention.

Case studies

Case study 1. Benthic community structure and environmental stability: Streams at Cass

A small number of investigations undertaken in South Island rivers have indicated that flow-mediated disturbances reduce benthic invertebrate density and diversity but that both recover quickly (Sagar, 1986; Scrimgeour et al., 1988; Scrimgeour and Winterbourn, 1989). In other words, the faunas of these rivers are resilient and some taxa may be resistant. In order to explore these relationships further, Death and Winterbourn (1995) examined invertebrate species richness and evenness in 10 streams and on the stony shore of Lake Grasmere where water chemistry was similar but flow patterns and channel stability differed. Stability was assessed with the multivariate index of Death and Winterbourn (1994) that incorporates a suite of hydrological and thermal variables, most of which were measured monthly at each site. In a parallel study, Death (1995) asked whether spatial patterns in benthic community structure were also habitat specific, rather than simply a product of stream stability.

The 10 first-to-third order streams and the wave-beaten lake shore were located near Cass on the eastern side of the Southern Alps. All sites were within an 18 km radius of each other, were exposed to similar climate, including rainfall, had moderately hard water, circum-neutral pH and low concentrations of plant nutrients. The streams all drained catchments supporting tussock grassland vegetation and/or stands of southern beech forest (*Nothofagus solandri* var. *cliffortioides*). The sites differed in their flow patterns (variability and proneness to flooding) as some originated from springs whereas others were fed mainly by runoff.

The invertebrate faunas of riffles were sampled in five consecutive seasons by individual stone sampling (15 stones in three size classes per site; Death, 1995) and abundances were expressed as numbers per unit surface area of stone as described by Wrona et al. (1986). A total of 185 species (42–92 per site) was obtained from the 11 sites, the number of species per site declining in a linear fashion as environmental stability decreased (Fig. 1). Interestingly, the numbers of species present at the most unstable sites were largely unaffected by disturbance events including major floods that resulted in large-scale scouring and deposition of bed materials.

In contrast to the highly disturbed sites, those of intermediate stability demonstrated a distinct reduction in species numbers in response to increasing disturbance frequency, but not necessarily disturbance intensity. Thus, increasing disturbance frequency seems to act on species richness by reducing the time available for recolonization following disturbance events.

The most stable streams not only had the greatest species richness but their faunal composition was clearly site-specific (Death, 1995). Unlike the less stable sites where leptophlebiid mayflies (*Deleatidium* spp.) and small chironomids dominated numerically, a gastropod *Potamopyrgus antipodarum* and a large chironomid *Maoridiamesa harrisi* were two of the commonest species in stable, spring-fed streams. The densities of these two algal grazers were strongly correlated with epilithic algal biomass (as chlorophyll a), i.e., food supply, which was less subject to physical removal where disturbance frequency and intensity were low.

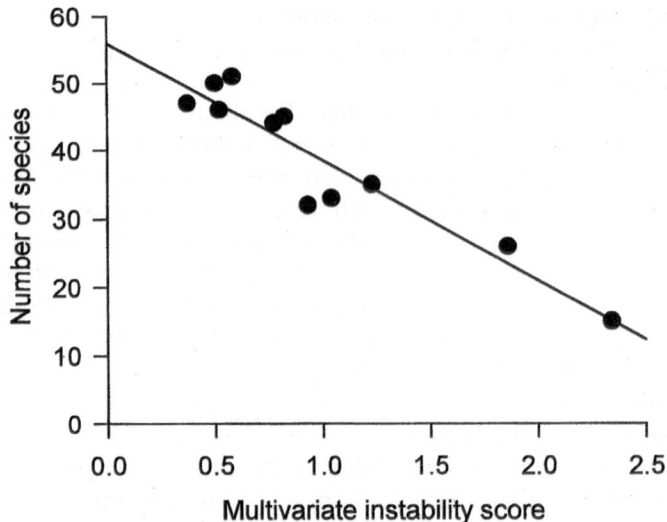

Figure 1. Relationship between the mean number of species collected at each of eleven Cass sites in five seasons and stream stability (the multivariate instability score of Death and Winterbourn, 1994). $r^2 = 0.78$. (after Death and Winterbourn, 1995).

The numerical dominance of *Deleatidium* in all but the most stable streams reflects the strong colonizing characteristics, behaviour, and generalised feeding and habitat requirements of the larvae. One feature that must enhance their success in flood-disturbed environments was demonstrated by Graesser (1987) who found that densities of drifting larvae, measured as numbers per volume of water filtered, declined when stream discharge increased. From this she inferred that the larvae respond to increases in flow by exhibiting refuge-seeking behaviours, either amongst larger, more stable substrata, or deeper within the stream bed.

In summary, comparative studies of riffle faunas of Cass Basin streams indicate that both species composition and abundances of taxa are related to stream stability including disturbance frequency, and to the intrinsic nature of the stream. More specifically, the data show that species richness was positively correlated with stability, and that faunal similarity was greatest between the most unstable streams. In contrast, site-specific factors have a much stronger influence on the nature of stream assemblages where disturbance frequency and intensity is low. These factors may include competitive displacement of species at very stable sites (Death and Winterbourn, 1995), and predation effects. Thus, McIntosh and Townsend (1994) demonstrated that predatory fish could have substantial impacts on the movement of invertebrates among patches in New Zealand streams, and Flecker and Townsend (1994) reported that trout predation had cascading effects on invertebrates and algae in experimental channels in a South Island mountain stream.

Case study 2. Forest stream faunas at Cass: Temporal variability

The studies of Death (1995) and Death and Winterbourn (1995) discussed above compared the benthic invertebrate communities of contrasting streams in a single year, and the only temporal comparisons were among successive seasons (not considered here). Although strong relationships were found between species richness, species composition and habitat stability, the question remains as to whether these relationships persist over longer periods of time. This question was addressed by sampling the riffle communities of three small beech forest streams at Cass in 5 consecutive years. Middle Bush Stream and Sugarloaf Stream drained adjacent catchments, whereas Craigieburn Cutting Stream was 12 km to the south.

Sampling of the benthic invertebrate fauna was undertaken in late March- early April (autumn) in each of the 5 years (1992–1996). Six quantitative samples were taken with a Surber sampler (0.1 m², 0.5 mm mesh) from riffles within the same reach on each occasion. Sampling followed the peak emergence and flight periods of some of the more abundant insects in these streams (Winterbourn, 1978) so that larvae hatching from eggs laid in the preceding summer months could be expected to represent a significant component of the faunas.

The results of this field study are of interest in several respects. First, despite the close proximity of the three streams and their physico-chemical similarity (Tab. 1), their faunas differed somewhat in species composition.

Table 1. Physico-chemical and faunal features of the three forest streams at Cass sampled in the autumn of 5 successive years, 1992–1996. Values for the physico-chemical factors are 5-year means. Sorensen's scores and values of Kendal's W are for within-stream comparisons over the 5 years. CV = coefficient of variation (%)

	Middle Bush	Sugarloaf Bush	Craigieburn Cutting
Discharge ($1s^{-1}$)	0.9	1.5	2.6
Conductivity ($\mu S_{25} cm^{-1}$)	104	95	56
Alkalinity (mg · l^{-1} CaCO₃)	44	42	24
Taxa collected each year (Mean and CV)	22 (9)	20 (24)	15 (17)
Total taxa per stream	35	35	30
Total invertebrates (m²) (Mean and CV)	676 (57)	796 (45)	700 (47)
Sorensen's score (Mean and CV)	0.73 (9)	0.65 (11)	0.56 (23)
Kendall's W (a) Top 10 taxa (b) Top 6 taxa	0.63 0.54	0.51 0.50	0.49 0.42
Top 3 taxa	*Deleatidium* Hydraenidae *Neppia*	*Deleatidium* Elmidae *Neppia*	*Deleatidium* *Neppia* Elmidae

Nevertheless, a species of *Deleatidium* (Ephemeroptera) was always the numerically dominant taxon, and subdominants in the streams were a flatworm *Neppia montana* and beetles of the families Hydraenidae and Elmidae (Tab. 1). Sorensen's similarity coefficients calculated from the full lists of species collected in the 4 years indicated that the two streams in closest proximity had the most similar faunas (Sorensen's score 0.80) and that average inter-year similarity exceeded 55% in all three streams (Tab. 1).

Second, despite differences in annual flow patterns (inferred from climate data recorded at the University of Canterbury biological station, Cass, where annual rainfall in the 5 years ranged from 1223–1708 mm, and in the month prior to sampling totalled 35–147 mm), all three streams exhibited high levels of faunal stability as indicated by Kendall's coefficient of concordance (W), a multiple correlation coefficient used to compare the rank order of abundance of the 10 most common taxa in each stream. This lack of strong between-year variation in the relative abundances of common taxa was despite the up to sixfold differences in absolute abundance of invertebrates over the 5 years (coefficients of variation 45–57%; Tab. 1). Interestingly, the data indicate that high- and low-abundance years did not coincide in the three streams (Kendall's W = 0.38), despite their presumably having similar discharge patterns and therefore comparable disturbance regimes. This suggests that factors other than physical disturbance must have an important role in regulating population size.

It is of interest to compare these findings with those of Richards and Minshall (1992) who examined the stability of species assemblages in Idaho (USA) headwater streams over a 5-year period. They sampled benthic faunas of five streams in catchments unaffected by wildfire (and five disturbed by fire but not considered here) once a year in late-summer (as at Cass) and evaluated the stability of species assemblages with Kendall's W as in the New Zealand study.

Richards and Minshall (1992) found that faunal persistence and the relative abundances of common species were remarkably stable over time. Thus, W calculated from the top 10 taxa (excluding Chironomidae and Oligochaeta as at Cass) ranged from 0.40–0.50 for their five reference streams, cf. 0.49–0.63 at Cass. They suggested that faunal stability may have been attributable to the relative predictability of the environment, for example the timing and magnitude of flow events, and inferred that a less stable community might be expected where flow and other physical events are less predictable – as for example at Cass. This is not borne out, however, by the New Zealand findings which indicate that the maintenance of stable/persistent stream faunal assemblages over time is not solely a function of environmental predictability.

One of the most characteristic features of running water ecosystems is that they are open systems (Fisher and Likens, 1973) depending to variable extents on external carbon sources at least as an energy supplement, and on the continual colonization of biota, most obviously of insects with

terrestrial adult (reproductive) stages. The importance of the latter as determinants and regulators of community structure demands greater attention
than it has received in the past and is an area in which numerous potentially
significant questions can be asked. For example: What factors stimulate the
females of particular species to oviposit in particular streams? Do adult
insects characteristically return to their larval stream to oviposit? Is larval
population size predominantly a function of the terrestrial reproductive
phase, and the instream phase merely "fine tuning")? How much oviposition (or how many ovipositing females) is necessary to maintain stream
insect populations at the levels characteristically seen? With respect to the
last of these questions, calculations made by Bunn (1995) for several
species of insect in an Australian stream indicate that the number of return-
ing adults needed may be very few.

To complicate the issue further, not all common stream invertebrates
have terrestrial adult stages, and of the nine most abundant taxa in the
three Cass streams three do not fall into that category. Thus, the flatworm
Neppia montana is fully aquatic throughout its life, an hydraenid beetle has
terrestrial larvae and pupae and an aquatic adult, whereas an elmid beetle
has aquatic larvae and adults and presumably a terrestrial pupa.

Regardless of the mechanisms that determine and regulate stream communities, the finding of a high level of stability in assemblage structure
under such diverse environmental conditions as those found at Cass and in
Idaho (Richards and Minshall, 1992) provides a very positive message with
regard to the representativeness of "one-off" (limited time) surveys as
reliable, medium-term indicators of stream fauna composition.

Case study 3. Catchment level disturbance: Forestry and stream communities

In many parts of New Zealand, native forest vegetation has been felled and
replaced with exotic forestry species (mainly *Pinus radiata*), or plantations
have been established on previously non-forested land. Such major changes
to catchment vegetation represent large scale ecological disturbances that
might be expected to have significant effects on running water ecosystems.
Thus, logging and afforestation affect a host of factors including runoff
characteristics of catchments, stream water temperature, riparian shading,
bank protection, instream cover, and the supply of dissolved and particulate
allochthonous matter that provides the major energy resource utilized by
forest stream communities (Campbell and Doeg, 1989).

In North Westland, fluvio-glacial terrace systems derived from deposits
left by late-Pleistocene glaciations are now used extensively as sites for
exotic plantation forestry. In high rainfall areas (>2200 mm y^{-1}), the soils
on many terraces are naturally infertile, extremely acid (pH may be <3.5),
and both vertical and lateral water movement through the soil is very slow.

This "swampy, acidic, barren type of land" (Hulme, 1984) is known locally as pakihi, and typically supports non-forest vegetation comprising manuka scrub (*Leptospermum scoparium*), rushes and ferns including bracken (*Pteridium esculentum*). Pakihi are prepared for sustainable exotic forestry by digging channels (V-blading) to improve drainage, and by the application of fertilizers (especially those containing phosphate), activities that can be expected to influence the hydrology and water chemistry of streams draining the affected catchments.

The effects of afforestation on the invertebrate faunas of pakihi streams was investigated in 1987 by Collier et al. (1989). Faunal and physico-chemical surveys were undertaken at 26 stream sites on two drainage networks, some in undisturbed native forest and others where exotic pine plantations had been established for 1–10 years at the time of the survey. Riparian vegetation at different sites consisted of indigenous forest, *Pinus radiata*, scrub, gorse (*Ulex europaeus*) and bracken. Invertebrates were sampled semi-quantitatively in a wide range of habitats at each site to maximise the number of taxa collected. Classification, ordination and correlation analyses were undertaken using presence-absence data.

A total of 83 invertebrate taxa were collected at the 26 sites. The Trichoptera, Plecoptera and Diptera contained most species, but the most widely distributed taxa were Chironomidae (not identified beyond family level) and the ubiquitous *Deleatidium,* recorded at 25 and 24 sites, respectively.

Classification of sites by TWINSPAN (Two-way Indicator Species Analysis, a divisive clustering technique based on reciprocal averaging; McCune and Mefford, 1995) using faunal presence-absence data, distinguished four site groups differing in mean stream water pH, water temperature, and channel width (Tab. 2). A significant correlation was found between pH and time since afforestation at the sites suggesting that catchment development had resulted in increased acidification of streams. In fact, development was most advanced in small headwater catchments where pH was naturally higher, and the results therefore provided no

Table 2. Mean stream water pH, channel width, November water temperature and rank order of catchment development of West Coast forest stream sites in four groups distinguished by TWINSPAN (Collier et al., 1989)

	TWINSPAN groups			
	A	B	C	D
pH	6.1	5.0	4.6	4.4
Channel width (m)	5.2	4.0	2.5	1.3
Temperature (°C)	15	15	22	28
Rank order*	1.0	1.0	3.3	4.3

* Ranks are 1 = native forest, 2 = pakihi and scrub, 3–5 = planted in *Pinus radiata* for 10, 5–6, and 1–2 years, respectively.

convincing evidence that the planting of pines contributed to the low pH recorded at these sites.

Nevertheless, catchment development for afforestation is a disturbance that affects the stream environment by increasing sedimentation, reducing shading, and increasing water temperature (Collier et al., 1989). Valentine (1995) also demonstrated that it affected stream channel morphology and both the occurrence and size distribution of woody debris. Wood was less abundant but more homogeneous in size where exotic forest was present as trees were young and of similar age to each other.

Valentine (1995) repeated the survey of Collier et al. in 1995 (i.e., 8 years later) and found that the relationships between time of catchment development and physico-chemical attributes of the streams reported earlier were still apparent but less marked. Except in the most acid headwater catchments, stream faunas were also more homogeneous among sites, with stream size and bed stability, but not time since afforestation, being implicated as primary variables influencing community structure.

The findings of the two pakihi stream surveys indicated that although catchment-scale disturbances had significant effects on stream environments, stream faunas within indigenous forest and plantation forest of different ages exhibited few differences in species composition among sites. Instead, the most pronounced differences in community structure were associated with extreme physico-chemical conditions (low pH and associated factors), not with afforestation. In Westland, the faunas of forest streams appear to include numerous habitat and food generalists with strong colonizing characteristics, features that appear to preadapt them to life in changing and changeable environments.

Case study 4. Recovery and community stability in afforested streams: The Maimai Experimental Area

The Maimai Experimental Area was established near Reefton, Westland, by the (then) Forest Research Institute in the 1970s. It incorporates eight small catchments (2.3–8.3 ha) originally in native beech-mixed podocarp-hardwood forest, six of which were clear-cut in different ways between 1976 and 1978. Some catchments were subsequently burnt before being planted in exotic conifers (*Pinus radiata*) and eucalypts (*Eucalyptus delegatensis*) in 1977–1980. Two catchments were retained as unmodified controls. The streams draining the catchments are small (mean width 0.7 m) and flood-prone, with beds dominated by cobbles, fine gravel and sand. Stream water is slightly acid (pH 5.0–6.5) and nutrient-poor (conductivity $<40\,\mu S_{25}\,cm^{-1}$). No fish are present.

The main purpose of the Maimai research programme is to evaluate the effects of forest management techniques on runoff, sediment loads and nutrient export (see Neary et al., 1978; Moore, 1989; Rowe et al., 1994). In

addition, the opportunity has been taken to monitor and compare the recovery and stability of the stream faunas over time as physical and vegetational changes have proceeded within the catchments. Results of some of
the earlier ecological studies on streams have been published by Winterbourn and Rounick (1985), Winterbourn (1986) and Rounick and Winterbourn (1986).

Preliminary collections of stream fauna were made in some of the
Maimai streams in 1976 and 1978, and more systematic sampling of four
of the streams was undertaken on 12 occasions between 1981 and 1995.
Field surveys were made in both summer (Nov–Jan) and autumn (April,
May) and have been analysed separately. The four "core" streams were an
unlogged control, two streams with riparian strips of indigenous vegetation, and one without a riparian strip (Tab. 2). Benthos was collected from
the same reach of each stream on all occasions by kick sampling and turning over stones into triangular nets (0.2 and 0.8 mm mesh). All collections
were made by the same operator (the writer) in as similar a manner as
possible.

Numerically dominant taxa in the Maimai streams are the mayflies
Deleatidium and *Zephlebia*, the stoneflies *Austroperla*, *Stenoperla* and
Spaniocerca, Elmidae and the amphipod *Paraleptamphopus* (Tab. 3). The
latter was the most abundant invertebrate in the control stream, which was
also the smallest of the four. Most of the species in all four streams can be
categorized as collector-browsers that feed on fine particulate detritus and

Table 3. Characteristics of the Maimai streams and faunal statistics for six autumn and six
summer surveys. Sorensen's scores are means of all year-year combinations. S = summer only,
A = autumn only

	Streams			
	208	209	214	215
Catchment area (ha)	3.84	8.26	4.62	2.64
Buffer strip	Present	Present	Absent	Forest control
Total taxa	42	50	46	41
Sorensen's scores (mean and CV)				
(a) Autumn	0.54 (19)	0.66 (9)	0.60 (13)	0.73 (10)
(b) Summer	0.58 (16)	0.68 (7)	0.67 (15)	0.66 (15)
Kendall's W				
(a) Autumn	0.22	0.49	0.32	0.62
(b) Summer	0.32	0.55	0.39	0.58
Top 3 taxa	*Austroperla* *Deleatidium* *Spaniocerca* (A) *Zephlebia* (S)	*Deleatidium* *Stenoperla* Elmidae (A) *Austroperla* (A)	*Deleatidium* *Stenoperla* *Austroperla*	*Paraleptamphopus* *Zephlebia* *Deleatidium*

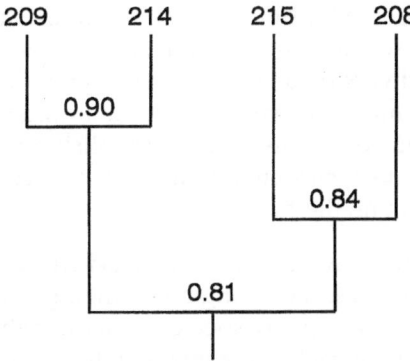

Figure 2. The degree of similarity of invertebrate faunas in Maimai streams 208, 209, 214 and 215 calculated using presence-absence of species, Sorensen's index and average linkage clustering. Full taxonomic lists derived from 12 surveys between 1981 and 1995 provide the data base (author's unpublished data).

epilithon although *Austroperla* larvae also feed as shredders and grazers of filamentous algae, and *Stenoperla* is a predator (Winterbourn and Rounick, 1985). Small numbers of obligate leaf shredders, caddisflies in the families Oeconesidae and Leptoceridae, occur in these streams which also support the predominantly detritivorous crayfish, *Paranephrops planifrons*. The total number of taxa recorded from the four streams ranged from 41–50, the complement of species in each being very similar as indicated by the high similarity coefficients (0.81–0.90) shown in Figure 2. Furthermore, stability of individual stream faunas over time, as indicated by Sorensen's index (0.54–0.73) was almost the same as that for the three beech forest streams at Cass (Tab. 1). Although a common pool of species was found in each stream, the relative abundances of the common species differed among them as did faunal persistence measured by Kendall's W (Tab. 3). Persistence measured in this way was greatest in the control stream (Stream 215) and the largest stream (Stream 209), and in no stream were the changes indicative of succession. When the findings of this study are considered in the context of catchment-level disturbance, several features stand out:

a) Although clearcutting resulted in large inputs of sediment to stream channels, increases in water temperature and light levels, and changes in flow patterns (Fahey and Rowe, 1992; Rowe et al., 1995), benthic faunas recovered rapidly (Winterbourn and Rounick, 1985).
b) The stream faunas present less than 2 years after planting of exotic trees had comparable species richness and dominance to those found on subsequent sampling days over the next 14 years.
c) Changes in forest type from native podocarp-hardwood-beech to exotic pines and eucalypts had no obvious effect on the composition of stream

faunas. Rather, observed differences among faunas of the four streams appeared to be associated with small differences in the sizes of the streams and the diversity of habitats they provide. Even substantial differences in the immediate stream-side environment, i.e., presence or absence of riparian forest, and the re-establishment of riparian cover (grasses, ferns, shrubs), particularly marked in Stream 214 (Fig. 3), had little apparent effect on the fauna.

Results of the Maimai study provide further evidence of the resilience of New Zealand stream faunas, the strong recolonizing abilities and non-specific habitat requirements of many species, and their ability to live in, and maintain populations in, unstable streams with highly variable flows.

Discussion

Stream environments may seem harsh to the human observer, but members of the aquatic biota have evolved morphological, behavioural, and life history adaptations that reduce stresses imposed upon them (Hynes, 1970). Thus, many members of the New Zealand stream fauna are clearly well adapted to life in variably disturbed, and seemingly harsh conditions. Their characteristics are products of their evolution under such conditions, and appear to have also pre-adapted them to human-induced disturbances associated with activities such as forestry and alluvial mining that have analogous effects.

Few New Zealand authors have considered the ecological characteristics of stream invertebrates in an explicitly evolutionary or biogeographic context. A notable exception is Towns (1983) who discussed the role of intraspecific competition and abiotic factors in the evolution of life-history patterns of Leptophlebiidae (Ephemeroptera), the mayfly family to which the ubiquitous, and highly successful *Deleatidium* belongs. He considered that the various life-history patterns seen within the family had evolved in response to abiotic factors (not spelt out) rather than as a competition-avoidance mechanism, and so sided with the prevalent view that in general New Zealand streams are, and by implication were, physically dominated systems (e.g., Winterbourn et al., 1981).

Poorly synchronized life histories are prevalent among the Leptophlebiidae (Towns, 1983; Collier and Winterbourn, 1990) and some other groups of New Zealand stream-dwelling insects (e.g., Winterbourn, 1978; Towns, 1981), and are also common in Australia and Chile, two other southern lands. All three countries share major families of stream-dwelling insects (e.g., Leptophlebiidae and Grypopterygidae) that demonstrate lack of seasonality in occurrence and growth of larvae and of adult emergence. Hynes and Hynes (1975) attributed these flexible characteristics of Australian grypopterygids to the harsh and unpredictable climate to which they

(A)

(B)

(C)

Figure 3. Maimai stream 214 showing the re-establishment of riparian vegetation in (A) 1982, (B) 1986, and (C) 1994. Native forest in this catchment was clearcut in 1977, and burnt and planted with *Pinus radiata* in 1978.

are exposed, and Towns (1983) noted that both Australian and New Zealand aquatic insect faunas have been subjected to wide climatic variations since their isolation from other continents. He also proposed that these had led to selection for ecological flexibility, one expression of which is the widespread occurrence of non-seasonal life cycles.

Although climatic variability has been invoked to account for the prevalence of ecological flexibility and life cycle asynchrony in New Zealand, Australia and Chile, the nature and periodicity of this variability (notably in rainfall patterns that influence discharge patterns) differs markedly between them. Thus, in much of New Zealand rainfall occurs year round, and storm events that are perceived to induce major disturbance events can occur in any month. In contrast, seasonal droughts and the drying up of streams are seen as the major disturbances in eastern Australia, while southern Chile has a rainy season when rapid and high discharge fluctuations occur (Gonser, 1995).

Regardless of the exact nature of the pressures that have led to the evolution of the New Zealand stream fauna as we see it today, it is clear that our streams are colonized by some very successful generalist species. The wide ranging distributions of taxa like *Deleatidium* in all kinds of stony rivers and streams is testament to this, as are the results of intensive field studies that have demonstrated the broad habitat tolerances of a range of common species (Graesser, 1987; Jowett and Richardson, 1990; Jowett et al., 1991; Stewart 1993). To give two examples, Jowett and Richardson (1990) found that only 25% of the variation in *Deleatidium* larval abundance was explained by velocity, depth and substrate composition in a North Island river riffle, and Graesser (1987) found that although species richness and abundance of invertebrates in four South Westland streams was greater on larger (and more stable) stones, neither was correlated with water velocity or depth.

Environmental variation in running water environments can be perceived as a continuum from harsh to benign (Peckarsky, 1983). Highly variable, and/or unpredictable flow regimes can be expected to have a dominant effect on ecological patterns (Resh et al., 1988, Poff and Ward, 1989), but within streams the impacts of flow perturbation may be reduced by the presence of refugia that confer resilience on the system, speed the recovery process, and provide a source of colonists (Townsend, 1989; Cobb et al., 1992a). These two factors (variable flows and refugia) can be integrated nicely if a patch dynamics approach is taken (Pringle et al., 1988), an approach that seems particularly appropriate with respect to South Island mountain streams, since it emphasizes the rapid colonizing ability of many stream-dwelling organisms, and the continually changing mosaic of local environmental conditions (Allan, 1995). However, as Allan points out it is the predictability of environmental conditions *in the aggregate* rather than in one place or time that ensures some regularity and persistence of communities. The case studies presented above, and other New Zealand studies, demonstrate the resilience and stability of benthic faunas in disturbance-prone running waters, in agreement with a patch dynamics framework. They also point to the strong colonizing capacities of New Zealand stream invertebrates (a requirement of the patch dynamics model), and the prevalence of habitat and trophic generalists. The Westland case studies demonstrate these attributes in the context of large-scale catchment disturbance. Not only was there rapid recolonisation following deforestation and habitat destruction by heavy sedimentation, but the composition and structure of the re-established communities appeared little affected by forest type. As such, they support an earlier New Zealand-wide survey encompassing four forest types (Rounick and Winterbourn, 1982), and a more intensive and localized study on the eastern side of the Southern Alps (Harding and Winterbourn, 1995).

Because streams and rivers are open systems, and because many of the invertebrates inhabiting them are insects with aerial or terrestrial adults, the colonizing abilities of adults are critical for the persistence of species populations. I suspect they may also play a much greater role in determining the size of larval populations than is generally recognized, a contention also alluded to by McElravy and Resh (1987) who commented that the separation of seasonal or annual variability from variability produced by perturbation is a major concern in aquatic biology. Schmidt et al. (1995) demonstrated widespread gene flow between drainages in Queensland, Australia, in a study of an unidentified but common species of *Baetis* (Ephemeroptera), and their results also suggest that the larval populations found at a particular site may be the offspring of few adults. Little is known about the biology of adults of common New Zealand stream insects, but the Cass studies suggest they may play an important role in habitat selection (species assemblages at stable sites were individualistic and differed from each other), and in determining baseline population levels. Wisseman and Anderson (1984) also emphasized that life stages other than larvae need to be studied to put larval data into a "life system" context, and importantly, that quantitative data are needed on mortality factors affecting *all* life stages. How this is done is another thing altogether, and represents a major challenge to stream ecologists concerned with population and community dynamics.

Although I have emphasized the potentially significant role of terrestrial and aerial stages in the population dynamics of aquatic insects, it must be remembered that many non-insect taxa in particular, are aquatic throughout their lives. Molluscs are either unrepresented or rare in the more unstable New Zealand streams (e.g., Death and Winterbourn, 1995), and are almost certainly slower colonisers than many insects because their life cycles include no terrestrial stage. Similarly, the turbellarian flatworm *Neppia montana* appears to be restricted to mountain streams with a high proportion of stable, substratum elements. In contrast, several crustaceans including harpacticoid copepods (Tank and Winterbourn, 1995) and the amphipod *Paraleptamphopus subterraneous* can be abundant in unstable streams, yet have no terrestrial life history stage. Their ability to use the hyporheic zone and perhaps deeper groundwaters too, is implicated in their success as colonists of streams whose surface sediments and water channels are subject to frequent scouring and movement. Interestingly, several species of *Spaniocercoides*, a genus of notonemourid stonefly that occurs in unstable, West Coast streams (Winterbourn and Ryan, 1994) also appears to have predominantly hyporheic larvae in addition to a winged adult.

Compared with our results from Cass where faunal persistence (and assemblage concordance) in three beech forest streams was high, Bunn et al. (1995) reported marked temporal variations in stream community composition in subtropical Australian streams over a 4-year period. Their calculations indicated that very small differences in the numbers of ovipositing, female insects could account for much of the spatial and temporal variation in assemblage composition. Although assemblage structure persisted over time at Cass, population densities differed between years and in different ways in the three streams studied. These findings provide little support for a "climatic" theory of population size regulation but refocus attention on the biological requirements and behaviour of the reproductive stages.

Resh et al. (1988) contended that physical disturbance may be a dominant determinant in stream ecology. Studies of physically disturbed and temporally variable South Island streams support this contention, and indicate that a well established and "well adapted" fauna occur in these harsh environments. While flow variability and the occurrence of discharge events resulting in bed disturbance occur unpredictably on scales of months and years, they can be seen as components of a predictably variable environment on a larger (evolutionary) time scale. Many characteristic features of present-day South Island stream faunas – life history flexibility, a lack of habitat specificity, strong colonizing ability, generalist food habits, are best explained in a historical context, or to quote Poff (1992) "a system incorporates its disturbance history." Without question, many South Island mountain streams are unstable, disturbed environments, but because of their evolutionary history, and depending on the time scale of interest, their faunas can be considered both stable and disturbed.

Acknowledgments
The development of my ideas about South Island stream communities has been greatly enhanced by the studies of a diverse group of students who have worked with me at the University of Canterbury. It gives me particular pleasure to acknowledge the contributions of Kevin Collier, Russell Death, Anne Graesser and David Valentine whose work I have drawn on most strongly here. Curt Lively and Angus McIntosh provided constructive comments on the manuscript and helped give it better balance. Lastly, I thank Colin O'Loughlin (formerly of the Forest Research Institute, Christchurch) and Lindsay Rowe (Landcare Research, Christchurch) for providing me with the opportunity to work in the Maimai Experimental Area for almost two decades.

References

Allan, J.D. (1995) *Stream Ecology. Structure and Function of Running Waters*. Chapman & Hall, London.

Anderson, N.H. (1992) Influence of disturbance on insect communities in Pacific Northwest Streams. *Hydrobiologia* 248:79–92.

Biggs, B.J.F. (1995) The contribution of flood disturbance, catchment geology and land use to the habitat template of periphyton in stream ecosystems. *Freshwater Biol.* 33:419–438.

Bunn, S.E., Hughes, J.M. and Marshall, C. (1995) Temporal patterns in a guild of algal grazers in subtropical rainforest streams. *Bulletin of the North American Benthological Society* 12:144.

Campbell, I.C. and Doeg, T.J. (1989) Impact of timber harvesting and production on streams: a review. *Australian Journal of Marine and Freshwater Research* 40:519–539.

Close, M.E. and Davies-Colley, R.J. (1990) Baseflow water chemistry in New Zealand rivers. *N. Z. J. Mar. Freshwater Res.* 24:319–342.

Cobb, D.G., Galloway, T.D. and Flannagan, J.F. (1992a) Effects of discharge and substrate stability on density and species composition of stream insects. *Can. J. Fisheries Aquat. Sci.* 49:1788–1795.

Cobb, D.G., Galloway, T.D. and Flannagan, J.F. (1992b) The effect on the Trichoptera of a stable riffle constructed in an unstable reach of Wilson Creek, Manitoba, Canada. *In:* C. Tomaszewski (ed.) *Proceedings of the 6th International Symposium on Trichoptera*, Adam Mickiewicz University Press, Poznan, pp 81–88.

Collier, K.J. (1987) Spectrophotometric determination of dissolved organic carbon in some South Island streams and rivers. *N. Z. J. Mar. Freshwater Res.* 21:349–351.

Collier, K.J. and Winterbourn, M.J. (1990) Population dynamics and feeding of mayfly larvae in some acid and alkaline New Zealand streams. *Freshwater Biol.* 23:181–189.

Collier, K.J., Winterbourn, M.J. and Jackson, R.J. (1989) Impacts of wetland afforestation on the distribution of benthic invertebrates in acid streams of Westland, New Zealand. *N. Z. J. Mar. Freshwater Res.* 23:479–490.

Connell, J.H. and Sousa, W.P. (1983) On the evidence needed to judge ecological stability or persistence. *Amer. Nat.* 121:780–824.

Cowie, B. (1985) An analysis of changes in the invertebrate community along a southern New Zealand montane stream. *Hydrobiologia* 120:35–46.

Death, R.G. (1995) Spatial patterns in benthic invertebrate community structure: products of habitat stability or are they habitat specific? *Freshwater Biol.* 33:455–467.

Death, R.G. and Winterbourn, M.J. (1994) Environmental stability and community persistence: a multivariate perspective. *J. N. Amer. Benthol. Soc.* 13:125–139.

Death, R.G. and Winterbourn, M.J. (1995) Diversity patterns in stream benthic invertebrate communities: the influence of habitat stability. *Ecology* 76:1446–1460.

Fahey, B.D. and Rowe, L.K. (1992) Land use impacts. *In:* M.P. Mosley (ed.): *Waters of New Zealand,* Longman Paul, Auckland, pp 265–284.

Fisher, S.G. and Likens, G.E. (1973) Energy flow in Bear Brook, New Hampshire: an integrative approach to stream ecosystem metabolism. *Ecol. Monogr.* 43:421–439.

Flecker, A.S. and Townsend, C.R. (1994) Community-wide consequences of trout introduction in New Zealand streams. *Ecol. Appl.* 4:798–807.

Gonser, T. (1995) Feeding strategies, population dynamics and competitive crunch periods of a Southern Chilean stream mayfly assemblage. *Bulletin of the North American Benthological Society* 12:144.

Gore, J.A., Kelly, J.R. and Yount, J.D. (1990) Application of ecological theory to determining recovery potential of disturbed lotic ecosystems: research needs and priorities. *Environ. Manage.* 14:755–762.

Graesser, A.K. (1987) Invertebrate drift in three flood-prone streams in South Westland, New Zealand. *Verhandlungen der Internationalen Vereinigung für Theoretische und Angewandte Limnologie* 23:1427–1431.

Graesser, A.K. (1988) *Physico-chemical Conditions and Benthic Community Dynamics in Four South Westland Streams*. Ph.D. thesis, University of Canterbury, Christchurch.

Greenland, D.E. (1977) Weather and climate at Cass. *In:* C.J. Burrows (ed.) *Cass*, Department of Botany, University of Canterbury, pp 93–116.

Griffiths, G.A. (1979) High sediment yields from major rivers of the western Southern Alps, New Zealand. *Nature* 282:61–63.

Griffiths, G.A. and McSaveney, M.J. (1983) Hydrology of a basin with extreme rainfalls – Cropp River, New Zealand. *New Zealand Journal of Science* 26:293–306.

Grimm, N.B. and Fisher, S.G. (1989) Stability of periphyton and macroinvertebrates to disturbance by flashfloods in a desert stream. *J. N. Amer. Benthol. Soc.* 8:292–307.

Harding, J.S. and Winterbourn, M.J. (1995) Effects of contrasting land use on physico-chemical conditions and benthic assemblages of streams in a Canterbury (South Island, New Zealand) river system. *N. Z. J. Mar. Freshwater Res.* 29:479–492.

Holling, C.S. (1973) Resilience and stability of ecological systems. *Annu. Rev. Ecol. Syst.* 4:1–23.

Hulme, K. (1984) *The Bone People.* Spiral Collective, New Zealand.

Hynes, H.B.N. (1970) *The Ecology of Running Waters.* Liverpool University Press, Liverpool.

Hynes, H.B.N. and Hynes, M.E. (1975) The life histories of many of the stoneflies (Plecoptera) of south-eastern mainland Australia. *Australian Journal of Marine and Freshwater Research* 26:113–153.

Jowett, I.G. and Richardson, J. (1990) Microhabitat preferences of benthic invertebrates in a New Zealand river and the development of instream flow-habitat models for *Deleatidium* spp. *N. Z. J. Mar. Freshwater Res.* 24:11–22.

Jowett, I.G., Richardson, J., Biggs, B.J.F., Hickey, C.W. and Quinn, J.M. (1991) Microhabitat preferences of benthic invertebrates and the development of generalized *Deleatidium* spp. habitat suitability curves, applied to four New Zealand rivers. *N. Z. J. Mar. Freshwater Res.* 25:187–200.

Lake, P.S. and Barmuta, L.A. (1986) Stream benthic communities: Persistent presumptions and speculations. *In:* P. De Deckker and W.D. Williams (eds): *Limnology in Australia,* CSIRO/Junk, Dordrecht, pp 263–276.

Linklater, W. (1995) Breakdown and detritivore colonization of leaves in three New Zealand streams. *Hydrobiologia* 306:241–250.

McCune, B. and Mefford, M.J. (1995) PC-ORD. Multivariate Analysis of Ecological Data, Version 2.0. MjM Software Design, Gleneden Beach, Oregon, USA.

McElravy, E.P. and Resh, V.H. (1987) Diversity, seasonality, and annual variability of caddisfly (Trichoptera) adults from two streams in the California Coast Range. *Pan-Pac. Entomol.* 63:75–91.

McIntosh, A. and Townsend, C.R. (1994) Interpopulation variation in mayfly anti-predator tactics: differential effects of contrasting predatory fish. *Ecology* 75:2078–2090.

Meffe, G.K. and Minckley, W.L. (1987) Persistence and stability of fish and invertebrate assemblages in a repeatedly disturbed Sonoran Desert stream. *Amer. Midland Naturalist* 117:116–191.

Menge, B.A. (1976) Organization of the New England rocky intertidal community: role of predation, competition, and environmental heterogeneity. *Ecol. Monogr.* 46:355–393.

Moore, T.R. (1989) Dynamics of dissolved organic carbon in forested and disturbed catchments, Westland, New Zealand I. Maimai. *Water Resour. Res.* 25:1321–1330.

Mosley, M.P. and Rowe, L.K. (1981) Low flow water chemistry in forested and pasture catchments, Mawheraiti River, Westland. *N. Z. J. Mar. Freshwater Res.* 15:307–320.

Neary, D.G., Pearce, A.J., O'Loughlin, C.L. and Rowe, L.K. (1978) Management impacts on nutrient fluxes in beech-podocarp-hardwood forests. *N. Z. J. Ecol.* 1:19–26.

Peckarsky, B.L. (1983) Biotic interactions or abiotic limitations? A model of lotic community structure. *In:* Fontaine, T.D. and Bartell, S.M. (eds): *Dynamics of Lotic Ecosystems,* Ann Arbor Science, Ann Arbor, pp 303–323.

Pfankuch, D.J. (1975) *Stream reach inventory and channel stability evaluation.* United States Forest Service, Region 1, Missoula, Montana.

Poff, N.L. (1992) Why disturbances can be predictable: a perspective on the definition of disturbance in streams. *J. N. Amer. Benthol. Soc.* 11:86–92.

Poff, N.L. and Ward, J.V. (1989) Implications of streamflow variability and predictability for lotic community structure: a regional analysis of stream flow patterns. *Can. J. Fisheries Aquat. Sci.* 46:1805–1818.

Pringle, C.M., Naiman, R.J., Bretschko, G., Karr, J.R., Oswood, M.W., Webster, J.R., Welcomme, R.L. and Winterbourn, M.J. (1988) Patch dynamics in lotic systems: the stream as a mosaic. *J. N. Amer. Benthol. Soc.* 7:503–524.

Quinn, J.M. and Hickey, C.W. (1990) Characterization and classification of benthic invertebrate communities in 88 New Zealand rivers in relation to environmental factors. *N. Z. J. Mar. Freshwater Res.* 24:387–410.

Resh, V.H., Brown, A.V., Covich, A.P., Gurtz, M.E., Li, H.W., Minshall, G.W., Reice, S.R., Sheldon, A.L., Wallace, J.B. and Wissmar, R.C. (1988) The role of disturbance in stream ecology. *J. N. Amer. Benthol. Soc.* 7:433–455.

Richards, C. and Minshall, G.W. (1992) Spatial and temporal trends in stream macroinvertebrate communities: the influence of catchment disturbance. *Hydrobiologia* 241:173–184.

Rounick, J.S. and Winterbourn, M.J. (1982) Benthic faunas of forested streams and suggestions for their management. *N. Z. J. Ecol.* 5:140–150.

Rounick, J.S. and Winterbourn, M.J. (1986) Stable carbon isotopes and carbon flow in ecosystems. *Bioscience* 36:171–177.

Rowe, L.K., Pearce, A.J. and O'Loughlin, C.L. (1994) Hydrology and related changes after harvesting native forest catchments and establishing *Pinus radiata* plantations. Part 1. Introduction to study. *Hydrol. Process.* 8:263–279.

Sagar, P.M. (1986) The effects of floods on the invertebrate fauna of a large unstable braided river. *N. Z. J. Mar. Freshwater Res.* 20:37–46.

Scarsbrook, M.R. and Townsend, C.R. (1993) Stream community structure in relation to spatial and temporal variation: a habitat templet study of two contrasting New Zealand streams. *Freshwater Biol.* 29:395–410.

Schmidt, S.K., Hughes, J.M. and Bunn, S.E. (1995) Gene flow among conspecific populations of *Baetis* sp. (Ephemeroptera): adult flight and larval drift. *J. N. Amer. Benthol. Soc.* 14:147–157.

Scrimgeour, G.J., Davidson, R.J. and Davidson, J.M. (1988) Recovery of benthic macroinvertebrate and epilithic communities following a large flood, in an unstable, braided, New Zealand river. *N. Z. J. Mar. Freshwater Res.* 22:337–344.

Scrimgeour, G.J. and Winterbourn, M.J. (1989) Effects of floods on epilithon and benthic macroinvertebrate populations in an unstable New Zealand river. *Hydrobiologia* 171:33–44.

Stewart, A. (1993) *Aluminium and pH Tolerance of Some New Zealand Stream Invertebrates.* M.Sc. Thesis, University of Canterbury, Christchurch.

Tank, J.L. and Winterbourn, M.J. (1995) Biofilm development and invertebrate colonization of wood in four New Zealand streams of contrasting pH. *Freshwater Biol.* 34:303–315.

Towns, D.R. (1981) Life histories of benthic invertebrates in a kauri forest stream in northern New Zealand. *Australian Journal of Marine and Freshwater Research* 32:191–211.

Towns, D.R. (1983) Life history patterns of six sympatric species of Leptophlebiidae (Ephemeroptera) in a New Zealand stream and the role of interspecific competition in their evolution. *Hydrobiologia* 99:37–50.

Townsend, C.R. (1989) The patch dynamics concept of stream community ecology. *J. N. Amer. Benthol. Soc.* 8:36–50.

Valentine, D.A. (1995) *The Effects of Forestry Operations and Catchment Development on Lotic Ecosystems in North Westland.* M. For. Sci. thesis, University of Canterbury, Christchurch.

White, P.S. and Pickett, S.T.A. (1985) Natural disturbance and patch dynamics: an introduction. *In:* S.T.A. Pickett and P.S. White (eds): *The Ecology of Natural Disturbance and Patch Dynamics,* Academic Press, New York, pp 3–9.

Winterbourn, M.J. (1978) The macroinvertebrate fauna of a New Zealand forest stream. *N. Z. J. Zool.* 5:157–169.

Winterbourn, M.J. (1986) Forestry practices and stream communities with particular reference to New Zealand. *In:* I.C. Campbell (ed.) *Stream Protection: The Management of Rivers for Instream Uses.* Chisholm Institute of Technology, Caulfield, pp 57–73.

Winterbourn, M.J. (1987) Invertebrate communities. *In:* A.B. Viner (ed.) *Inland Waters of New Zealand,* D.S.I.R., Wellington, pp 167–190.

Winterbourn, M.J. (1995) Rivers and streams of New Zealand. *In:* C.E. Cushing, K.W. Cummins and G.W. Minshall (eds) *River and Stream Ecosystems,* Ecosystems of the World 22, Elsevier, Amsterdam, pp 695–716.

Winterbourn, M.J. and Collier, K.J. (1987) Distribution of benthic invertebrates in acid, brown water streams in the South Island of New Zealand. *Hydrobiologia* 153:277–286.

Winterbourn, M.J. and Rounick, J.S. (1985) Benthic faunas and food resources of insects in small New Zealand streams subjected to different forestry practices. *Verhandlungen der Internationalen Vereinigung für Theoretische und Angewandte Limnologie* 22:2148–2152.

Winterbourn, M.J. and Ryan, P.A. (1994) Mountain streams in Westland, New Zealand: benthic ecology and management issues. *Freshwater Biol.* 32:359−373.

Winterbourn, M.J., Rounick, J.S. and Cowie, B. (1981) Are New Zealand stream ecosystems really different? *N. Z. J. Mar. Freshwater Res.* 15:321−328.

Wisseman, R.W. and Anderson, N.H. (1984) Mortality factors affecting Trichoptera eggs and pupae in an Oregon Coast Range watershed. *In:* J.C. Morse (ed.) *Fourth International Symposium on Trichoptera,* series Entomologica Volume 30, Junk, The Hague, pp 455−460.

Wrona, F.J., Calow, P., Ford, I., Baird, D.J. and Maltby, L. (1986) Estimating the abundance of stone-dwelling organisms: a new method. *Can. J. Fisheries Aquat. Sci.* 43:2025−2035.

Evolutionary Ecology of Freshwater Animals
ed. by B. Streit, T. Städler and C. M. Lively
© 1997 Birkhäuser Verlag Basel/Switzerland

Complex trophic interactions in freshwater benthic food chains

C. Brönmark, J. Dahl and L. A. Greenberg

Department of Ecology, Lund University, Ecology Building, S-22362 Lund, Sweden

Summary. Complex trophic interactions play an important role in structuring benthic communities in lakes and streams. In this review of freshwater, benthic food chains, a variety of different types of complex interactions were identified, including trophic cascades, interaction modifications and indirect commensalisms. The strength of these interactions varied greatly among studies, depending on habitat complexity, the type and intensity of environmental stresses, the ways in which prey defended themselves from predators, and the degree of trophic omnivory in the food chain. Areas that deserve further study include the effect of trophic omnivory on interaction strength and indirect effects of piscivores in benthic communities. Future studies should include whole-system, long-term manipulations to deepen our understanding of the role of complex interactions in freshwater communities.

Introduction

In recent years, many theoretical and empirical studies have recognised the importance of complex trophic interactions for structuring freshwater communities (see e.g., Kerfoot and Sih, 1987; Carpenter, 1988; Carpenter and Kitchell, 1993). By complex interactions we refer to sequences of biotic interactions that functionally link species (Neill, 1988). The defining feature of a complex or indirect interaction is that there is an intermediary species, mediating the interaction between two species. Much of our initial knowledge of complex or indirect interactions in freshwaters has arisen from studies carried out in pelagic habitats. This is probably because pelagic habitats are relatively easy to sample and manipulate, the communities themselves are relatively simple, and traditionally, many ecologists have worked in this habitat (Lodge et al., 1988). As Lodge et al. (1988) point out, complex interactions in other freshwater habitats should be studied as the pelagic zone is not an isolated system; instead it has functional links with other habitats, littoral and profundal habitats in lakes as well as with stream habitats. Further, studies of other habitats may reveal how complex interactions resemble and differ between habitats as well as uncover important unrecognized factors that affect complex interactions, and that would be difficult to identify if studies were restricted to one habitat.

The benthic habitat is characteristically very patchy, with large variation in the type of structure and its complexity. Structure of the habitat arises not only from the type of physical bottom, whether it be sand or cobble, but also from the organisms covering the bottom, such as algae and macrophytes, which provide structure and food for other organisms. Benthic habitats are characterized as being species-rich habitats, with many species exhibiting trophic omnivory. Given the physical and biological diversity found in benthic habitats, it will not be surprising if more intricate complex interactions are uncovered. Here, we review the state of our knowledge of complex interactions in benthic habitats in lakes and streams. Specifically, we discuss the different types of interactions that have been reported and the factors that affect the strength of these interactions. Clearly, this research area is only in its initial stage, especially in streams where the study of complex interactions has not received as much attention as in lakes.

Types of interactions

There are many different possible indirect interactions, and it is impossible to classify them so that this is fully appreciated. Nevertheless a discussion of indirect interactions is facilitated by having some descriptors of different indirect interactions. Wootton (1994) describes some of the more commonly studied types of indirect effects, which we adopt for our discussion as well. Below we briefly describe these.

Perhaps the most commonly studied type of indirect interaction is that of *exploitative competition* (Fig. 1(A)). Although often not regarded as an indirect interaction, exploitative competition involves two species that indirectly affect another by their direct effects on the abundance of a shared food resource. Exploitative competition has been discussed extensively in the literature, and, thus, it will not be treated further in this review.

A *trophic cascade* is an indirect interaction characteristic of linear food chains. In this interaction, a predator A has an indirect positive effect on species C by reducing the abundance of B (Fig. 1(B)). If the indirect positive effect on species C is mediated through a change in the behaviour of species B, the effect may be referred to as a behavioural cascade.

Apparent competition is an indirect interaction involving two prey species that share a common predator (Fig. 1(C)). An increase in the abundance of prey species B leads to an increase in the predator through a numerical response, and this elevates attack rates on prey species C, resulting in reduced densities of C. Few experimental studies of this kind of interaction exist (e.g., Schmitt, 1987), and we are unaware of any in freshwaters. Consequently, we do not discuss this interaction further in this review.

Another type of indirect interaction is *indirect commensalism* (Fig. 1(D)). In this interaction a consumer species (A) that is specialized on one of two competing resource organisms (B) will benefit a consumer (E) that is specialized on the other resource organisms (D). Examples of this interaction come from studies of interactions between grazers and algae in marine and freshwater habitats.

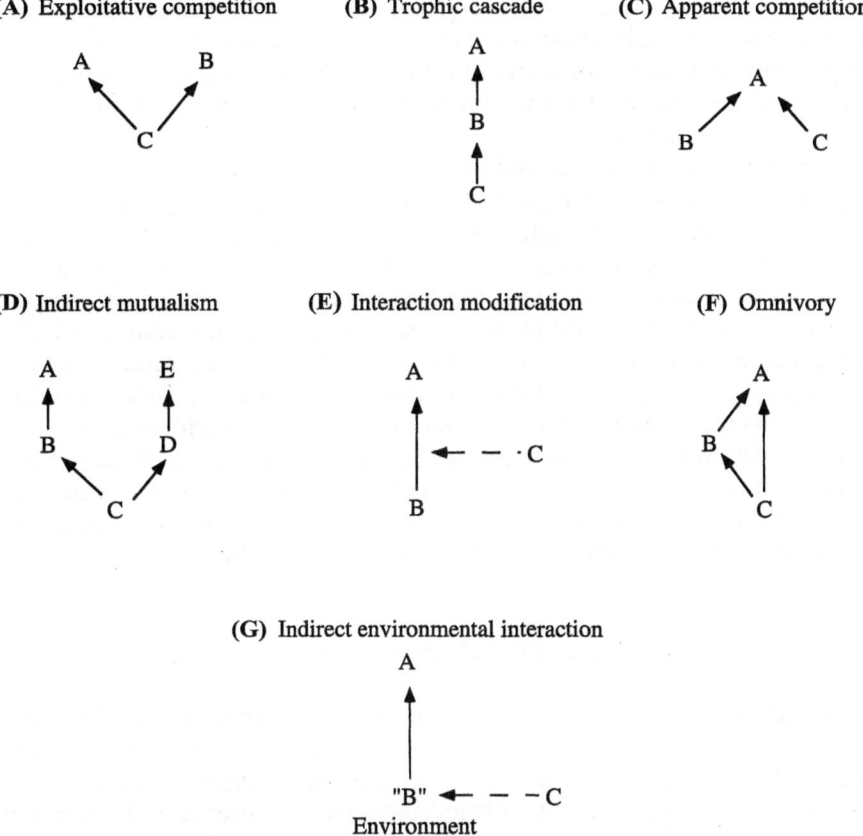

Figure 1. Some commonly observed types of complex interactions. Solid arrows show energy flows, whereas dotted arrows show modification of an interaction. Modified after Wootton (1994).

Interaction modifications is yet another type of indirect effect. In these cases, one species modifies the interaction between other species, often through changes in behaviour (Fig. 1(E)). For example, interpredator aggression may reduce consumption of a prey species (Soluk, 1993), and, further, the behaviour of one predator may enhance prey availability to another predator (Soluk and Collins, 1988a).

Cascading trophic interactions

In 1960, Hairston, Smith and Slobodkin proposed a theory (referred to as HSS) regarding the importance of food chain interactions for community structure. According to HSS, which was developed from observations of trilevel trophic terrestrial systems, predators regulate herbivores, freeing primary producers from herbivore regulation. Instead, primary producers are regulated by their resources. HSS and its extensions (e.g., Fretwell, 1987) claim that the factor (competition or predation) regulating a specific population depends on its position in the food chain and the number of trophic levels in the chain.

In freshwaters, the most notable examples of trophic cascades come from experimental studies of pelagic food chains in lakes (see e.g., Henrikson et al., 1980, Shapiro and Wright, 1984, Kerfoot and Sih, 1987, Carpenter, 1988; Carpenter and Kitchell, 1993). Typically, an increase in the abundance of the top trophic level, the piscivores, results in a decrease in planktivorous fish followed by an increase in large, cladoceran zooplankton and a decrease in phytoplankton biomass. Thus, a manipulation of the top trophic level cascades all the way down to the primary producers, the phytoplankton. Until recently, few studies had been performed on food chains involving freshwater benthic organisms, but recent studies have shown that manipulations of top trophic levels may cause trophic cascades in these food chains as well. Below we summarise the results of studies of trophic cascades in benthic food chains in lakes and streams.

Trophic cascades in benthic food chains in lakes

A number of recent studies have demonstrated the importance of indirect interactions in freshwater benthic food chains in lake littoral systems. Interestingly, in all these studies the herbivore and primary producer levels consist of snails and periphytic algae, respectively. Patterns of distribution and abundance of freshwater snails have traditionally been attributed to abiotic factors, especially calcium, which is needed for shell construction and maintenance (e.g., Hubendick, 1947; Økland, 1990). On a regional scale, calcium availability most certainly affects snail distributions, but recent studies suggest that biotic interactions become important as soon as calcium levels are > 5 mg/l (Aho, 1984; Lodge et al., 1987). Lodge et al. (1987) developed a conceptual model in which predation was viewed as the major factor structuring assemblages of freshwater snails in large, permanent ponds and lakes, and recent empirical studies have supported this result (e.g., Weber and Lodge, 1990; Merrick et al., 1991; Brönmark et al., 1992; Martin et al., 1992).

Periphytic algae, i.e., algae that grow on solid substrates such as stones, macrophytes etc., are the major food source of freshwater snails, and many studies have shown that snails may have strong effects on the biomass, productivity and species composition of periphyton (see Brönmark, 1989 for a review). Consequently, changes in predation pressure on snails may have profound effects on periphyton assemblages. Recently, the density of molluscivorous fish or crayfish has been manipulated in field experiments.

Fish-snail-algae

Many fish species, including cichlids, centrarchids, and cyprinids, have evolved morphological adaptations that enable them to feed on strong-shelled molluscs. For example, the pumpkinseed sunfish (*Lepomis gibbosus*) is a widely distributed centrarchid in temperate lakes of North America, specializing on snails as an adult (Sadzikowski and Wallace, 1976; Mittelbach, 1984; Osenberg and Mittelbach, 1990). Pumpkinseeds are morphologically adapted to crush molluscan shells, using their strong jaw muscles and pharyngeal teeth (Lauder, 1983). The effect of pumpkinseed sunfish on snail assemblages and periphyton was tested in enclosure/exclosure experiments in two northern Wisconsin, USA lakes (Brönmark et al., 1992). Predation by pumpkinseed sunfish dramatically reduced the biomass of snails (Fig. 2) as well as caused a change in snail species composition from large, thin-shelled to small, thick-shelled species (Klosiewski, 1991). The reduced biomass of snails resulted in an increased accumulation of periphyton biomass on artificial substrates (Fig. 2). In cages without pumpkinseeds there was a high biomass of snails and low biomass of periphyton. Although seemingly a classic example of trophic cascades, the increase in periphyton biomass may have been due to nutrient enhancement mediated via fish excretion or fish churning up the sediments (Threlkeld, 1987). This alternative explanation was tested and refuted by comparing the biomass of periphyton in cages with the non-molluscivore yellow perch (*Perca flavescens*) with cages containing pumpkinseeds and cages lacking fish. Brönmark et al. (1992) found that the density of snails was not affected by perch, and further, periphyton biomass in yellow perch cages was identical to biomass in fish exclosures. Thus, changes in periphyton biomass were not an effect of fish presence *per se*, but rather due to an indirect trophic cascade induced by a molluscivorous fish.

Strong, indirect effects due to predation by sunfish were also reported by Martin et al. (1992). They conducted an enclosure experiment to test the effects of predation by large and small sunfish, especially redear sunfish (*Lepomis microlophus*), in the littoral macroinvertebrate community. They predicted that large redear would reduce the density of large invertebrates and snails, whereas small sunfish would affect smaller macroinvertebrates such as trichopterans and chironomids but not snails (see Stein et al., 1984; Mittelbach, 1984). However, they found a strong effect of sunfish, independent of size, on the biomass of snails. It was suggested that the effect of small sunfish on snails was due to their consumption of juvenile, newly hatched snails (< 1 mm), which prevented recruitment of snails to larger sizes. During the second summer they noted that cages with sunfish had more periphyton than cages without fish, suggesting that redear sunfish had a similar cascading effect as the pumpkinseed sunfish. The strong effect of small sunfish on snails and periphyton suggested that a predator may have a strong, long-term effect on interactions in a food chain through a short-term predation event, influencing the recruitment of a

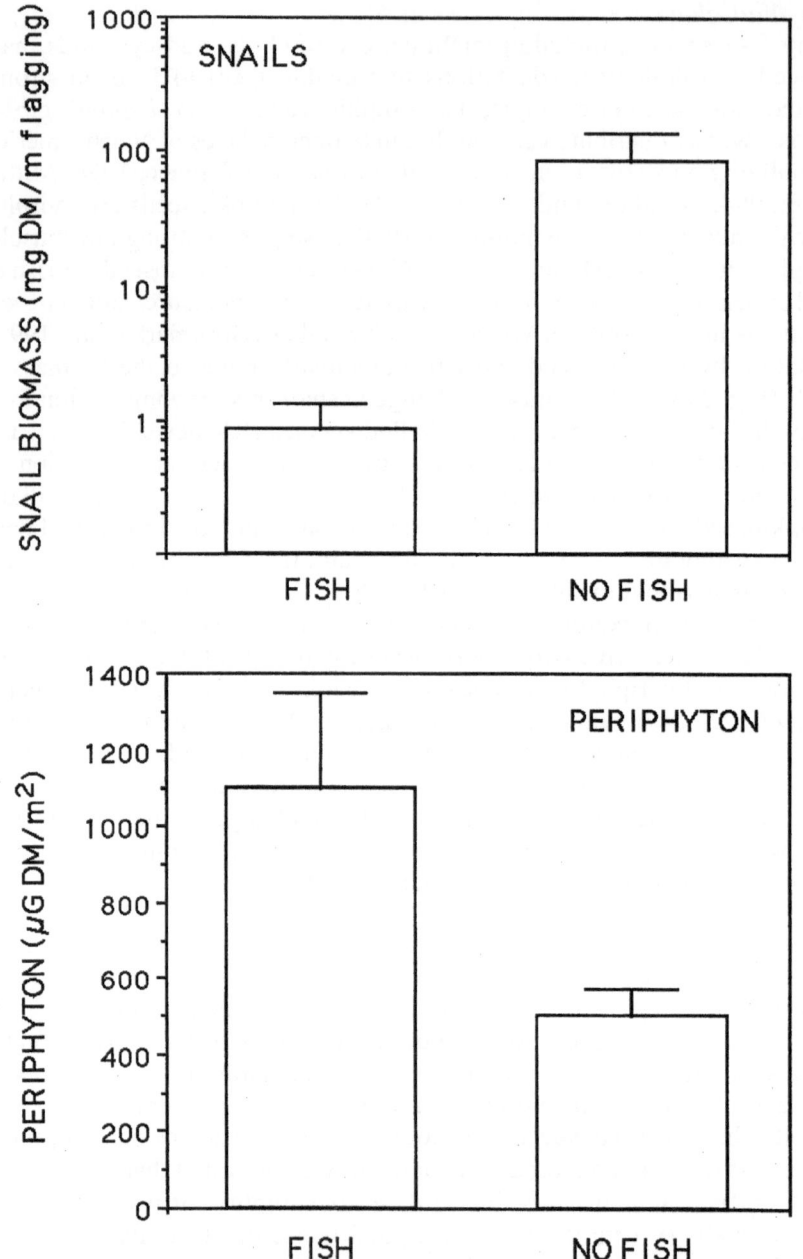

Figure 2. Final biomass of snails on plastic flagging tape (top panel) and periphyton (bottom panel) in cages with pumpkinseed sunfish (FISH) and in cages without fish (NO FISH). Data from Brönmark et al. (1992).

dominant herbivore (adult snails) that it does not normally include in its diet.

The sunfishes in the studies above are highly specialized snail predators, and thus strong trophic cascade effects on snails and periphyton were expected. A generalist is less likely to produce strong cascading effects. In a recent enclosure study, Brönmark (1994) investigated the effects of two benthivorous fish, tench (*Tinca tinca*) and perch (*Perca fluviatilis*). Earlier studies had suggested that both these species were generalist foragers, feeding on zooplankton as juveniles and then switching to a broad range of benthic macroinvertebrates as they grew larger, with perch becoming piscivorous at sizes > 150 mm (Kennedy and Fitzmaurice, 1970; Johansson and Persson, 1986). Tench have morphological adaptations (molariform pharyngeal teeth) to crush molluscan shells, but previous studies of tench diets suggested that snails comprised only a small proportion of their diets. Perch rarely consume snails. Consequently, Brönmark (1994) predicted that cascading effects induced by tench would be weak in comparison to specialists such as sunfishes, and the only effects perch might have would be mediated by its effects on non-molluscan herbivores. However, Brönmark (1994) found that tench had a strong, direct effect on the biomass of snails (Fig. 3), and this in turn resulted in higher periphyton biomass. Perch, on the other hand, had no effect on snails, other benthic macroinvertebrates or periphyton biomass. The results suggest that tench is a specialised molluscivore with a strong preference for snails. This is further supported by gut content analyses of tench that were introduced to a previously fishless pond with high densities of snails, which showed that tench fed almost exclusively on snails at these high availabilities (Miner and Brönmark, unpublished data). Why then do earlier diet data based on gut content analyses indicate that tench is a generalist forager that should have a minor impact on snail-periphyton interactions? This is probably because tench had reduced snail populations, forcing them to include non-molluscan prey in their diet. In parallel, Osenberg et al. (1992) found that in a lake with a high density of pumpkinseeds, only a small proportion of snails (<3%) were included in pumpkinseed diets, whereas in a lake with a low density of pumpkinseeds, gut contents consisted predominately of snails (64–95%). Thus, prey selectivity of a predator as estimated by gut content data do not necessarily indicate the strength of direct and indirect interactions.

The empirical tests of food chain theory in benthic food chains in lakes described above and in streams (see below) clearly show that cascading trophic interactions may be of great importance, affecting biomass, density and species composition of lower trophic levels. However, these studies do not evaluate cascading effects in food chains containing piscivores, i.e., a fourth trophic level (but see Power, 1990, 1992). Further, the generality of the results can be questioned as the experiments were performed in relatively small cages. This is a common method for manipulating the density

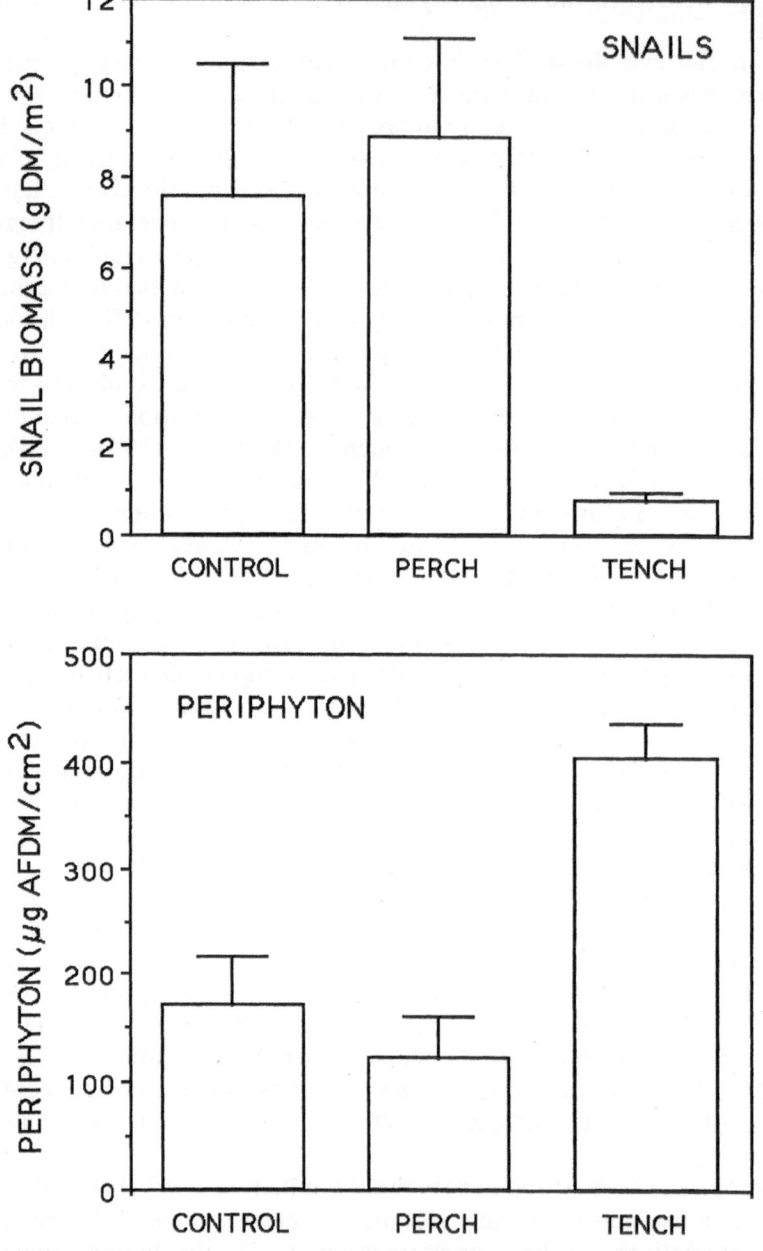

Figure 3. Final biomass of snails (top panel) and periphyton (bottom panel) in cages placed in an eutrophic pond in southern Sweden. Control cages had no fish, whereas the other cages were stocked with either perch or tench. Data from Brönmark (1994).

of predators in the field, and it has many advantages but also limitations (e.g., Diamond, 1986; Frost et al., 1988; Cooper and Barmuta, 1993; Neill, 1994). Small cages are easy to manipulate and replicate but results may be confounded by edge effects, replicates being poorly matched with regards to spatial heterogeneity, the behaviour of organisms may be affected within cages, etc. Cage effects may intensify or conceal the effect of the experimental manipulation. An increase in the scale of the experimental unit results in increased generality of the results, but also incurs considerable logistical problems and astronomical project budgets. Thus, it is hard to replicate and statistically evaluate large scale, whole system experiments. The temporal scale of manipulative experiments may also set constraints on the generality of results. Most studies which have shown strong cascading effects are performed over a rather short period (weeks to months, but see Brönmark et al., 1992), but on a longer time scale changes in species composition and size structure of prey assemblages, plasticity of prey defences and changes in prey behaviour may all ameliorate interaction strength between top predators and lower trophic levels. An alternative to experimental manipulations is to follow a system over a long period, preferably a period extending over several generations of the top predator. Mittelbach et al. (1995) performed a long-term study of the dynamics of a pelagic food chain consisting of piscivorous fish, planktivorous fish, zooplankton and phytoplankton, and were able to show that changes in piscivore abundance due to winterkill and later reintroduction resulted in patterns in biomass and size structure of lower trophic levels that were predicted from small-scale experiments and food chain theory. Another alternative to manipulative experiments is the "natural snapshot experiment" (Diamond, 1986), where a large number of natural systems in which the predator and prey have coexisted for many generations are surveyed, and patterns in distribution of biomass among food chains is compared to predictions from theory or from results from small-scale experiments.

Recently, Brönmark and Weisner (1996) surveyed 44 ponds in southern Sweden to test whether differences in fish community structure would affect the distribution of biomass among lower trophic levels of a benthic food chain, according to predictions from food chain theory. The ponds were divided into three different categories: ponds without fish, ponds with molluscivorous fish but no piscivorous fish, and ponds with piscivorous pike and/or perch. In ponds without fish and with molluscivores (two and three trophic level systems, respectively) the patterns among tropic levels were consistent with trophic cascade theory. Ponds without fish had a high density and biomass of snails and a low biomass of periphyton, whereas ponds with molluscivorous fish had a low density and biomass of snails and high periphyton biomass (Fig. 4). However, in ponds with four trophic levels, i.e., ponds that also had piscivores, the distribution of biomass among trophic levels deviated from predictions from theory. Ponds with piscivores had low densities of molluscivores and high densities of snails,

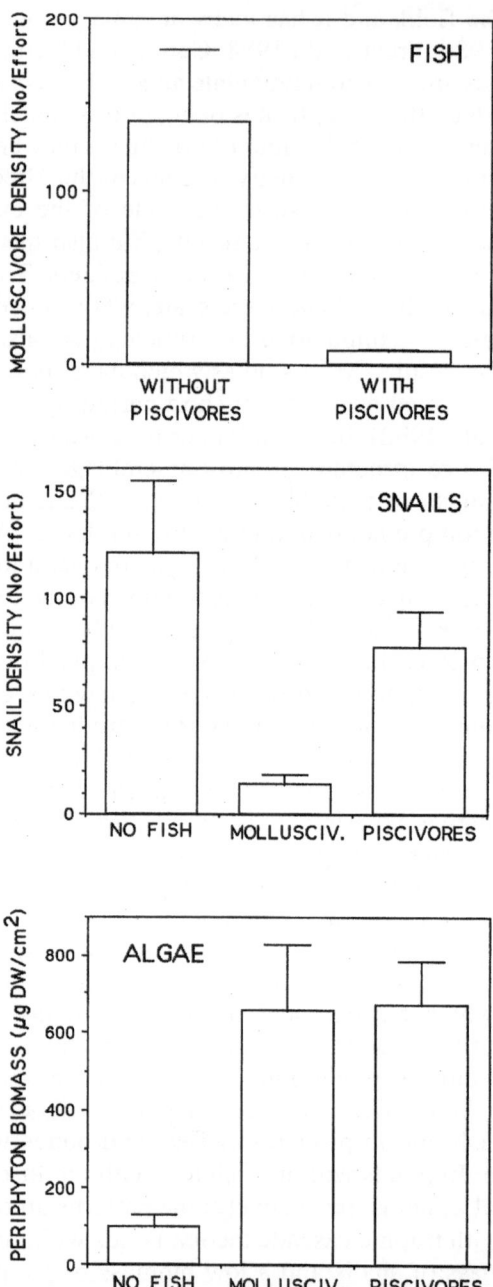

Figure 4. The density (catch per unit effort) of molluscivorous fish (top panel), snails (numbers per sweep net sample; middle panel) and the biomass of periphyton (bottom panel) in ponds in southern Sweden. Ponds were classified according to fish community structure, i.e., ponds without fish, ponds with molluscivorous fish and ponds that had both molluscivores and piscivores. Data from Brönmark and Weisner (1996).

whereas periphyton did not differ from ponds with three trophic levels (with molluscivorous fish only). Thus, in this benthic, freshwater food chain there was a decoupling of the cascading trophic interactions. This was probably due to the effect of large tench and crucian carp, which were invulnerable to predation by gape-limited piscivores (see below), and on the species composition of the snail assemblage in the ponds with piscivores. In these ponds, snail assemblages were dominated by small, detritivorous planorbid snails rather than large, periphyton-feeding, lymneid snails.

Crayfish-snail-algae
Crayfish are widespread omnivores, feeding on snails and other macro-invertebrates, macrophytes, periphyton and detritus. Omnivory is expected to reduce interaction strength in trophic cascading interactions as direct effects may be counteracted by indirect effects. For example, predation by crayfish may reduce the density of snails and thereby have a positive, indirect effect on periphyton growth as predicted by trophic cascades. However, crayfish may also have a direct, negative effect on periphyton through grazing and, further, they may reduce surface area available for colonization by periphyton by feeding on macrophytes. Hence, it is not intuitively obvious how changes in crayfish abundance would affect lower trophic levels.

Recently, Lodge and co-workers investigated the effects of crayfish on benthic macroinvertebrates, macrophytes and periphyton. In a correlative study in Trout Lake, Wisconsin, USA, they observed that in areas where crayfish were abundant, snails were scarce and cobbles had a thick cover of periphyton, whereas areas lacking crayfish had little periphyton and snails were abundant. They interpreted these distribution patterns as highly suggestive of strong cascading effects of crayfish predation (Weber and Lodge, 1990). To test this idea, Lodge et al. (1994) performed a field experiment with crayfish enclosures and exclosures and found that crayfish had dramatic effects on macrophytes and snails. There was a 100-fold decrease in snail densities and a 90% reduction of total macrophyte shoot densities in the presence of crayfish. The biomass of periphyton per unit surface area was greater in crayfish enclosures than in exclosures, indicating that the indirect effect of crayfish through decreases in snail grazing pressure is stronger than the direct effect of periphyton grazing by crayfish. The negative effect of crayfish on macrophytes may be due to direct consumption or indirectly through increased shading by an accumulating periphytic cover in the absence of snails. However, the occurrence of many floating fragments of submerged macrophytes suggested that direct consumption was the most important mechanism affecting macrophyte decline in crayfish enclosures. The results of this experiment show that even though crayfish were omnivorous they may still have strong, cascading effects on lower trophic levels.

Trophic cascades in benthic food chains in streams

Many studies on consumer-resource interactions have been performed in streams (see e.g., Allen 1994), but few studies have considered more than two trophic levels at a time. Many consumer-resource studies have found strong interactions and eventually it may be shown that these effects cascade down to lower trophic levels. This lack of studies in running waters contrasts with lakes, especially in pelagic food chains, where a large number of studies have shown strong cascading effects. The few studies showing cascading effects do not necessarily mean that trophic cascades are less likely to occur in running waters. It might only indicate that questions about cascading effects in running waters have not been asked until recently.

Most studies of trophic cascades in running waters have only considered three trophic levels. In several of these studies fishes are the main herbivore. Power and co-workers examined interactions between piscivores, grazing fishes and algae in a series of studies in Panama and Oklahoma streams. In Panama, Power (1984 a,b,c) found that the distribution of the armoured catfish in stream pools was constrained by depth- and size-specific predators. Large grazing fishes avoided shallow waters where wading and diving predators occurred, resulting in bands of algae being maintained along shallow stream margins. Small grazing fishes avoided deep water where swimming predators occurred. The density of small grazers, however, was too low to deplete algae along the shallow river margins. These body size-depth interactions explained variation in algal standing crops along depth gradients (Power, 1987). In a similar study in a Oklahoma prairie-margin stream, Power et al. (1985) tested whether largemouth bass (*Micropterus salmoides*) were responsible for differences in pool-to-pool distribution of an algal-grazing minnow (*Campostoma anomalum*) and attached algae (predominately *Spirogyra* sp. and *Rhizoclonium* sp.). They removed bass from a pool, divided the pool longitudinally into two sections, and added *Campostoma* to one side. Over the following 5 weeks, standing crops of algae decreased significantly on the *Campostoma* side but increased on the control side. In two nearby, unmanipulated *Campostoma* pools, three bass were added to each pool. Numbers of grazing *Campostoma* declined due to emigration and predation, whereas the standing crop of algae increased after bass addition.

In another three level system, where invertebrates instead of fish were the main herbivores, Bechara et al. (1992) found strong effects of brook trout (*Salvelinus fontinalis*). Water from a boreal stream was led to artificial stream channels which either had no fish or brook trout. Invertebrates were allowed to colonise during a 3-week period prior to introduction of brook trout. They found that size selective predation by brook trout on the herbivorous mayfly *Baetis* and the omnivorous caddisfly *Psychoglypha subborealis*, had positive effects on periphyton biomass. The increase in periphyton biomass was mostly due to trout's consumption of the grazer, *Baetis*.

Complementary cage experiments, involving both the inclusion and exclusion of selected invertebrates from periphyton-covered surfaces, demonstrated that *Baetis* could reduce periphyton biomass.

So far, only one study has examined cascading interactions in benthic food chains with four trophic levels, i.e., with piscivorous fish as the top predator. Power and co-workers studied interactions in a food chain involving piscivores, benthivorous fish, grazing invertebrates and algae (Fig. 5) in small enclosures in the Eel River, California, USA (Power, 1990, 1992 a, b). Although not piscivores in the classical sense, juvenile roach (*Hesperoleucas symmetricus*) and steelhead (*Oncorhynchus mykiss*) were

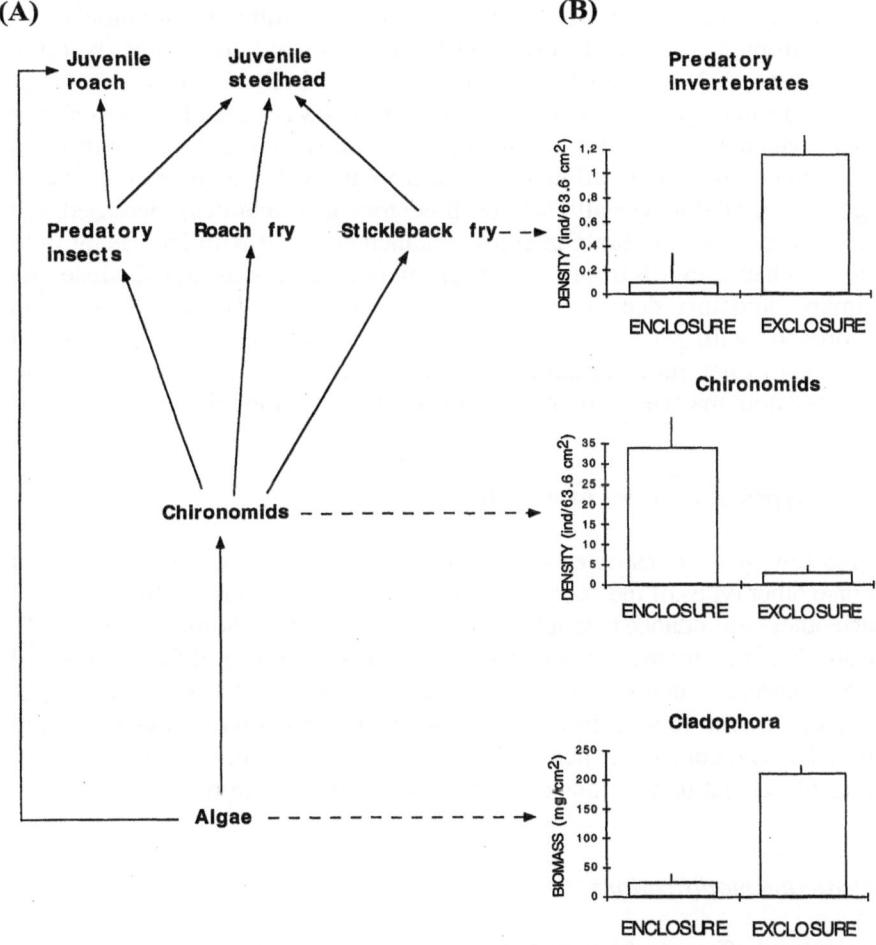

Figure 5. Food chain interactions in pools in a California stream (A), densities of predatory insects and chironomids, and algal biomass water 5 weeks in enclosures with juvenile roach and steelhead compared to exclosures without juvenile fish (B). Modified from Power (1990).

the top predators in the system, feeding on fish fry. Juvenile roach and steelhead reduced abundances of various predatory invertebrates and fish fry (roach and stickleback *Gasterosteus aculeatus*), which resulted in an increase in abundance of algivorous chironomid larvae (*Pseudochironomus richardsoni*). This, in turn, resulted in a decrease in biomass of the algae, *Cladophora* and *Nostoc*. Further analyses revealed that the strong trophic cascades were associated with biota from boulder-bedrock substrates, but not with gravel-dwelling biota (Power, 1992 b).

Environmental productivity has been incorporated into later extensions of HSS-theory (e.g., Fretwell, 1977; Oksanen et al., 1981). According to theory food chain length is constrained by environmental productivity. Moreover, theory also predicts how changes in productivity affect biomass among trophic levels in food chains of different lengths. For example, increasing productivity in three trophic level systems should result in higher biomass of the top trophic and primary producer levels. In stream enclosures, Wootton and Power (1993) tested the effect of productivity in food chains of different lengths. Primary production in stream enclosures was varied by altering light levels, which was achieved by covering enclosures with screens that differed in amount of light penetration. In a three trophic level system, the biomass of algae and predators (small fish and predatory invertebrates) increased, but the biomass of invertebrate grazers remained constant with increasing light levels. When a fourth trophic level, piscivorous fish, was added, algae and primary predators decreased, but grazers increased. The results were in accordance with predictions of classic food chain theory, suggesting that responses of natural communities to changes in environmental productivity and predators may be adequately described by these models.

Other types of indirect interactions

Up to now we have restricted our discussion to trophic cascades. There are of course other types of indirect interactions where two species indirectly affect each other's abundance through an intermediate species. Some involve modifications of the interaction between two species as a result of the presence of a third species (interaction modification). Others involve facilitative relationships among four or more species in which competitive relationships are altered by two consumer species. Below we describe examples of these other types of indirect interactions in benthic, freshwater communities.

Interaction modifications

Behavioral effects in food chains
Predator mediated changes in habitat use may affect the diet of intermediate consumers and thus change the strength and direction of inter-

actions in benthic food chains. Perhaps the best example from lakes comes from a series of studies on centrarchid fishes in North America. The piscivorous largemouth bass (*Micropterus salmoides*) affects habitat use by and interspecific interactions between juvenile bluegill (*Lepomis macrochirus*) and pumpkinseed sunfishes (*Lepomis gibbosus*). In the presence of bass, juvenile sunfishes leave the pelagic zone, restricting themselves to the vegetated littoral zone where they feed on benthic macroinvertebrates (e.g., Werner et al., 1983). Mittelbach (1988) suggested that this refuge-seeking behaviour of juvenile sunfish causes intense interspecific food competition, with subsequent effects on the benthic invertebrates. In a field experiment where he stocked cages in the littoral zone with sunfish at densities commonly found in lakes containing bass, he found that sunfish had a strong negative effect on the density of large invertebrates, resulting in a smaller average size of macroinvertebrates. Thus, the presence of largemouth bass may have strong, indirect effects on benthic macroinvertebrates, mediated through a behavioral shift in habitat use and thereby diet of the intermediate consumer, juvenile sunfish.

Centrarchid bass also have strong, indirect effects in streams. Power and co-workers found that bass (*M. salmoides* and *M. dolomieui*) had strong, indirect effects on periphyton mediated through stonerollers (*Campostoma*) via both trophic cascades (described above) and behavioural shifts in habitat use (Power and Matthews, 1983; Power et al., 1985). Through a series of experimental manipulations, they found that *Campostoma* reacted to the presence of bass by both moving from the deeper parts of pools to the shallow edges and by emigrating from the pools. These behavioural responses resulted in an increase in algal biomass and a change in algal species composition and suggest that bass effects may cascade down to periphyton through both lethal and behavioural effects on herbivores.

Indirect effects of largemouth bass on crayfish and their prey have also been studied (Hill and Lodge, 1995). In this case the bass were too small to consume crayfish. Nevertheless, the presence of bass increased crayfish mortality, possibly due to an increase in crayfish aggression when competing for shelter. The percent cover of macrophytes was higher in bass treatments, which was suggested to be due to higher mortality and/or reduced activity levels of crayfish. Further studies of this food chain are needed to determine how larger bass, which are able to consume crayfish, affect food chain dynamics.

Interactions between two predators and one prey species
Soluk and Collins (1988a) discuss three types of interactions between predators: neutral, where predators do no affect each other's prey consumption, negative, where total prey consumption is less than the sum of prey consumption by each of the two predators, and positive, where total prey consumption is greater than neutral values. For negative predator inter-

actions, removal of one of the predators may not reduce prey consumption due to compensation from the other predator. Removal of one of the predators for a positive predator interaction should decrease prey consumption.

Effects of more than one predator on prey species are difficult to predict based on functional response curves generated from single predator experiments (Soluk, 1993). Soluk and Collins (1988a) reported a positive predator interaction or facilitation between sculpin (*Cottus bairdi*) and stonefly (*Agnetina capitata*) predators feeding on the mayfly *Ephemerella subvaria*. The number of mayflies consumed by sculpins and stoneflies when together was nearly twice as many as the sum of their consumption when alone. The stonefly caused *Ephemerella* to increase its use of the tops and sides of stones, thereby increasing their accessibility to sculpins (Soluk and Collins, 1988b). Interestingly, when *Baetis* was the prey, the predator interaction was negative due to *Baetis* fleeing the stonefly by leaving the substrate, thereby being inaccessible to both predators (Soluk and Collins, 1988b).

Interactions between one predator and two prey species
Many examples of prey selection are examples of indirect interactions. This is because a predator's preference for one prey species has repercussions for the other prey species. For example, Greenberg et al. (1995) found that pike (*Esox lucius*) or zander (*Stizostedion lucioperca*) consumed the same number of crucian carp (*Carassius carassius*) whether or not rudd (*Scardinius erythrophtalmus*), an alternative prey fish, were present, whereas these piscivores ate fewer rudd when rudd were together with crucian carp than when alone (Fig. 6). The increased survival of rudd in the presence of crucian carp and the lack of an effect of rudd on crucian carp survival could be described as an example of an indirect commensalism. In this case, crucian carp's greater vulnerability to piscivores was attributed to it being a relatively poor schooler and a slow swimmer.

Huang and Sih (1990, 1991) studied interactions in outdoor swimming pools between the green sunfish (*Lepomis cyanellus*), the small-mouthed salamander (*Ambystoma barbouri*) and isopods (*Lirceus fontinalis*), three common inhabitants of stream pools in Kentucky, USA. In early spring, the sunfish can consume newly hatched salamanders and isopods but salamanders cannot consume isopods (Huang and Sih, 1990). Salamanders caused female isopods to leave a refuge, resulting in increased fish predation on them. The presence of isopods outside of the refuge increased fish activity, which signalled salamanders to remain in the refuge. This provided an indirect positive effect on salamander survival. Interestingly, the nature of this interaction shifted in late spring, when salamanders could consume isopods. At this time, fish benefitted isopod survival by reducing salamander predation on isopods (Huang and Sih, 1991). Salamander predation was reduced due to increased segregation between isopods and salamanders and reduced activity by salamanders.

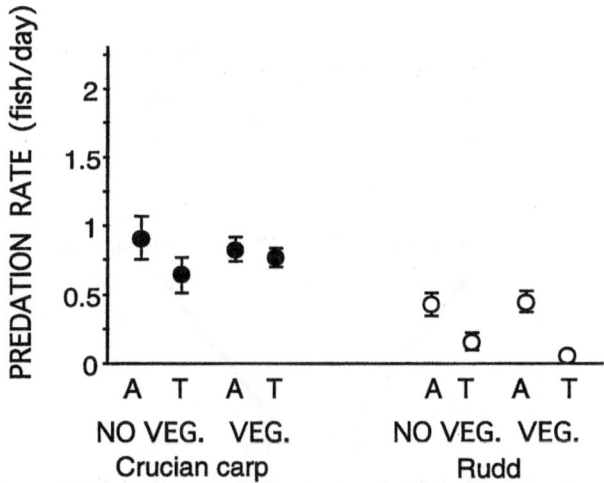

Figure 6. Number of crucian carp (filled circles) and rudd (open circles) missing per day when prey species were alone (A) and together (T). Results are shown for northern pike in the presence and absence of artificial vegetation. Error bars represent one standard error.

Indirect commensalism

Vandermeer (1980) suggested a scenario involving two consumers and two resource organisms. In a situation where the two resource organisms compete, a consumer that specialises on one of the resource organisms indirectly benefits a consumer that has specialised on the other resource. This interaction was defined as "indirect mutualism" (Fig. 1 (E)). Dethier and Duggins (1984), studying interactions between grazing snails, macroalgae and microalgae in the marine intertidal zone, later defined "indirect commensalism" for situations with asymmetric competition between resource organisms and generalist consumers without food preferences. Brönmark et al. (1991) suggested that such facilitative interactions may be operating in a system with tadpoles, snails and two different algal food resources (Fig. 7). Tadpoles and snails, which are common in small ponds without vertebrate predators, graze on periphytic algae and are thus potential resource competitors. However, the relationship between tadpoles and snails is more complicated than so. Tadpoles are efficient foragers on periphytic microalgae but they avoid the larger filamentous green algae, *Cladophora*. Snails, on the other hand, prefer periphytic microalgae, but may also include *Cladophora* in their diet. *Cladophora*, which is competitively dominant over periphytic mircroalgae, reduces availability of phosphorus to periphytic microalgae, thereby limiting their growth. Brönmark et al.'s (1991) field experiment showed that tadpoles had a negative effect on snail growth and egg production due to resource depression (microalgae), whereas snails feeding on *Cladophora* had a positive effect on

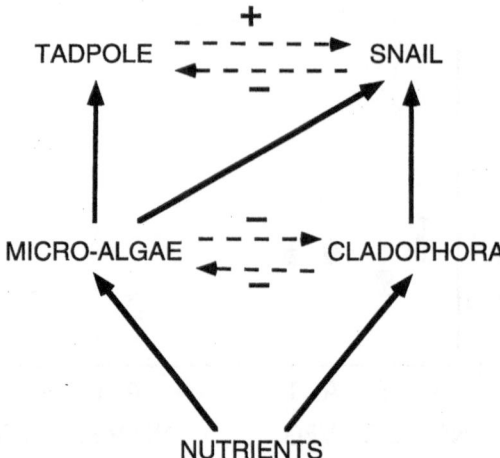

Figure 7. Interactions between tadpoles, snails, their algal food resources and nutrients. Arrows between trophic levels denote consumption, whereas broken arrows denote competitive (negative sign) or facilitative (positive sign) interactions. From Brönmark et al. (1991).

length of tadpole larval period, tadpole mass and tadpole growth rate. This was because snail grazing on *Cladophora* increased availability of nutrients for the microalgae, resulting in increased availability of micro-algae for the tadpoles.

Indirect commensalism reported in streams have involved various grazers together with *Cladophora* and microalgae. For example, crayfish, where they occur, appear to be keystone predators with far-reaching direct and indirect effects on other organisms. In Augusta Creek, Michigan, USA, crayfish were effective at cropping *Cladophora* in deep water (20–50 cm) and in areas with current velocities < 50 cm s⁻¹ (Hart, 1992; Creed, 1994). Exclusion of the competitively dominant *Cladophora* facilitated develop-ment of epilithic diatoms as Brönmark et al. (1991) observed in their tadpole-snail experiment above. Thus, the presence of crayfish could in-directly favour epilithic diatoms, which in turn had positive effects on sessile grazing insects such as the trichopterans, *Leucotrichia* and *Psycho-mia*. The absence of crayfish favoured *Cladophora*, which in turn favoured other benthic invertebrates such as the mayfly *Stenonema* and the stoneflies *Taeniopteryx* and *Isoperla*. However, the absence of crayfish from high velocity areas did not automatically lead to *Cladophora* establishment. This was probably due to priority effects being important in determining whether *Cladophora* or microalgal lawns became established (Hart, 1992). Presumably, if grazers such as *Leucotrichia* and *Psychomia* could arrive at these areas first, they could maintain the microalgal lawns.

Feminella and Resh (1991) also described a similar situation involving competition between the competitive dominant, *Cladophora*, and micro-algae in a stream. In this case, the caddisfly *Gumaga nigricula* preferred

Cladophora and the caddisfly *Helicopsyche borealis* preferred microalgae, and both species were ineffective at reducing the biomass of their non-preferred algae. In early summer, *G. nigricula* facilitated the caddisfly *Helicopsyche borealis* by selectively grazing *Cladophora* and thereby favouring the development of *H. borealis's* preferred food resource, microalgae. Interestingly, by late summer when *Cladophora* abundance was low, the mutualistic relationship between *H. borealis* and *G. nigricula* appeared to shift to asymmetrical competition, favouring *H. borealis*.

Predators affect physical environment

Predators may also indirectly affect another species by modifying its physical environment (Fig. 1 (G)). Flecker (1992) found that sediment processing by fish decreased abundance of many insect taxa, presumably by a number of different mechanisms. Power (1990) found algal enhancement due to incidental sediment removal during grazing by armoured catfish (*Ancistrus*). However, at higher densities of armoured catfish, algae were simply depleted as direct algal grazing effects became more limiting than sedimentation. Gelwick and Matthews (1992) found that the density of nontanypodine chironomids and the proportion of fine particulate organic matter were greater in pools grazed by stoneroller (*Campostoma anomalum*) than in nongrazed pools. They suggested that this may be due to facilitation of collector-gatherers (chironomids) by reducing algal matter to sizes usable by them via either faecal production by the fish or by mechanical fragmentation of the algae. The growth or survival of collector-gatherers may also be facilitated by invertebrate shredders in general as has been suggested by many authors, but as Heard and Richardson (1995) point out, this deserves more critical study.

Factors affecting the interaction strength in benthic food chains

Recently, Strong (1992) questioned the generality of the cascading trophic interaction concept. He argued that strong effects of trophic cascades are prominent only in species-poor aquatic communities with a few strong key species at the predator and herbivore level and with algae at the bottom of the food chain. In more speciose systems, top-down effects should be buffered by, for example, a high degree of temporal and spatial heterogeneity and omnivory. However, benthic systems in streams and lakes are often characterized by a high spatial heterogeneity with species-rich food webs where omnivory commonly occurs (e.g., Lodge, et al. 1988; 1994, Diehl, 1992, 1995). And yet, as described above, a number of studies have shown that cascading trophic interactions may be of importance in structuring freshwater, benthic food chains. In addition to habitat heterogeneity and omnivory a number of other factors may reduce the strength of trophic cascades and other indirect interactions. Efficient prey defence adaptations should reduce the effect of predators on prey population dynamics, thereby decreasing the strength of indirect interactions (Neill, 1994). Anti-predator adaptations include both constitutive defences that have evolved in response to long-term changes in predation pressure, as well as predator-induced phenotypic changes in morphology and behaviour. Environmental stress may also prevent predator populations from reaching densities where they limit prey populations, thus reducing the importance of complex interactions (e.g., Menge and Sutherland, 1987; Lodge et al., 1987; Morin and Lawler, 1995). Below we will discuss how omnivory, habitat structure, environmental stress and prey defence adaptations may affect the strength of indirect interactions in benthic food chains.

Habitat complexity

Theoretical and experimental studies have shown that increasing habitat heterogeneity reduces the strength of predator-prey interactions and allows coexistence of predator and prey by providing refugia for prey organisms. Although there is a large literature on the effects of habitat heterogeneity on predator-prey interactions, few studies so far have explicitly tested how indirect interactions are affected by changes in habitat heterogeneity. However, differences in structural complexity between habitats or between systems that affect the strength of direct consumer-resource interactions should also affect the strength of indirect interactions. Below we summarize the state of our knowledge about the effects of habitat heterogeneity on predation using examples from two trophic level systems.

In ponds and lakes, the littoral zone generally has the highest structural complexity. Along windswept shores this complexity derives from coarse substrates. On less exposed sites, where sediments accumulate, submerged and emergent vegetation provide structural complexity. Stands of freshwater macrophytes house a high diversity and biomass of benthic macroinvertebrates (e.g., Gerkin, 1962; Dvorac and Best, 1982; Sozska 1975), and experimental studies have shown that increasing macrophyte densities results in higher diversity and biomass of macroinvertebrates (e.g., Gilinsky, 1984; Diehl, 1992). This is because macrophytes provide refugia from predation, as well as a surface for epiphytic algae, a food source for many grazers.

While submerged and emergent vegetation often give rise to structural complexity in lakes, coarse substrates generally have the same function in streams. As in lakes, the strongest effects of fish predation on benthic organism occur in habitats of low structural heterogeneity. For example, short-term predation trials (24 h) conducted in artificial systems showed that the mottled sculpin (*Cottus bairdi*) was less effective at consuming the stonefly *Hesperoperla pacifica*, the mayfly *Ephemerella grandis* and the caddis-fly *Rhyacophila vaccua* over cobble and pebble bottoms than over sandy bottoms, where mortality was 95–100% (Brusven and Rose, 1981). Similarly, longer term studies of the effects of fish predation on benthic communities have generally reported strongest effects when conducted in habitats with relatively low complexity (Gilliam et al., 1989; Schlosser and Ebel, 1989; Power, 1990, 1992b; Bechara et al., 1992, 1993). Studies that directly compare habitats with low and high complexity have also reached the same conclusion (e.g., Brusven and Rose, 1981; Fraser and Cerri, 1982; Power, 1992a; Bechara et al., 1992; Holomuzki and Stevenson, 1992). Holomuzki and Stevenson (1992) studied the impact of longear sunfish (*Lepomis megalotis*) and green sunfish (*Lepomis cyanellus*) on a benthic community in simple bedrock and more complex stony habitats in an ephemeral stream. They found moderate effects of fish predation in the simple habitat but no effects in the more complex habitat. In another study, Power (1992a) found that predation by American roach (*Hesperoleucas symmetricus*) and steelhead trout (*Oncorhyncus mykiss*) produced strong effects on biota associated with boulder-bedrock substrates (simple substrates), but few effects on biota associated with the more complex gravel. The results from these studies suggest that coarse substrates, which offer increased structural complexity and thereby refuges for benthic invertebrates, may mitigate indirect effects of fish on other trophic levels.

Not all studies that have varied structural complexity have reported indirect effects of fish predation. Dudgeon (1993) found no effects of substrate complexity on the loach's (*Orenectes platycephalus*) ability to reduce benthic prey abundance. However, this may be an artefact of his experimental set-up. Loaches could easily move the tile fragments, which provided substrate complexity, and thereby consume concealed prey that under natural conditions should have been more difficult to locate. Bechara et al. (1993) found that brook trout (*Salvelinus fontinalis*) affected benthic invertebrates over relatively complex pebble substrates. However, the effects were restricted to drifting and epibenthic prey; there were no effects on infauna. Thus, brook trout only affected invertebrates on top of pebbles, microhabitats of low structural complexity.

Similar results were found by Dahl and Greenberg (unpublished data). They studied the influence of habitat on interactions between brown trout and benthic invertebrates in Snällerödsån Creek in southern Sweden. A stream section was divided into sections using metal sheets and steel grids, and trout were placed in the enclosures for 1 month. Three habitat treatments were tested: a shallow, sandy habitat, a deep habitat containing a mixture of large and small cobbles and a deep habitat with large cobbles. They found that habitat type had weak effects on interactions between trout and benthic invertebrates as few taxa were affected. The authors suggested that the lack of effect might have been due to the large amount of terrestrial prey consumed by trout (30–40% of their total diet biomass) and suggested that this may be one of the reasons why the effects of fish on benthic invertebrates have been so variable. A meta-analysis, i.e., a statistical analysis of standardized comparisons of data from published studies (Gurevitch and Hedges 1993), using studies in which fish predators were manipulated (n = 10 studies) in natural systems showed that strictly benthic-feeding species (e.g., sculpins and creek chub) had moderate effects on benthic prey, whereas drift feeders (salmonids) had no effect (Dahl and Greenberg 1996). Consequently, trophic cascading effects in running waters should be more likely to occur in streams where benthic feeders are the dominant predators.

The above studies of habitat complexity in streams have considered fish-invertebrate interactions. Habitat complexity also has strong effects on interactions between piscivorous and benthic feeding fishes, and this should have consequences for lower trophic levels. Fraser and Cerri (1982) manipulated adult piscivorous creek chubs (*Semotilus atromaculatus*) and juvenile blacknose dace (*Rhinichtys atratulus*) and juvenile creek chubs in semi-natural artificial streams. They found that both prey species actively avoided compartments in the streams that contained piscivores, but structural complexity and time of day affected the strength of the avoidance response. Fraser and Emmons (1984) found that when physical structure in predator compartments was low, dace avoided predators at night but not during the day, regardless of predator density. When structure was increased, avoidance of predators at night depended on both the amount of structure and the density of predators. Fraser and Emmons (1984) suggested such predator-induced alterations of habitat use could have consequences for lower trophic levels.

The type and magnitude of structural complexity varies within streams, indicating that interaction strength may be habitat dependent. Streams are often divided, based on water depth and current speed, into habitats such as riffles, runs and pools. Coarse substrates are typical for riffles and runs, whereas finer substrates are usually associated with pools. The general pattern of larger fish, many of which are piscivorous, being restricted to deeper habitats (pools) and smaller insectivorous and algivorous fish to shallower habitats (riffles), suggests that trophic cascades or indirect interactions in general are likely to depend on habitat. The generally higher structural complexity in riffles and runs may weaken interaction strength. Schlosser and Ebel (1989) showed in outdoor artificial stream experiments that cyprinids reduced invertebrate abundance in pools but had little effect in shallow riffle and raceway habitats. They attributed this differential effect to differences in habitat complexity.

Omnivory

An omnivore can be defined as a consumer that uses resources from different trophic levels. The simplest case of omnivory involves a basic resource, an intermediate forager that consumes the resource, and a top forager that feeds on both the resource and the intermediate forager (Fig. 1(F)). The concept of cascading trophic interactions relies on strong interactions in linear food chains with distinct trophic levels, and, thus, a high prevalence of omnivory should obscure patterns in food chains as predicted from food chain theory (see Strong 1992). Populations in food chains with omnivory may interact through direct predation, indirect cascade effects, competition, apparent competition and/or mutual enhancement (e.g., Diehl 1993). Traditionally, theoretical and empirical studies have argued that omnivory is rare in natural food webs (Pimm, 1980), but recent studies have shown that omnivores are common in many systems (e.g., Polis, 1991; Diehl 1993). Menge and Sutherland (1987) argued in a conceptual model of food chain interactions, based on empirical studies of marine intertidal systems, that omnivory is important. Their model predicts that primary producers in food chains with a high degree of omnivory are controlled by consumers, whereas traditional food chain theory predicts that control of primary producers depends on the number of trophic levels in the food chain.

Diehl (1993) reviewed studies where the density of a top omnivore had been manipulated. In aquatic systems, these food chains typically consisted of a top consumer (omnivorous fish), intermediate consumers (macroinvertebrate predators, e.g., odonates, chironomids, salamanders, megalopterans), and resource organisms (herbivorous or detritivorous invertebrates, e.g., chironomids, microcrustaceans, amphipods). Generally, top consumers had a negative effect on intermediate consumers, whereas the effect of the top consumer on the basic resource populations varied widely from negative effects, no effects, to positive effects. Diehl suggested that relative body size differences between trophic levels affected the strength of the direct and indirect interactions. In food chains where intermediate consumers and resource organisms were relatively similar in size and both considerably smaller than the top consumer, as is the case in many freshwater food chains (fish-predatory invertebrates-non-predatory invertebrates), top consumers had relatively strong direct effects on intermediate consumer and basic resource levels, whereas indirect effects were relatively weaker.

None of the studies reviewed by Diehl involved primary producers. In fact, we are only aware of one study including primary producers. Lodge et al. (1994) found strong cascading effects of the omnivorous crayfish on snails and algae, suggesting that omnivory not always results in weak indirect interactions.

In streams, most communities are generally characterized by high connectance (a high fraction of possible trophic links are realized) (Rundle and Hildrew 1992), and this may explain why examples of indirect effects in running waters are rare. Stream fishes are usually omnivores, consuming predatory invertebrates, non-predatory invertebrates and terrestrial prey. When fish selectively prey on large, predatory invertebrates (e.g., Crowder and Cooper 1982, Hambrigth et al. 1986, Power 1992b), smaller, non-predatory invertebrates may be released from predation (Power 1992b). The fact that invertebrate predators can be consumed by fish and also compete with fish for smaller non-predatory invertebrates may explain the commonly observed lack of an effect of fish on the biomass of invertebrate prey (i.e., indirect effects mask direct effects of fish predation). Enclosures without fish are not predator-free, and consumption of invertebrates by predatory invertebrates may compensate for the absence of fish. Moreover, competition between fish and invertebrate predators is likely asymmetrical, at least for salmonids, as they may complement their benthic diet by consuming drifting invertebrates of terrestrial origin (Bridcut and Giller 1995, Dahl and Greenberg, unpublished data).

Despite the existence of many papers measuring effects of predation from either vertebrate or invertebrate predators (e.g., Allan, 1982; Peckarsky, 1985; Malmqvist and Sjöström, 1987; Cooper et al., 1990), few studies have directly compared the effects of vertebrate and invertebrate predators and the consequences of these interactions for benthic communities. Wooster (1994) reviewed 20 studies that either considered the effects of invertebrate predators on invertebrate prey density or fish predators on invertebrate prey density. He found that predatory invertebrates had on average significantly stronger effects on invertebrate prey density than fish. Wooster (1994) suggested that the reason for this difference was related to differences in prey emigration responses to vertebrate and invertebrate predators. Dahl and Greenberg (1997) directly compared the effects of vertebrate and invertebrate predators on benthic communities within the same system and found that invertebrate predators at natural densities could consume prey at the same rate as fish under some circumstances. They measured consumption rates of brown trout (*Salmo trutta*) and three densities of leeches (*Erpobdella octaculata*) on two prey types (*Gammarus pulex* and *Baetis rhodani*) and two types of substrate, fine gravel and fine gravel with cobbles. Trout had higher foraging rates over fine gravel bottoms than over cobble bottoms, whereas leeches were more effective over cobble bottoms. Prey type also affected foraging rates, with trout consuming more *Baetis* than *Gammarus*, and leeches consuming more *Gammarus* than *Baetis*. A direct comparison of foraging rates of trout and leeches when alone showed that trout consumed more prey than the highest density of leeches, except when *Gammarus* occurred over cobble bottoms; then trout consumed prey at the same rate as the lowest leech density. When leeches and trout were together, trout foraging was unaffected by leeches, but leech foraging was affected by trout when feeding on *Baetis* but not *Gammarus*. The effect of trout on leeches was probably due to behavioral effects on leeches as only one leech was consumed by trout. The results suggest that predation from predatory invertebrates on benthic invertebrates could be of the same magnitude as predation from vertebrate predators, at least for some prey species.

Environmental stress

In Menge and Sutherland's (1987) model of community regulation, environmental stress plays an important role in structuring communities, and thereby affects the nature and strength of complex interactions. They describe two types of environmental stress, those where mechanical forces act directly on organisms, referred to as physical stresses, and those influencing biochemical reactions in organisms referred to as physiological stresses (Menge and Sutherland, 1987). In freshwater systems, mechanical stress involves movement of water and objects in the water, and is a commonly occurring type of stress in streams. Physiological stress involves factors such as temperature and light, which directly affects rates of biological processes, and is relevant in both lentic and lotic environments.

The most characteristic feature of streams is that water flow varies both spatially and temporally, producing environmental stress. The nature of this variation sets the framework for the types and strengths of complex interactions in benthic stream communities. Under highly variable, unpredictable flow regimes lotic systems are dominated by physical stresses, whereas more predictable flow regimes should allow the expression of complex interactions, and even physiological stresses may be important here. Poff and Ward (1989) classified lotic systems based on the degree of intermittence, flood frequency, flood predictability and flow predictability. Essentially, they suggested that biotic interactions are more likely to be important in streams with low intermittency and flood frequency and with high flood and flow predictability. It is in these types of streams then that complex interactions should occur.

Unfortunately, little empirical work has been conducted looking at the relationships between physical stress and complex interactions. Schlosser and Angermeier (1990), who observed reductions in densities of benthic invertebrates and age 0+ fish after spring floods, suggested future research should be directed at flow variability. A few studies have looked at how flow affects competition between two species, i.e., not a complex interaction. For example, Greenberg (1988) found that during two years of low flow and high temperatures, the fish *Etheostoma rufilineatum* restricted *E. simoterum* to stream margins through interference, whereas during a year of high discharge and low temperature, when the density of *E. rufilineatum* was lower, there was no interaction. Similarly, year-to-year variation in the timing and intensity of winter storms affected the relative importance of interspecific competition and disturbance for two filter-feeding invertebrates, *Hydropsyche oslari* and *Simulium virgatum* (Hemphill and Cooper 1983, Hemphill 1991).

As stated above, few studies have treated the effects of flood disturbance on complex interactions in benthic stream communities. An exception is described in Power (1992b). She found that variable floods could produce great variation in the dynamics and strength of top-down effects. In one year, a June flood severely reduced stands of *Cladophora*, removing a potential refuge for chironomids from fish predators, and thus no trophic cascades occurred. In another year lacking this late spring flood, *Cladophora* attained high biomass, protected chironomids from fish, and resulted in strong top-down effects. McAuliffe (1984) found that the caddisfly *Leucotrichia pictipes* was an aggressive competitive dominant, monopolising space over other sessile and mobile invertebrates. However, physical disturbance could alter this monopolization of space. At low flow, in shallow areas where stones may be temporarily exposed, short-lived sessile invertebrates were favoured. *Leucotrichia* density was also reduced on stones that overturned during high flows, favouring other invertebrate taxa.

Flooding is an extreme flow event, but even more modest changes in flow may affect complex interactions in benthic communities. In outdoor artificial stream experiments, Schlosser (1995) found that the presence of fish predators had a proportionately greater effect on invertebrate colonisation under elevated flows, especially in pools. Although the mechanism for this difference is unclear, this indicates that flow may be able to modify complex interactions. Extremely low flows or droughts may also affect complex interactions, both by concentrating animals but also trough increased physiological stress due to increased temperatures and decreased oxygen concentrations. A number of studies have reported lethal effects of low flows, although the consequences of this for complex interactions have not been studied (Matthews, 1982).

Physiological stress such as light may limit primary production and thereby affect interaction strength in streams (Hill and Harvey, 1990; Hill et al., 1995). Feminella et al. (1989) found that periphyton biomass was negatively correlated to canopy cover, and they were unable to demonstrate a significant effect of grazers in a highly shaded stream. Similarly, Hill and Harvey (1990) found no indirect effects of creek chubs and snails on periphyton density in a shaded stream. In contrast, Rosemond et al. (1993) could not find that biotic interactions were stronger in stream sections with high light levels relative to stream sections with low light levels. In addition to photosynthetically active radiation, ultraviolet radiation, which has deleterious effects on both algae and invertebrates, may also affect trophic interactions in shallow freshwater habitats. In outdoor artificial flume experiments Bothwell et al. (1994) found that middle ultraviolet radiation (UV-B: 320–400 nm) was more harmful to larval chironomids than algae, resulting in an increase in diatom biomass. Bothwell et al. (1994) speculated that the harmful effects of UV-B on herbivores might have strong repercussions for higher trophic levels.

Prey defences

Effective prey defences tend to decouple predator-prey dynamics and decrease the strength of interactions, both direct and indirect, in food chains. Prey have evolved many morphological, chemical and behavioural defences against predation in freshwaters. Calcium carbonate and chitinous covering as well as spines protect organisms such as snails, crayfish and fish from predation (Klosiewski, 1991, Lodge et al., 1994, Hoogland et al., 1957). Chemical defences (toxins) make many algae, such as bluegreens, unpalatable to herbivores, and many macroinvertebrates in the littoral zone, including notonectids, dytiscids and gyrinids, have glands that produce distasteful secretions that ward off attacking predators (Scrimshaw and Kerfoot, 1987). Behavioral responses to predators include changes in activity, habitat shifts and evasive movements away from predators (e.g., drifting stream invertebrates).

In freshwater systems, fish are often the top predator. Most fish are gape-limited predators; that is they must consume prey whole. Consequently there exists an upper prey size above which prey are no longer susceptible to predation, i.e., there is an absolute size refuge. Hambright et al. (1991) suggested that absolute size refuges may have important consequences for interactions in food chains. In pelagic food webs, gape-limited piscivores produce prey populations dominated by large, deep-bodied planktivores (Hambright et al. 1991, Hambright 1994). These large individuals may continue feeding on zooplankton, resulting in a decoupling of trophic interactions. In benthic food chains similar effects of piscivores may occur. Brönmark and Weisner (1996) surveyed 44 ponds in southern Sweden and found no evidence that the biomass of the primary producer, periphyton, was related to piscivore density (see above; Fig. 4). In ponds with pisci-

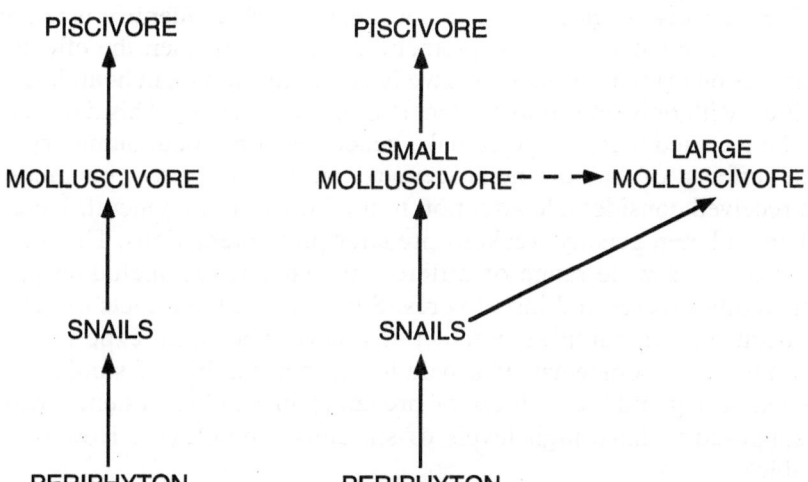

Figure 8. Trophic interactions in a benthic freshwater food chain. To the left a food chain lacking size refugia, whereas in the food chain to the right large molluscivores have reached an absolute size refuge from predation by piscivores, decoupling the cascade effect of piscivores. The broken arrow denotes recruitment of small molluscivores to the larger stage.

vores, populations of molluscivorous fish were dominated by large individuals (Brönmark et al., 1995), which selectively reduced the density of
large effective periphyton-grazing snail species, resulting in a snail community dominated by small, thick-shelled, mainly detritivorous, snail
species. Thus, changes in size-structure caused by top predators resulted in
a decoupling of cascading trophic interactions (Fig. 8).

Conclusions and future directions

We tend to think of the study of complex interactions as a relatively new
research area, with its early beginnings in 1960 with the seminal work of
Hairston, Smith and Slobodkin (Hairston et al., 1960). However, ideas about
complex indirect interactions with often non-intuitive effects were discussed for bird-insect-crop interactions over 100 years ago by Lorenzo Camerano (Camerano, 1880). This underscores that the role of complex interactions in communities is not a new research area, rather that there has been
a renewed interest in the area. The question then is how far have we come
and where should we focus our research efforts in the immediate future?

Clearly, complex interactions are important in freshwater, benthic communities, even if these communities are typically species-rich, with many
of the species being omnivorous and occurring in habitats with high levels
of structural complexity. Food chain theory points out that a high level of
omnivory should weaken interaction strength. However, the mere presence
of omnivory does not guarantee that this will be so. Factors such as the relative size of interacting species and the degree of omnivory are important to
consider. Clearly we need more experimental work to identify when and
how much omnivory weakens interaction strength. Further, the effects of
omnivores on primary producers have been poorly studied in benthic communities, with only one study to date (Lodge et al., 1994). This deficiency
must be rectified before we can make generalizations about omnivory.

The effects of habitat heterogeneity on predator-prey interactions
have received considerable attention in the literature. In general, increasing habitat heterogeneity weakens predator-prey interactions. This seems
to be true for a wide range of different predator types, including piscivores, molluscivores and insectivores. Surprisingly, the effects of habitat
heterogeneity on complex interactions have not been experimentally studied. It is somewhat of a paradox that a number of studies have
revealed strong indirect effects of predators in benthic systems, which
are supposed to have high levels of structural complexity. How is this
possible?

Curiously, most studies of food chain dynamics in benthic communities
have ignored piscivores. This contrasts strongly with pelagic systems in
which many studies have shown strong effects of piscivores on other
trophic levels. Why has this predator type been ignored in benthic systems?

There is no apparent reason why their importance should be less in benthic systems than in pelagic ones.

Most studies of complex interactions in benthic systems have used enclosures of various types, many of which have been quite small in size. This, of course, may lead to biases in our interpretations of results. Because enclosures often have been small rather high densities of organisms have been used, which may exaggerate effects of predators. This is further complicated by the fact that enclosures affect immigration and emigration of organisms, making it difficult to identify mechanisms behind observed effects. Clearly we must increase the scale of our experiments, both spatially but also temporally. We know very little about the long-term effects of complex interactions on community structure and function. We implore researchers in this field to employ whole-system, long-term manipulations in their work as this will better enable us to identify the role of complex interactions in freshwater communities.

References

Aho, J. (1984) Relative importance of hydrochemical and equilibrial variables on the diversity of freshwater gastropods in Finland. *In:* A. Solem and A.C. Van Bruggen (eds): *Biogeographical Studies on Non-marine Mollusca.* E.J. Brill, Leiden, The Netherlands, pp 198–206.

Allan, J.D. (1982) The effects of reduction in trout density on the invertebrate community of a mountain stream. *Ecology* 63:1444–1455.

Allan, J.D. (1994) *Stream Ecology: Structure and Function of Running Waters.* Chapman & Hall, London, UK.

Bechara, J.A., Moreau, G. and Planas, D. (1992) Top-down effects of brook trout (*Salvelinus fontinalis*) in a boreal forest stream. *Can. J. Fish. Aquat. Sci.* 49:2093–2103.

Bechara, J.A., Moreau, G. and Hare, L. (1993) The impact of brook trout (*Salvelinus fontinalis*) on an experimental stream community: the role of spatial and size refugia. *J. Anim. Ecol.* 62:451–464.

Bothwell, M.L., Sherbot, D.M.J. and Pollock, C.M. (1994) Ecosystem response to solar ultra-violet-B radiation: influence of trophic-level interactions. *Science* 265:97–100.

Bridcut, E and Giller, P. (1995) Diet variability in brown trout (*Salmo trutta*) sub-populations and individuals. *Can. J. Fish. Aquat. Sci.* 52:2543–2552.

Brönmark, C. (1989) Interactions between epiphytes, macrophytes and freshwater snails: a review. *J. Mollus. Stud.* 55:299–311.

Brönmark, C. (1994) Effects of tench and perch on interactions in a freshwater, benthic food chain. *Ecology* 75:1818–1824.

Brönmark, C. and Weisner, S.E.B. (1996) Decoupling of cascading trophic interactions in a freshwater, benthic food chain. *Oecologia* 108:534–541.

Brönmark, C., Rundle, S.D. and Erlandsson, A. (1991) Interactions between freshwater snails and tadpoles: competition and facilitation. *Oecologia* 87:8–18.

Brönmark C., Klosiewski S.P. and Stein R.A. (1992) Indirect effects of predation in a freshwater, benthic food chain. *Ecology* 73:1662–1674.

Brönmark, C., Paszkowski, C.A., Tonn, W.M. and Hargeby, A. (1995) Predation as a determinant of size structure in populations of crucian carp (*Carassius carassius*) and tench (*Tinca tinca*). *Ecology of Freshwater Fish* 4:85–92.

Brusven, M.A. and Rose, S.T. (1981) Influence of substrate composition and suspended sediment on insect predation by the torrent sculpin *Cottus rhotheus. Can. J. Fisheries Aquat. Sci.* 38:1444–1448.

Camerano, L. (1880) On the equilibrium of living beings by means of reciprocal destruction. Reprinted in: S. Levin (ed.): *Lecture Notes in Biomathematics 100,* pp. 360–380.

Carpenter, S.R. (1988) (ed.) *Complex Interactions in Lake Communities*. Springer-Verlag, New York, USA.

Carpenter S.R. and Kitchell J.F. (1993) (eds) *The Trophic Cascade in Lakes*. Cambridge University Press, Cambridge.

Cooper, D.D., Walde, S.J. and Peckarsky, B.L. (1990) Prey exchange rates and the impact of predators on prey populations in streams. *Ecology* 71:1503–1514.

Cooper, S.C. and Barmuta, L.A. (1993) Field experiments in biomonitoring. *In:* D.M. Rosenberg and V.H. Resh (eds) *Freshwater Biomonitoring and Benthic Macroinvertebrates*. Chapman and Hall, New York, pp 399–441.

Creed, R.P. (1994) Direct and indirect effects of crayfish grazing in a stream community. *Ecology* 75:2091–2103.

Crowder, L.B. and Cooper, W.E. (1982) Habitat structural complexity and the interactions between bluegills and their prey. *Ecology* 63:1802–1813.

Dahl, J. and Greenberg, L.A. (1996) Impact on stream benthic prey by benthic vs. drift feeding predators: a meta-analysis. *Oikos* 77:177–181.

Dahl, J. and Greenberg, L.A. (1997) Foraging rates of a vertebrate and an invertebrate predator in stream enclosures. *Oikos* 78:459–466.

Dethier, M.N. and Duggins, D.O. (1984) An "indirect commensalism" between marine herbivores and the importance of competitive hierarchies. *Amer. Nat.* 124:205–219.

Diamond, J. (1986) Laboratory experiments, field experiments, and natural experiments. *In:* J. Diamond and T.J. Case (eds): *Community Ecology*. Harper and Row, New York, pp 3–22.

Diehl, S. (1992) Fish predation and benthic community structure: the role of omnivory and habitat complexity. *Ecology* 73:1646–1661.

Diehl, S. (1993) Relative consumer sizes and the strengths of direct and indirect interactions in omnivorous feeding relationships. *Oikos* 68:151–157.

Diehl, S. (1995) Direct and indirect effects of omnivory in a littoral lake community. *Ecology* 76:1727–1740.

Dudgeon, D. (1993) The effects of spate-induced disturbance, predation and environmental complexity on macroinvertebrates in a tropical stream. *Freshwater Biol.* 30:189–197.

Dvorac, J. and Best, E.P.H. (1982) Macroinvertebrate communities associated with the macrophytes of Lake Vechten: structural and functional relationships. *Hydrobiologia* 95:115–126.

Feminella, J.W. and Resh, V.H. (1991) Herbivorous caddisflies, macroalgae, and epilithic microalgae: dynamic interactions in a stream grazing system. *Oecologia* 87:247–256.

Feminella, J.W., Power, M.E. and Resh, V.H. (1989) Periphyton responses to invertebrate grazing and riparian canopy in three northern coastal California streams. *Freshwater Biol.* 22:445–457.

Flecker, A.S. (1992) Fish trophic guilds and the structure of a tropical stream: weak direct vs. strong indirect effects. *Ecology* 73:927–940.

Fraser, D.F. and Cerri, R.D. (1982) Experimental evaluation of predator-prey relationship in a patchy environment: consequences for habitat use patterns in minnows. *Ecology* 63:307–313.

Fraser, D.F. and Emmons, E.E. (1984) Behavioral response of blacknose dace (*Rhinichys atratulus*) to varying densities of predatory creek chub (*Semotilus atromaculatus*). *Can. J. Fish. Aquat. Sci.* 41:364–370.

Fretwell, S.D. (1987) Food chain dynamics: the central theory of ecology? *Oikos* 50:291–301.

Frost, T.M., DeAngelis, D.L., Bartell, S.M., Hall, D.J. and Hurlbert, S.H. (1988) Scale in the design and interpretation of aquatic community research. *In:* S.R. Carpenter (ed.): *Complex Interactions in Lake Communities*. Springer-Verlag, New York, pp 229–258.

Gelwick, F.P. and Matthews, W.J. (1992) Effects of an algivorous minnow on temperate stream ecosystem properties. *Ecology* 73:1630–1645.

Gerking, S.D. (1962) Production and food utilization in a population of bluegill sunfish. *Ecol. Monogr.* 32:31–78.

Gilinsky, E. (1984) The role of fish predation and spatial heterogeneity in determining benthic community structure. *Ecology* 65:455–468.

Gilliam, J.F., Fraser, D.F. and Sabat, A.M. (1989) Strong effects of foraging minnows on a stream benthic invertebrate communities. *Ecology* 70:445–452.

Greenberg, L.A. (1988) Interactive segregation between the stream fishes *Etheostoma simoterum* and *E. rufilineatum*. *Oikos* 51:193–202.

Greenberg, L.A., Paszkowski, C.A. and Tonn, W.M. (1995) Effects of prey species composition and habitat structure on foraging by two functionally distinct piscivores. *Oikos* 74:522–532.

Gurevitch, J. and Hedges, L.V. (1993) Meta-analysis: combining the results of independent experiments. *In:* S.M. Scheiner and J. Gurevitch (eds) *Design and Analysis of Ecological Experiments.* Chapman and Hall, New York, London. pp 378–398.

Hairston, N., Smith, F.E. and Slobodkin, L. (1960) Community structure, population control, and competition. *Amer. Nat.* 94:421–425.

Hambright, K.D. (1994) Morphological constraints in the piscivore-planktivore interaction: Implications for the trophic cascade hypothesis. *Limnol. Oceanogr.* 39:897–912.

Hambright, K.D., Trebatosky, R.J., Drenner, R.W. and Kettle, D. (1986) Experimental study on the impact of bluegill (*Lepomis macrochirus*) and largemouth bass (*Micropterus salmoides*) on pond community structure. *Can. J. Fish. Aquat. Sci.* 43:1171–1176.

Hambright, K.D., Drenner, R.W., McComas, S.R. and Hairston, Jr. N.G. (1991) Gape-limited piscivores, plantivore size refuges, and the trophic cascade hypothesis. *Hydrobiologia* 121:389–404.

Hart, D.H. (1992) Community organization in streams: the importance of species interactions, physical factors, and chance. *Oecologia* 91:220–228.

Heard, S.B. and Richardson, J.S. (1995) Shredder-collector facilitation in stream detrital food webs: is there enough evidence. *Oikos* 72:359–366.

Hemphill, N. (1991) Disturbance and variation in competition between two stream insects. *Ecology* 72:864–872.

Hemphill, N. and Cooper, S.D. (1983) The effect of physical disturbance on the relative abundances of two filter-feeding insects in a stream. *Oecologia* 58:378–382.

Henrikson, L., Nyman, H., Oscarson, H. and Stenson, J. (1980) Trophic changes, without changes in the external nutrient loading. *Hydrobiologia* 68:257–263.

Hill, A.M. and Lodge, D.M. (1995) Multi-trophic level impact of sublethal interactions between bass and omnivorous crayfish. *J. N. Amer. Benthol. Soc.* 14:306–314.

Hill, W.R. and Harvey, B.C. (1990) Periphyton responses to higher trophic levels and light in a shaded stream. *Can. J. Fish. Aquat. Sci.* 47:2307–2314.

Hill, W.R., Ryon, M.G. and Schilling, E.M. (1995) Light limitation in a stream ecosystem: responses by primary producers and consumers. *Ecology* 76:1297–1309.

Holomuzki, J.R. and Stevenson, R.J. (1992) Role of predatory fish in community dynamics of an ephemeral stream. *Can. J. Fish. Aquat. Sci.* 49:2322–2330.

Hoogland, R., Morris, D. and Tinbergen, N. (1957) The spines of sticklebacks (*Gasterosteus* and *Pygosteus*) as means of defence against predators (*Perca* and *Esox*). *Behaviour* 10:205–236.

Huang, C. and Sih, A. (1990) Experimental studies on behaviorally-mediated, indirect interactions through a shared predator. *Ecology* 71:1515–1522.

Huang, C. and Sih, A. (1991) Experimental studies on direct and indirect interactions in a three trophic-level stream system. *Oecologia* 85:530–536.

Hubendick, B. (1947) Die Verbreitungsverhältnisse der limnischen Gastropoden in Südschweden. *Zoologiska Bidrag, Uppsala* 24:415–559.

Johansson, L. and L. Persson (1986) The fish community of temperate, eutrophic lakes. *In:* B. Riemann and M. Søndergaard (eds): *Carbon Dynamics of Eutrophic, Temperate Lakes: The Structure and Functions of the Pelagic Environment.* Elsevier Scientific Publishers, Amsterdam, The Netherlands, pp 237–266.

Kennedy, M. and P. Fitzmaurice (1970) The biology of Tench, *Tinca tinca* (L.) in Irish waters. *Proceedings of the Royal Irish Academy* 69:31–82.

Kerfoot, W.C. and Sih, A. (1987) (eds) *Predation. Direct and Indirect Impacts on Aquatic Communities.* University Press of New England, Hanover and London

Klosiewski SP (1991) Selective predation by pumpkinseed sunfish and its influence on snail assemblage structure. *Dissertation.* Ohio State University, Columbus, Ohio, USA.

Lauder, G.V. (1983) Functional and morphological bases of trophic specialization in sunfishes (Teleostei, Centrarchidae). *J. Morphol.* 178:1–21.

Lodge, D.M., Brown, K.M., Klosiewski, S.P., Stein, R.A., Covich, A.P., Leathers, B.K. and Brönmark, C. (1987) Distribution of freshwater snails: spatial scale and the relative importance of physicochemical and biotic factors. *Amer. Malacol. Bull.* 5:73–84.

Lodge, D.M., Barko, J.W., Strayer, D., Melack, J.M., Mittelbach, G.G., Howarth, R.W., Menge, B. and Titus, J.E. (1988) Spatial heterogeneity and habitat interactions in lake communities.

In: S.R. Carpenter (ed.): *Complex Interactions in Lake Communities*. Springer-Verlag, New York, pp 181–209.

Lodge, D.M., Kershner, M.W. and Aloi, J. (1994) Effects of an omnivorous crayfish (*Orconectes rusticus*) on a freshwater littoral food web. *Ecology* 75:1265–1281.

Malmqvist, B. and Sjöström, P. (1987) Stream drift as a consequence of disturbance by invertebrate predators. *Oecologia* 74:396–403.

Martin, T.H., Crowder, L.B., Dumas, C.F. and Burkholder, J.M. (1992) Indirect effects of fish on macrophytes in Bays Mountain Lake: evidence for a littoral trophic cascade. *Oecologia* 89:476–481.

Matthews, W.J., Surat, E. and Hill, L.G. (1982) Heat death of the orangethroat darter, *Etheostoma spectabile* (Precidae), in a natural environment. *Southwest. Naturalist* 27:216–217.

McAuliffe, J.R. (1984) Competition for space, disturbance, and the structure of a benthic stream community. *Ecology* 65:894–908.

Menge, B.A. and J.P. Sutherland (1987) Community regulation: Variation in disturbance, competition, and predation in relation to environmental stress and recruitment. *Amer. Nat.* 130:730–757.

Merrick, G.W., Hershey, A.E. and McDonald, M.E. (1991) Lake trout (*Salvelinus namaycush*) control of snail density and size distribution in an arctic lake. *Can. J. Fish. Aquat. Sci.* 48: 498–502.

Mittelbach, G.G. (1984) Predation and resource partitioning in two sunfishes (Centrarchidae). *Ecology* 65:499–513.

Mittelbach, G.G. (1988) Competition among refuging sunfishes and effects of fish density on littoral zone invertebrates. *Ecology* 69:614–623.

Mittelbach, G.G., Turner, A.M., Hall, D.J., Retting, J.E. and Osenberg, C.W. (1995) Perturbation and resilience: A long term, whole-lake study of predator extinction and reintroduction. *Ecology*: 2347–2360.

Morin, P.J. and Lawler, S.P. (1995) Food web architecture and population dynamics: Theory and empirical evidence. *Annu. Rev. Ecol. Syst.* 26:505–529.

Neill, W.E. (1988) Complex interactions in oligotrophic lake food webs: responses to nutrient enrichment. *In:* S.R. Carpenter (ed.): *Complex Interactions in Lake Communities*. Springer-Verlag, New York, pp 31–45.

Neill, W.E. (1994) Spatial and temporal scaling and the organization of limnetic communities. *In:* P.S. Giller, A.G. Hildrew and D.G. Raffaelli (eds): *Aquatic Ecology. Scale, Pattern and Process*. Blackwell Scientific Publications, Oxford, pp 189–231.

Økland, J. (1990). *Lakes and Snails*. Universal Book Services/Dr. W. Backhuys, Oegstgeest, The Netherlands.

Oksanen, L., Fretwell, S.D., Arruda, J. and Niemela, P. (1981) Exploitation ecosystems in gradients of primary productivity. *Amer. Nat.* 118:240–261.

Osenberg, C.W. and Mittelbach G.G. (1990) The effects of body size on the predator-prey interaction between pumpkinseed sunfish and gastropods. *Ecol. Monogr.* 59: 405–432.

Osenberg, C.W., Mittelbach, G.G. and Wainwright, P.C. (1992) Two-stage life histories in fish: the interaction between juvenile competition and adult performance. *Ecology* 73:255–267.

Peckarsky, B.L. (1985) Do predaceous stoneflies and siltation affect the structure of stream insect communities colonizing enclosures? *Can. J. Zool.* 63:1519–1530.

Pimm, S.L. (1982) *Food Webs*. Chapman and Hall, London.

Poff, N.L. and Ward, J-V. (1989) Implications of streamflow variability and predictability for lotic community structure: a regional analysis of streamflow patterns. *Can. J. Fish. Aquat. Sci.* 46:1805–1818.

Polis, G.A. (1991) Complex trophic interactions in deserts: An empirical assessment of food chain theory. *Amer. Nat.* 138:123–155.

Power, M.E. (1984a) Depth distributions of armored catfish: Predator induced resource avoidance? *Ecology* 65:523–528.

Power, M.E. (1984b) Habitat quality and the distribution of algae-grazing catfish in a Panamanian stream. *J. Anim. Ecol.* 53:357–374.

Power, M.E. (1984c) The importance of sediment in the feeding ecology and social interactions of an armored catfish, *Ancistrus spinosus*. *Environ. Biol. Fish.* 10:173–181.

Power, M.E. (1987) Predator avoidance by grazing fishes in temperate and trophical streams: importance of stream depth and prey size. *In:* W. Kerfoot and A. Sih (eds): *Predation: Direct and Indirect Impacts on Aquatic Communities.* University Press of New England Hanover, New Hampshire, pp 333–351.

Power, M.E. (1990) Resource enhancement by indirect effects of grazers: armored catfish, algae, and sediment. *Ecology* 71:897–904.

Power, M.E. (1990) Effects of fish in river food webs. *Science* 250:811–814.

Power, M.E. (1992a) Top-down and bottom-up forces in food webs: do plants have primacy? *Ecology* 73:733–746.

Power, M.E. (1992b) Habitat heterogeneity and the functional significance of fish in river food webs. *Ecology* 73:1675–1688.

Power, M.E. and Matthews, W.J. (1983) Algae-grazing minnows (*Campostoma anomalum*), piscivorous bass (*Micropterus spp.*), and the distribution of attached algae in a small prairie-margin stream. *Oecologia* 60:328–332.

Power, M.E., Matthews, W.J., and Stewart, A.J. (1985) Grazing minnows, piscivorous bass, and stream algae: dynamics of strong interaction. *Ecology* 66:1448–1456.

Power, M.E., Stewart, A.J. and Matthews, W.J. (1988) Grazer control of algae in an Ozark mountain stream: effects of short-term exclusion. *Ecology* 69:1894–1898.

Rosemond, A.D., Mulholland, P.J. and Elwood, J.W. (1993) Top-down and bottom-up control of stream periphyton: effects of nutrients and herbivores. *Ecology* 74:1264–1280.

Rundle, S.D. and Hildrew, A.G. (1992) Small fish and small prey in the food webs of some southern English streams. *Hydrobiologia* 25:25–35.

Sadzikowski, M.R. and D.C. Wallace (1976) A comparison of the food habits of size classes of three sunfishes (*Lepomis macrochirus Rafinesque, L. gibbosus (Linnaeus)* and *L. cyanellus Rafinesque*). *Amer. Midland Naturalist* 95:220–225.

Schlosser, I.J. (1995) Dispersal, boundary processes, and trophic-level interactions in streams adjacent to beaver ponds. *Ecology* 76:908–925.

Schlosser, I.J. and Angelmeyer, P.A. (1990) The influence of environmental resource variability, abundance, and predation on juvenile cyprinid and centrachid fishes. *Polskie Archiwum Hydrobiologii* 37:265–284.

Schlosser, I. and Ebel, K. (1989) Effects of flow regime and cyprinid predation on a headwater stream. *Ecol. Monogr.* 59:41–57.

Schmitt, R.J. (1987) Indirect interactions between prey: apparent competition, predator aggregation, and habitat segregation. *Ecology* 68:1887–1897.

Scrimshaw, S. and Kerfoot, C.W. (1987) Chemical defenses of freshwater organisms: beetles and bugs. *In.:* C.W. Kerfoot and A. Sih (eds): *Predation: Direct and Indirect Impacts on Aquatic Communities.* University Press of New England Hanover, New Hampshire, pp 240–263.

Shapiro, J. and Wright, D.I. (1984) Lake restoration by biomanipulation. *Freshwater Biol.* 14:371–383.

Soluk, D.A. (1993) Multiple predator effects: predicting combined functional response of stream fish and invertebrate predators. *Ecology* 74:219–225.

Soluk, D.A. and Collins, N.C. (1988a) Synergistic interactions between fish and stoneflies: facilitation and interference among stream predators. *Oikos* 52:94–100.

Soluk, D.A. and Collins, N.C. (1988b) Balancing risks? Responses and non-responses of mayfly larvae to fish and stonefly predators. *Oecologia* 77:370–374.

Soszka, G.J. (1975) The invertebrates on submerged macrophytes in three Masurian lakes. *Ekologika Polska* 23:393–415.

Stein, R.A., Goodman, C.G. and Marshall, E.A. (1984) Using time and energetic measures of cost in estimating prey value for fish predators. *Ecology* 65:702–715.

Strong, D.R. (1992) Are trophic cascades all wet? Differentiation and donor-control in speciose ecosystems. *Ecology* 73:747–754.

Threlkeld, S.T. (1987) Experimental evaluation of trophic-cascade and nutrient mediated effects of planktivorous fish on plankton community structure. *In:* C.W. Kerfoot and A. Sih (eds): *Predation: Direct and Indirect Impacts on Aquatic Communities.* University Press of New England, Hanover, New Hampshire, USA. pp 161–173.

Vandermeer, J. (1980) Indirect mutualism: variations on a theme by Stephen Levine. *Amer. Naturalist* 116:441–448.

Weber, L.M. and Lodge, D.M. (1990) Periphytic food and predatory crayfish: relative roles in determining snail distribution. *Oecologia* 82:33–39.

Werner, E.E. Gilliam, J.F., Hall, D.J. and Mittelbach, G.G. (1983) An experimental test of the effects of predation risk on habitat use in fish. *Ecology* 64:1540–1548.

Wooster, D. (1994) Predator impacts on stream benthic prey. *Oecologia* 99:7–15.

Wootton, J.T. (1994) The nature and consequences of indirect effects in ecological communities. *Annu. Rev. Ecol. Syst.* 25:443–466.

Wootton, J.T. and Power, M.E. (1993) Productivity, consumers, and the structure of a river food chain. *Proc. Natl. Acad. Sci. USA* 90:1384–1387.

Evolutionary Ecology of Freshwater Animals
ed. by B. Streit, T. Städler and C. M. Lively
© 1997 Birkhäuser Verlag Basel/Switzerland

Prey dispersal and predator impacts on stream benthic prey

D. E. Wooster, A. Sih and G. Englund

Center for Ecology, Evolution and Behavior, T. H. Morgan School of Biological Sciences, University of Kentucky, Lexington, KY 40506-0225, USA

Summary. The impact that predators have on local invertebrate prey density in streams varies between different types of predators as well as across different types of prey. Insights into mechanisms generating this variation can be gained by considering dispersal responses of prey to predators. If predators induce increases in prey dispersal rates then local prey density will decline. In contrast, if predators suppress prey dispersal rates then this will tend to increase prey density. In this chapter we review evidence revealing that prey dispersal responses to predators are important in driving patterns in predator impact on local prey density in streams. Future studies on predator impact on local prey density should examine the dispersal responses of prey to determine the degree to which dispersal generates patterns of predator impact on local prey density. Finally, we show that the spatial scale at which a study is conducted will have a strong influence on the degree to which prey dispersal influences predator impact on prey density. Small-scale studies will reflect mostly prey dispersal effects. In contrast, large-scale studies will reflect mostly direct (consumptive) effects of predators on prey density.

Introduction

Predators often have important effects on prey behavior (Sih, 1987; Sih and Wooster, 1994; Lima and Dill, 1990), population dynamics (Taylor, 1990; Murdoch, 1994) and community structure (Sih et al., 1985; Menge et al., 1994). In this chapter, we focus on the influence of predators on prey density, which we refer to as "predator impact". Predator impacts vary. A given predator can have little or no effect on some prey, strong negative effects on other prey, and paradoxically, positive effects on other prey. Different predators have different impacts on a given prey community. Predator impacts appear to vary across community types (e.g., marine vs. freshwater, or lotic vs. lentic; Sih et al., 1985). A central issue in ecology is thus to understand the variation in predator impacts on prey.

Predator-prey patterns can depend on spatial patchiness. That is, if we view the world as a system of connected patches, both predator and prey densities and predator impacts typically vary among patches. To understand patterns of predator impact among such patches, it is necessary to look at both predator-prey interactions within patches and movements of predators and prey among patches (Taylor, 1990; Cooper et al., 1990; Murdoch, 1994; Sih and Wooster, 1994). Perceived predator impacts can then depend on spatial scale. Strong predator impacts within a given patch need not indicate strong impacts at a global level (e.g., even if predators drive prey extinct within patches, movements of prey among patches might allow long-term persistence of prey; Taylor, 1990). Conversely, weak predator impacts within a given patch might correspond to strong predator impacts at a global level (see section *Effects of scale on prey movements and predator impacts* in this chapter).

To understand mechanisms underlying variation in predator impacts it is often useful to look at predator and prey behaviors. For example, if predators actively prefer to attack and consume some prey over others, this could explain stronger predator impacts on these preferred prey. Alternatively, weaker predator impacts on some prey could be explained by those prey showing more effective antipredator behaviors than other prey (Sih and Moore, 1990). Again, because overall predator impacts often depend on spatial patchiness, relevant behaviors could include both predator and prey behaviors within patches (that govern predation rates within patches) and predator and prey dispersal among patches.

The importance of predators in determining the densities of prey in streams has been the subject of considerable debate. Some studies have shown significant effects of predators on prey densities in streams (e.g., Walde and Davies, 1984; Peckarsky, 1985; Schlosser and Ebel, 1989; Feltmate and Williams, 1989; Gilliam et al., 1989; Lancaster, 1990; Holomuzki and Stevenson, 1992; Sih et al., 1992; Bechara et al., 1993; Forrester, 1994), while other studies have shown little or no effects (e.g., Allan, 1982; Flecker and Allan, 1984; Culp, 1986; Reice and Edwards, 1986; Lancaster et al., 1991). Here, we discuss our ideas on how prey behavior within patches, and in particular, prey movements among patches might influence patterns of predator impacts in streams. Specifically, we discuss:

1) A recent review that used a meta-analysis to quantify overall patterns of predator impact on local (within-patch) prey density in streams, with a particular emphasis on differences between predator impacts caused by vertebrate (primarily, fish) versus invertebrate (primarily, stonefly) predators (Wooster, 1994). Interestingly, the review showed that, on average, invertebrate predators appear to have stronger impacts on local prey density than vertebrate predators.

2) Two recent models (Cooper et al., 1990; Sih and Wooster, 1994) that generated predictions on how prey movements among patches (hereafter, referred to as "prey exchange") and alterations in prey-dispersal tendencies in response to predators should influence predator impacts. The latter model (Sih and Wooster, 1994) suggested a hypothesis to explain differences in predator impact between vertebrate and invertebrate predators.

3) Another recent literature review that partially tested the hypothesis discussed in (2) by looking at differences in prey behavioral responses to vertebrate and invertebrate predators (Wooster and Sih, 1995).

4) Insights that emerge from more detailed analyses of some of the few studies that include data on predation rates, prey exchange, prey

emigration responses to predators and predator impacts. These studies allow us to evaluate the relative importance of different mechanisms in explaining variation in predator impacts in these specific systems.

5) Ideas on the relationship between prey movement rates and predator impacts on prey density. We develop and test three hypotheses on this relationship through a series of new meta-analyses.

6) A model that addresses how spatial scale should govern the way prey exchange and prey emigration responses to predators influence predator impacts on prey (Englund and Olsson, 1996; Englund, in press).

7) Finally, we make suggestions for future study.

Patterns of predator impact in streams: Vertebrate vs. invertebrate predators

Dozens of studies have experimentally manipulated predator numbers to quantify effects of predators on prey densities in streams. Most of these studies have used cages to enclose or exclude predators. Some have relied on natural experimental units (e.g., distinct pools in a stream with alternating pools and riffles). A typical design compares prey densities in cages with versus without predators, or compares prey densities in cages with predators to cageless controls in areas without predators, or vice versa.

As noted, experimental studies in streams have yielded conflicting results (see earlier references). To look for generalities across studies, Wooster (1994) performed a literature review using meta-analytical techniques to draw inferences from 20 studies involving 35 comparisons of prey densities in the presence versus absence of predators. Meta-analysis is a statistical methodology designed to test for trends across multiple, independent studies (Hedges and Olkin, 1985; Gurevitch and Hedges, 1993; Arnqvist and Wooster, 1995). Meta-analysis has been used extensively in medicine (Shapiro, 1985; Mann, 1990) and education (Rosenthal, 1984; Hedges and Olkin 1985), but only recently has meta-analysis been used in ecology (Arnqvist and Wooster, 1995). The main advantages of meta-analysis are that it provides a quantitative summary of research domains and specific questions in ecology, it is a relatively powerful method of analysis, and it does not rely on comparing the statistical results of independent studies as do "vote-counting" reviews (Arnqvist and Wooster, 1995).

The goals of Wooster's (1994) meta-analysis were to test whether: 1) on average, predator impacts on local (i.e., within-patch) prey density in streams are significantly greater than zero; 2) vertebrate and invertebrate predators differ in average predator impacts; and 3) variations in experimental conditions (cage size, cage mesh size, duration of experiments) explain some of the variation in predator impacts.

On average, predators had small to moderate effects on prey density, with a mean effect size of -0.39 (95%) confidence interval: -0.23 to -0.55;

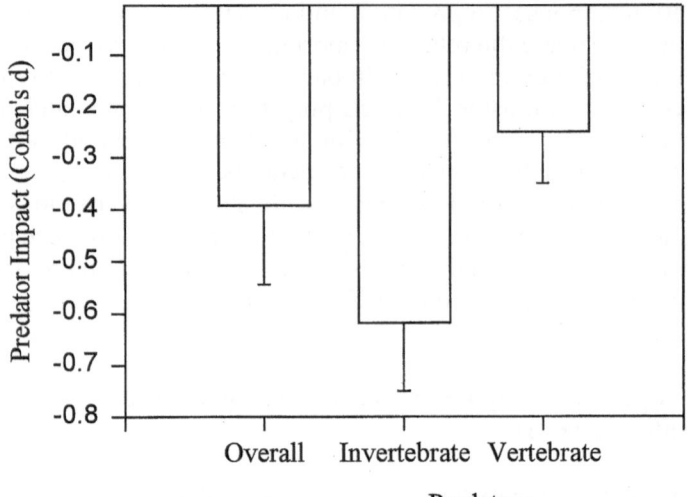

Figure 1. The impact that all predators, predatory invertebrates and predatory vertebrates have on stream invertebrate prey. Cohen's d values are a measure of the number of standard deviation units that the predator-treatment mean prey density is from the predator-free treatment mean prey density. For "overall," the bar indicates the mean predator effect (both predatory vertebrates and invertebrates combined), and the vertical line represents the 95% confidence interval. For "predatory invertebrates" and "predatory vertebrates," the bars indicate the mean effect size and the vertical lines represent one standard deviation (from Wooster, 1994).

Fig. 1). That is, on average, predators reduced prey density by 0.39 standard deviations below the mean prey density in predator-free areas. Invertebrate predators had on average, a significantly greater impact on prey density than vertebrate predators ($p < 0.05$; Fig. 1). Variation in the duration of experiments and in cage mesh size did not explain a significant portion of the variation in the magnitude of predator impacts (Wooster, 1994). Enclosure size did influence predator impact; smaller enclosures tended to show larger predator impacts.

Perhaps the most interesting outcome of the meta-analysis was that vertebrate predators generally had weaker effects on prey than did invertebrate predators. This is surprising because individual vertebrate predators (mostly, fish) have much higher consumption rates than invertebrate predators (mostly, stoneflies). Of course, invertebrate predators are often found at much higher densities than vertebrate predators; thus it is not clear how the two types of predators compare in total predation potential. Nonetheless, it is worth asking: why do vertebrate predators typically have weaker effects than invertebrate predators?

First, it is possible that the differences are merely artifactual. Experimental studies on the impacts of predatory vertebrates typically run for longer periods than those on predatory invertebrates (Wooster, 1994). However, no significant relationship was found between the duration of experiments and the observed effect sizes. Experiments with vertebrate predators also used significantly larger mesh sizes than experiments with invertebrate predators. Larger mesh sizes should allow greater prey movement rates in and out of cages as well as allowing access to cages by some invertebrate predators. Cooper et al. (1990) suggested that larger mesh sizes are associated with reduced predator impacts as the result of rapid prey immigration into cages. Wooster's (1994) review, however, found no significant relationship between mesh size and predator impact.

Finally, because vertebrate predators in streams are typically much larger than invertebrate predators, studies on vertebrates have, on average, used considerably larger cages (median $= 1.5$ m^2 for vertebrates, and 0.035 m^2 for invertebrates). Predator impacts tend to be greater in smaller cages (Wooster, 1994). Thus the stronger effects of invertebrate predators (as compared to vertebrate predators) on stream prey are confounded by differences in mean cage size.

Wooster (1994) also discussed three ecological reasons why invertebrate predators might have stronger effects on prey density than vertebrate predators. First, the availability and effectiveness of prey refuges might differ, particularly in stony streams. Prey can hide from vertebrate predators in the interstices among and below stones. For example, Power (1992) found no effect of fish (roach, *Hesperoleucas symmetricus*, and steelhead, *Oncorhynchus mykiss*) on prey densities in gravel, but strong effects on boulders that provide little refuge. In contrast, gravel and cobble do not provide good refuges for prey from smaller, invertebrate predators that can follow prey into interstices (Michael and Culver, 1986; Soluk and Collins, 1988b; Fuller and Rand, 1990). Although a few studies support the importance of refuge in explaining patterns of predator impact in streams (Schlosser and Ebel, 1989; Power, 1992), there is a need for further experimental studies that manipulate both predators and refuge availability in order to quantify effects of refuge on the relative impacts of different types of predators on prey.

The relatively weak impact of vertebrate predators on stream prey might also be explained by indirect positive effects of vertebrate predators on herbivorous invertebrate prey, through negative effects of vertebrate predators on invertebrate predators. That is, if vertebrates depress the numbers of invertebrate predators, then this should free herbivorous invertebrates from predation by predatory invertebrates. If intermediate predators potentially have strong effects on some prey (in the absence of top predators), then the presence of top predators can actually enhance the density of those prey (Sih et al., 1985). A few studies in streams include data that support this scenario (Bechara et al., 1992), and if it is often important, we would expect vertebrate predators to have stronger negative effects on predatory invertebrates than on herbivorous invertebrates. Wooster's (1994) meta-analysis, however, failed to support this pattern. Predatory vertebrates do not, on average, have stronger effects on predatory invertebrates than on herbivorous invertebrates. To more directly test the impact that vertebrate predators have on invertebrate predators, experimental manipulations are needed in which vertebrate and invertebrate predator densities are altered in the presence and absence of the other type of predator.

Finally, variation in predator impacts in streams might be explained partially by variation in prey emigration behavior. Specifically, differences in the impact of vertebrate and invertebrate predators in streams might be explained by differences in prey emigration responses to the two types of predators. This idea and tests of it are discussed in the following sections.

In sum, experimental studies in stream communities show that: 1) on average, predators have significant, but weak effects on prey densities; and 2) invertebrate predators typically have stronger effects on local prey densities than do vertebrate predators. Explaining the latter phenomenon is a major issue in stream ecology.

Prey exchange and predator impacts on prey: Two models

As noted earlier variation in prey exchange rates among patches can explain some of the observed variation in predator impacts in streams. In this section, we discuss two models that address this issue (Cooper et al., 1990; Sih and Wooster, 1994).

In a landmark paper in stream ecology, Cooper et al. (1990) presented various lines of evidence that prey exchange between patches can explain variation in local predator impacts on prey density in streams. Their central idea is that prey immigration can swamp local predator impacts. That is, in a patchy world where some patches have predators, but others do not, immigration of prey from predator-free patches into patches with predators can swamp predator impacts on local prey density. Even if predators consume prey at a high rate, if prey immigrate at a sufficient rate, then prey that have been consumed are continually replaced by new prey. The net effect might be little or no apparent predator impact.

One line of support for the "predator swamping" hypothesis is a simple, mathematical model (Cooper et al., 1990). Imagine a scenario with predators enclosed in cages placed in a stream that otherwise has no predators (as in many experimental studies). The goal of the model is to contrast prey densities in predator patches (N_p) versus the predator-free background ("controls," N_c). For simplicity, assume that N_c its constant over time (e.g., predation within cages has essentially no effect on prey density outside cages, or prey recruitment replaces predation), but that N_p can change.

Changes in N_p are caused by predation *per se*, and by prey immigration and emigration. In predator patches, the total proportion of prey killed by all predators per unit time is A; total predation rate is thus AN_p. Prey immigrate into predator patches at a rate that is proportional to prey density outside predator patches and prey emigrate out of predator patches at a rate that is proportional to prey density within predator patches. The model of Cooper et al. (1990) assumes that per capita exchange rates (E) are independent of the presence of predators. Thus total immigration into predator patches is EN_c, while total emigration out of predator patches is EN_p. Putting these all together, the rate of change in prey density in predator patches, $dN_p/dt = EN_c - N_p (A+E)$. By setting this derivative to zero, we solve for the equilibrium prey density in predator patches (recall that N_c is assumed to be constant). At equilibrium:

$$N_c/N_p = (A+E)/E. \tag{1}$$

N_c/N_p can be viewed as a measure of predator impact; larger values of N_c/N_p indicate greater predator impact. If $N_c/N_p = 1$, then predators have no impact on prey density. If $N_c/N_p > 1$, then predators reduce prey density, while if $N_c/N_p < 1$, then predators enhance prey density. Equation (1) predicts that an increase in prey movement rates (E) results in decreased predator impact (Fig. 2; $E_p = E_c$). The logic underlying this prediction is the predator swamping hypothesis discussed above.

A key assumption required by the above scenario and one that is often violated (see section *Prey emigration and predator impacts* for references) is that prey do not alter their per capita emigration rates in response to the presence of predators. If prey change their emigration rates in response to the presence of predators, will that influence the impact that predators have on local prey density?

To account for predator effects on prey emigration behavior, Sih and Wooster (1994) retained the scenario modeled by Cooper et al. (1990), but allowed E_p and E_c to differ, where E_p and E_c are defined as *per capita* emigration rates out of predator and predator-free patches, respectively. The equation for prey population dynamics in predator patches is then: $dN_p/dt = E_c N_c - N_p (A+E_p)$. At equilibrium

$$N_c/N_p = (A+E_p)/E_c. \tag{2}$$

Overall predator impact ($\log_e (N_c/N_p)$) depends on A/E_c (the sum of predation per se discounted by prey emigration rate out of predator-free areas) and E_p/E_c (the prey emigration behavioral response to predators). Figure 2 shows predicted relationships between prey exchange rates (mean of E_p and E_c) and predator impact for different prey emigration behaviors. As noted earlier, if prey do not alter their per capita emigration rates in response to predators (i.e., if $E_p = E_c$), then increased prey exchange swamps local predator impact. However, if prey increase their *per capita* emigration rates in response to predators ($E_p > E_c$), then this effect tends to enhance predator impact. If $E_p/E_c \gg 1$, then predators should significantly reduce prey density, even if predators consume no prey ($A = 0$).

Conversely, if predators cause prey to decrease their *per capita* emigration rates (i.e., if $E_p/E_c < 1$), then predator impact will be reduced (Fig. 2).

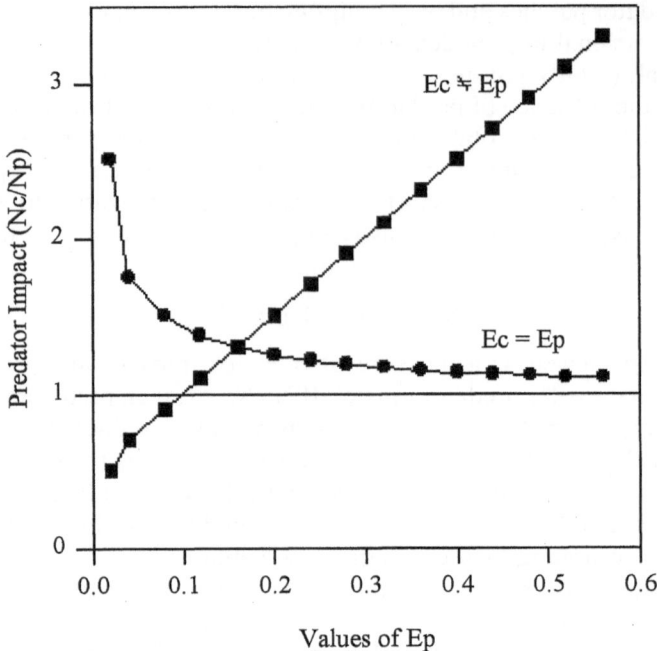

Figure 2. The influence of predator-induced changes in prey emigration rates on relationships between prey exchange and predator impact. Ep and Ec are per capita emigration rates out of predator and control (predator-free) patches, respectively. Predator impact is the ratio Nc/Np, where Nc and Np are prey densities in the control and predator patches, respectively. When Nc/Np > 1, predator presence causes a decline in local prey density. When Nc = Np, predators have no effect on local prey density. When Nc/Np < 1, predator presence causes an increase in local prey density. Note that when Ec = Ep, predator impact approaches one (indicating no effect of predators on prey density) as emigration rate increases (i.e., predator impact is swamped by prey movement). For all curves, the predation rate, A, is 0.06; for the Ec ≠ Ep line, Ec is 0.2.

Interestingly, the result can be higher prey density in patches with pred-ators, compared with predator-free controls. Although the notion that prey might be more abundant in patches with more predators might seem para-doxical, in a review of experimental studies of predation, Sih et al. (1985) found that in streams, about 40% of all prey taxa were indeed more abun-dant in sites with more predators (as compared to sites with fewer or no predators). This "unexpected effect" (cf. Sih et al., 1985) is usually attrib-uted to indirect effects of predators on some prey through either key-stone predator or intermediate predator effects (see Brönmark et al., this volume, for a review in freshwater systems). Few studies, however, include experimental evidence directly addressing the mechanism underlying this "unexpected effect." Predator-suppressed prey emigration is an alternative hypothesis to explain this phenomenon.

In sum, simple models show that: 1) if prey do not alter their *per capita* emigration rates in response to the presence of predators, then increased prey exchange should tend to swamp predator impacts; 2) however, if prey increase their *per capita* emigration rates in response to predators (i.e., if prey disperse as a means of escaping predators), then this behavior tends to increase local predator impact; while 3) if predators suppress prey *per capita* emigration rates, then predator impacts will be reduced, and can result in higher prey densities in patches with more predators.

Prey emigration behavior and predator impacts: Vertebrate vs. invertebrate predators

In the second section, we noted that in streams, predatory invertebrates tend to have stronger effects than predatory vertebrates on prey density. Can the ideas presented in the above section explain this pattern? That is, can variation in prey emigration responses to these two types of predators help explain differences in predator impact? In this section, we outline the idea and summarize the results of a relevant literature review (Wooster and Sih, 1995).

The mechanism discussed in the above section can explain differences in predator impact caused by vertebrate and invertebrate predators if prey tend to increase their *per capita* emigration rates in response to invertebrate predators, while decreasing their *per capita* emigration rates in response to vertebrate predators. To test this hypothesis, we reviewed the results of 17 experimental studies involving 97 comparisons of *per capita* emigration rates of stream prey in the presence versus absence of predators (Wooster and Sih, 1995).

About half of all prey taxa surveyed showed a significant change in *per capita* emigration rate in response to the presence of predators (51 out of 97, or 53%). The tendency for prey to show some emigration response (either increased or decreased emigration, as opposed to no change) to predators did not differ between vertebrate and invertebrate predators ($G = 1.54$; $p > 0.20$). Prey, however, tended to show qualitatively different responses to vertebrate versus invertebrate predators ($G = 20.83$; $p < 0.001$). In most instances, prey increased their drift in response to invertebrate predators, whereas they often decreased their drift in response to vertebrate predators (Tab. 1).

Why do prey show different emigration responses to vertebrate versus invertebrate predators? Sih and Wooster (1994) suggested that prey emigration responses to predators should depend on the relative costs and benefits of two alternative, prey behavioral responses to predators: hiding in refuge within a predator patch versus attempting to escape a predator patch by dispersing. A key factor in decision making should be the risk associated with attempting to emigrate out of a predator patch. Predation risk associated with emigration in response to less dangerous predators (i.e., predators that have a lower probability of detecting and capturing active prey) should be relatively low. In contrast, the risk associated with

Table 1. A comparison of the number of increases and decreases in prey drift and activity brought about by invertebrate and vertebrate predators

Prey behavior	Type of response	Predatory invertebrates		Predatory vertebrates	
		n	p	n	p
drift	increase	31	0.001	5	0.05
	decrease	3		12	
activity	increase	not enough data		2	<0.001
	decrease			19	

Comparisons were made using *G*-tests (*p*).

attempting to disperse out of patches with more dangerous predators might be prohibitively high. Instead of dispersing, the better option for prey might be to reduce activity and hide. The result would then be a paradoxical suppression of emigration out of predator patches (as compared to predator-free patches). Sih and Wooster (1994) termed this a "paradox of danger"; more dangerous predators might actually cause lower local predator impacts than less dangerous predators through a reduction of prey activity and thus reduced prey *per capita* emigration.

To test explicitly whether prey typically reduce their activity in response to the presence of vertebrate predators, we reviewed eight experimental studies involving 34 comparisons of stream prey activity in the presence versus absence of vertebrate predators (Wooster and Sih, 1995). In most cases, prey showed some significant response to predators (21/34 or 62%). In almost all instances, prey decreased their activity in response to the presence of vertebrate predators (Tab. 1); prey very rarely increased their activity in the presence of predatory vertebrates. Our suggestion is that this reduction in prey activity contributes to a tendency to reduce prey emigration in the presence of vertebrate predators.

The "paradox of danger" can explain differences in prey emigration response to vertebrate versus invertebrate predators. Stonefly nymphs (the most common predatory invertebrates in our survey) use tactile and hydrodynamic cues to detect and attack prey (Molles and Pietruszka, 1983; Sjöström, 1985; Peckarsky and Wilcox, 1989). Attack success is usually low. In particular, drifting prey are rarely attacked. Consequently, an effective response to these predators is to release from the substrate and enter the drift. In contrast, many fish (the most common predatory vertebrate in our survey) use visual cues to attack prey that are either exposed on the substrate or drifting (Allan, 1981; Brusven and Rose, 1981). Entering the drift or being active on the substrate are therefore very risky behaviors. The best response to drift-feeding fish (e.g., salmonids and many cyprinids) might be to hide in refuge and emigrate only at times when drift is relatively less risky (at night; Allan, 1978). However, some fish forage at night using mechano-sensory or olfactory cues to detect prey on top as well as within the substrate (e.g., *Rhinichthys cataractae*, Beers and Culp, 1990; *Galaxias vulgaris*, McIntosh and Townsend, 1994); we predict that the best response to these fish will be similar to the response to invertebrate predators, increasing drift rates in an effort to escape these predators. Indeed, this appears to be the response that mayfly nymphs exhibit to nocturnally foraging dace, *Rhinichthys cataractae* (Culp et al., 1991).

These results suggest that: 1) the assumption that $E_p = E_c$ is often invalid; and 2) differences in prey emigration responses to vertebrate and invertebrate predators can help to explain differences in predator impact caused by these two types of predators.

Variation in predator impacts: One predator, different prey

The above section looked at how one major pattern in streams – differences in the impacts of two kinds of predators on prey – might be explained by prey emigration behaviors. We next use the theoretical view developed above to examine another important pattern of predator impact: differences among prey in predator impacts caused by a given predator. We illustrate our approach by using data from a recent, experimental study on the effects of trout on an assortment of invertebrate prey (Bechara et al., 1993). We also discuss some other studies that include relevant data on predator impacts and prey exchange rates.

Equation (2) above predicts that predator impacts (as measured by N_c/N_p) should be linearly related to: 1) total predator attack rates (A); 2) the inverse of prey *per capita* emigration rates in the absence of predators ($1/E_c$); and 3) prey emigration responses to predators (E_p/E_c). These three predictions correspond to three mechanisms that can, in theory, determine variation in predator impacts: variation in predation rates *per se*, variation in the swamping effect of prey exchange, and variation in prey emigration behavior. Equation (2) thus suggests that to examine the relative importance of these mechanisms in explaining differences in predator impacts on different prey, one can run univariate linear regressions using relative predator impact (here, N_c/N_p) as the dependent variable, and A, $1/E_c$, and E_p/E_c as independent variables.

The study of Bechara et al. (1993) on effects of brook trout (*Salvelinus fontinalis*) on macroinvertebrate prey in experimental streams in Quebec, Canada, generated an excellent data set for doing this analysis. Their study compared benthic densities and drift rates of 15 prey taxa in the presence versus absence of trout. Relative E_p and E_c were estimated by dividing drift densities by benthic densities. Relative attack rates on different prey were estimated from fish stomach contents. Elsewhere, we present details on our re-analyses of their data (Sih et al., manuscript); here, we summarize the major results.

Using all 15 prey taxa, E_p/E_c was highly significant and positively related to local predator impact (PI = $0.30 + 0.79$ E_p/E_c; d.f. = 13, $p < 0.001$). Variation in prey emigration behavior explained 74% of the variation in predator impacts. In contrast, neither relative attack rate ($r^2 = 0.02$, d. f. = 13, $p = 0.49$) nor the inverse of prey exchange ($r^2 = 0.002$, d. f. = 13, $p = 0.89$) showed a significant relationship with predator impact.

We gained further insight by dividing the 15 prey taxa into two groups: taxa with high ($E_c > 0.1$, $N = 5$) versus low ($E_c < 0.04$, $N = 10$) inherent prey exchange rates (E_c appeared to be bimodally distributed with no prey taxa showing E_c between 0.036 and 0.108). We did this because we reasoned that if a prey type rarely disperses (low E_c), then changes in prey emigration rates in response to the presence of predators might have little effect on predator impact. In contrast, prey emigration behavior is more likely to have important effects on predator impact for prey that show high emigration rates.

The high and low prey exchange groups differed significantly in prey exchange rates (E_c and E_p), but not in predator impact, nor in prey emigration behavior (E_c/E_p). Neither group showed a consistent effect of predators on prey density (i.e., for both groups, N_c/N_p did not differ significantly from one). To evaluate mechanism governing variation in predator impacts, we ran separate sets of regressions (y = PI; x = A, $1/E_c$ or E_p/E_c) on the two groups of prey.

As predicted, for high prey exchange taxa, E_p/E_c was strongly correlated with local predator impact (PI = $0.32 + 0.91$ E_p/E_c; $r^2 = 0.995$; d. f. = 3, $p < 0.001$). In contrast, for low prey exchange taxa, E_p/E_c was not significantly correlated with predator impact ($r^2 = 0.02$, d. f. = 8, $p = 0.97$). In both groups, neither $1/E_c$ nor A were significantly related to predator impact (all $p = 0.20$). Overall, our analysis of Bechara et al.'s (1993) data suggests that in their system, an important factor explaining variation in predator impacts is variation among prey types in their emigration responses to predators. This was particularly true for prey taxa that inherently show relatively high prey exchange rates.

Forrester (1994), who studied effects of brook trout on prey (five species of mayflies), similarly found that variation in prey emigration behavior (measured as *per capita* drift rates from areas without trout) provided the best explanation for differences in predator impacts. The ranking of relative predator impacts on five prey taxa were as follows: *Baetis* > *Paraleptophlebia* > *Ephemerella* = *Eurylophella* = *Stenonema* = 0. The only two prey taxa that had their densities significantly reduced by trout both showed a significant tendency to increase their *per capita* drift rates in the presence of trout (*Baetis* and *Paraleptophlebia*). In contrast, the three prey taxa that showed no density response to trout did not alter their *per capita* drift rates in response to trout. Variation in attack rates did not explain variation in predator impacts; attack rates on the five prey taxa did not differ significantly. In direct opposition to the hypothesis that prey exchange swamps predator impact, prey with the highest drift propensities in the absence of predators (*Baetis*) showed the highest predator impact, while the densities of prey that showed the lowest drift propensities (*Eurylophella* and *Stenonema*) were not significantly affected by the presence of predators.

Few other studies include sufficient data for all the variables (A, E_c, E_p, predator impact) necessary to do the analysis shown above. Several other studies focusing on predator impacts on one or a few prey taxa, however, include enough data to evaluate the relative importance of predation *per se* versus predator-induced emigration in determing predator impacts. Reviews of these studies concur that predator-induced emigration often plays a major role in generating significant predator impacts on local prey density (Sih et al., 1992; Englund, in press. Although more studies are needed to yield strong generalities, existing data suggest that variations in prey emigration responses to predators have important effects on the impact that predators have on the density of different prey within small patches. However, at larger scales, prey emigration responses to predator presence can have different effects on predator impact on prey density (see section *Effects of scale on prey movements and predator impacts*).

Relationship between prey movement rates and predator impact

The examples given above provide strong evidence that prey emigration responses to predators can play an important role in generating predator impact. Is this a general pattern across streams? Do prey generally increase their movement rate in the presence of predators resulting in large impacts of predators on prey density? We envision three ways in which prey movement rate might influence predator impact on prey density (Fig. 3). The top path suggests that prey with high movement rates will have large predator impacts because of high encounter rates between predators and those prey. We refer to this as the "attack rate" hypothesis. The middle path suggests that inherent prey movement rates are positively correlated with the response of prey to the presence of predators in terms of changes in movement rate. Prey that are adapted for rapid movement (e.g., *Baetis* mayfly larvae) are expected to exploit that ability to escape areas with predators. Increases in movement rates in response to predators will result in large local predator impacts on prey density even in the absence of any consumption of prey by predators (Sih and Wooster, 1994). The work of Forrester (1994) suggests that this pathway is important for mayflies in a New England stream. We call this the "activity response" hypothesis. The final pathway is a direct link from prey movement rates to predator impacts on prey density. Rapid emigration from and immigration into areas with predators are expected to mask the effects that predators have on prey den-

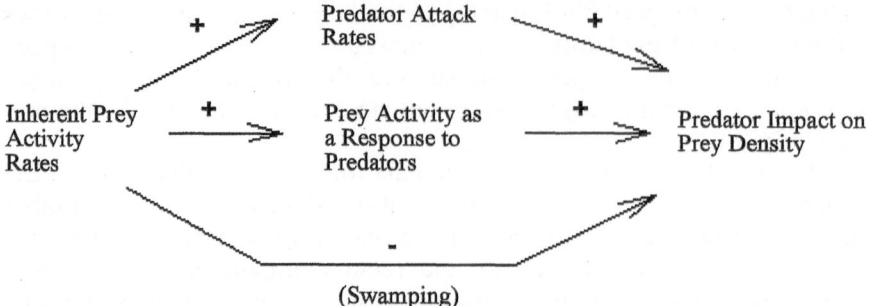

Figure 3. Possible pathways that inherent prey activity rates might influence predator impacts on prey density; + and − indicate positive or negative relationships between two variables.

sity at a small scale (Flecker, 1984; Cooper et al., 1990). Thus a negative relationship is expected between prey movement rate and predator impact on prey density. This pathway is referred to as the "swamping" hypothesis.

To examine these three hypotheses we conducted a series of meta-analyses. Data were compiled from published studies on stream insect colonization rates, predator preference for different prey, prey activity responses to predators, and predator impact on local prey density. Initially, we attempted to conduct the analyses on individual taxa; however, we found very few prey taxa for which all four measurements (colonization rate, predator preference, activity response to predators, and predator impact) had been made. In addition, for those prey taxa with some data on all four measurements there was generally little data on one or two of the measurements for a given prey taxon. Thus, analyses on individual taxa would suffer from low power. To increase power we placed prey taxa into groups following Merritt and Cummins (1984) who described and classified freshwater insects into eight groups based on adaptations to living in water. These adaptations should influence relative prey movement rates in streams. Our analysis is restricted to stream insect prey that we were able to classify into one of Merritt and Cummins' (1984) groups. We found data for insects belonging to four movement groups: swimmers, clingers, sprawlers, and burrowers (Tab. 2).

Our estimates of inherent movement rates were determined by estimating *per capita* colonization rates of individual prey taxa that could be classified into a movement group. Data were taken from published studies designed to examine the colonization rate of stream invertebrates onto natural substrates placed in replicate trays or baskets (data from Allan, 1975; Townsend and Hildrew, 1976; Ciborowski and Clifford, 1984; Peckarsky, 1986; Shaw and Minshall, 1980; Clifford et al., 1992; Schlosser, 1992). In each study, replicate trays were placed onto the streambed and subsampled over time. *Per capita* colonization rates were calculated by dividing the number of individuals of a given taxon colonizing trays in

Table 2. The four movement groups used to examine the relationship between movement rates and predator impact on prey density (from Merritt and Cummins, 1984)

Movement group	Description	Examples
swimmers	adapted for short burst of swimming	mayfly nymphs in *Baetidae, Siphlonuridae*
clingers	morphological/behavioral adaptations for attachment to substrate surface	mayfly nymphs in *Heptageniidae;* caddisfly larvae in *Hydropsychidae*
sprawlers	found in slow-flowing areas; crawl on surface of silt-covered substrate	mayfly nymphs in *Caenidae;* stonefly nymphs in *Leuctridae*
burrowers	dig discrete burrows in fine sediments	midge larvae in *Chironominae*

24 h by the number in the trays at the end of the experiment (duration of experiments ranged from 10 to 64 days). To determine if the four movement groups differed in colonization rates we conducted an ANOVA on arcsine transformed per capita colonization rates.

Predator impact on the density of the four movement groups was taken from studies that manipulated predator density in enclosures placed within or next to natural streambeds. All studies compared the densities of prey in predator-containing cages to that in predator-free enclosures (data from Dudgeon, 1993; Forrester, 1994; and studies used in Wooster, 1994). For each prey taxon (that could be classified into a movement group) within each study we calculated a standardized measure (Cohen's d) of predator impact on that taxon's density. Cohen's d is frequently used in meta-analysis and a variety of statistical tests have been developed to compare values of Cohen's d among different classes (Gurevitch et al., 1992; Gurevitch and Hedges, 1993; Cooper and Hedges, 1994). Cohen's d was calculated as

$$d = \frac{N_p - N_c}{s} Jm,$$ (3)

where N_p is prey density in the predator-containing enclosure, N_c is prey density in the predator-free enclosure, s is a measure of the standard deviation pooled across both predator and predator-free treatments, and Jm is a sample-size correction factor that approaches one as sample size increases. This analysis was designed to determine if the movement groups differed in the impact that predators had on their local density. To examine differences in predator impact on the density of the four movement groups, we conducted an ANOVA-like analysis on the Cohen's d values following the methods of Gurevitch et al. (1992).

Following similar methods, we conducted a meta-analysis on the movement responses of prey (e.g., drifting, crawling, swimming) to predator presence (data from Walton, 1980; Williams and Moore, 1982, 1985; Williams, 1986; Feltmate, 1987; Hershey, 1987; Malmqvist and Sjöström, 1987; Kohler and McPeek, 1989; Benton and Pritchard, 1990; Lancaster, 1990; Bechara et al., 1993; Ode and Wissinger, 1993; Forrester, 1994; Tikkanen et al., 1994). Cohen's d values were calculated by subtracting the movement rate in the absence of predators from the movement rate in the presence of predators and dividing by the standard deviation pooled across both treatments. This analysis was designed to determine if the four prey groups responded to predators by changing their movement rates and if the groups differed in the degree and direction (increasing movement versus decreasing movement) of their response.

Predator preference for prey belonging to the four different groups was taken from studies that calculated selectivity/electivity indices (e.g., Manly's α; Chesson, 1983) of stream predators consuming invertebrate prey, or that provided appropriate data that allowed calculation of Manly's α (data from Novak and Estes, 1974; Siegfried and Knight, 1976; Kovalak, 1978; Tippets and Moyle, 1978; Johnson, 1983; Cooper, 1984; Peckarsky, 1985; Fuller and Hynes, 1987; Walde and Davies, 1987; Williams, 1987; Angradi and Griffith, 1990; Fuller and Rand, 1990; Peckarsky et al., 1990;

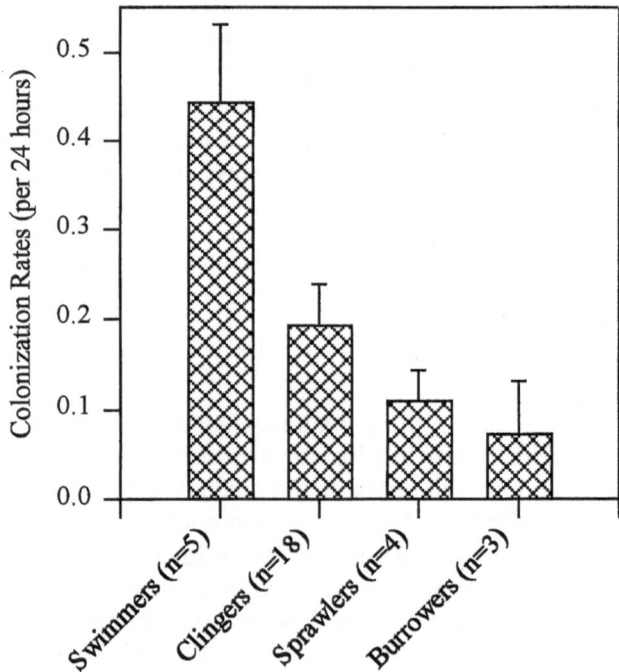

Figure 4. Twenty-four hour *per capita* colonization rates of four movement groups of stream insects. The bars represent the mean values and the vertical lines represent one standard error. Sample sizes are given next to the name of each group.

Bechara et al., 1993). Since selectivity indices are sensitive to the number of prey examined, we ranked the prey taxa within each study such that the most preferred taxa received a value of one and the least preferred taxa received a value of zero. We conducted an ANOVA on arcsine transformed preference ranks for the four movement groups.

Our analysis of colonization rates revealed that the four movement groups had different colonization rates and that swimmers have a much higher movement rate than the other three groups ($F = 3.96$, d.f. = 3, $p = 0.019$; Fig. 4). Both the attack rate hypothesis and the activity response hypothesis predict that predators should have the largest impact on the density of swimmers and the smallest impact on the density of burrowers. In contrast, the swamping hypothesis predicts that predators should have the smallest impact on the density of swimmers and the largest impact on the density of burrowers. An examination of predator impacts on the densities of the movement groups revealed that predators have the largest impact on swimmers (Qb = 9.51, d.f. = 3, $p < 0.025$; Fig. 5).

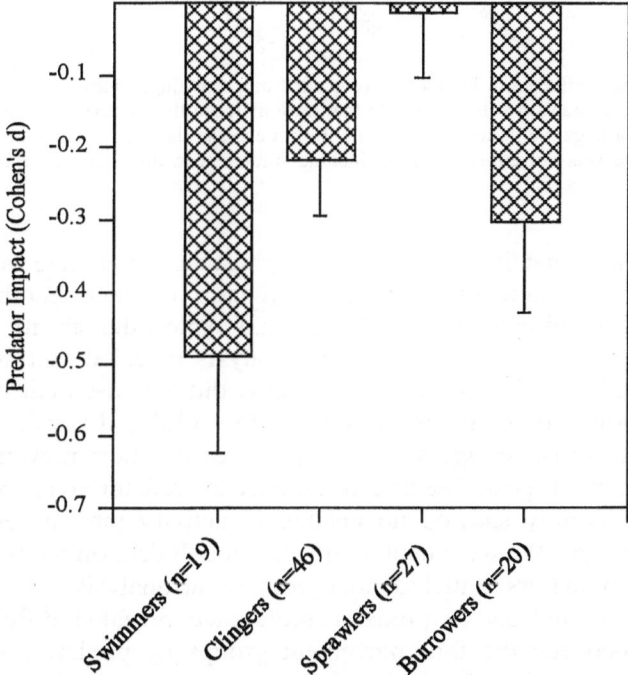

Figure 5. Predator impacts on the densities of four invertebrate taxa. Cohen's d was used as the index of predator impact (see the sixth section for calculation of Cohen's d). A negative value indicates that prey density was reduced in arenas with predators relative to predator-free arenas. The bars represent the mean predator impact and the vertical lines represent one standard deviation. Sample sizes are given next to the name of each group.

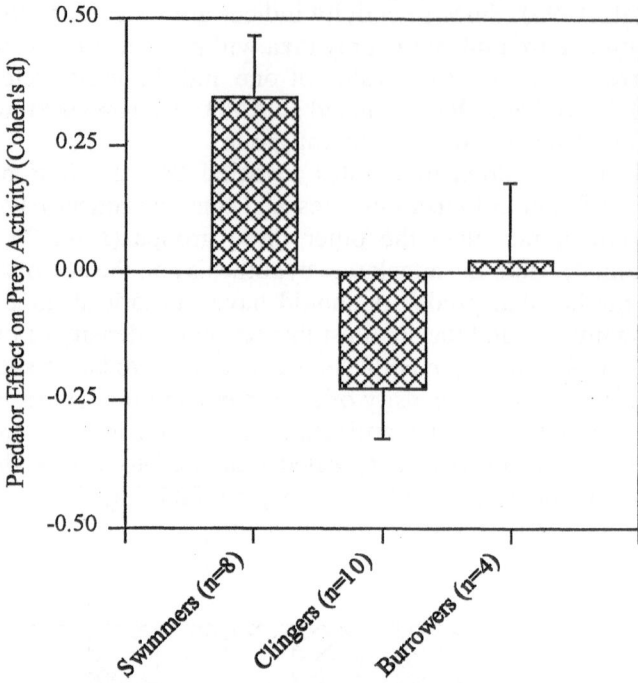

Figure 6. Predator effects on the activity (e.g., drifting, crawling, swimming rates) of insect prey. A positive Cohen's d value indicates that prey increase their activity in the presence of predators, and a negative value indicates that prey decrease their activity in the presence of predators. There was not enough data to include sprawlers in this analysis. Bars, lines and sample sizes as in Fig. 5.

We postulated that the group with the highest movement rate would show an increase in movement in response to predator presence. In other words, animals with good movement ability should exploit that ability to escape areas with predators (see above). Our analysis revealed that swimmers, the group with the highest colonization rate, did increase their movement rate in response to predator presence ($Qb = 13.07$, d.f. $= 2$, $p < 0.005$; Fig. 6). In contrast, clingers generally decreased their movement rate, either hiding in refuge or freezing in response to predator presence. Finally, burrowers generally showed no change in activity rate in response to predators (Fig. 6). We were unable to find enough data on the response of sprawlers to predators to include that group in the analysis.

Finally, our analysis of predator preference revealed differences in preference between the four movement groups by predators ($F = 4.59$, d.f. $= 3$, $p = 0.004$; Fig. 7). Both swimmers and burrowers had the largest preference values, suggesting a higher predator consumption rate on taxa belonging to those classes.

We developed three hypotheses about the relationship between prey movement rates and predator impact on prey density (Fig. 3). Our results

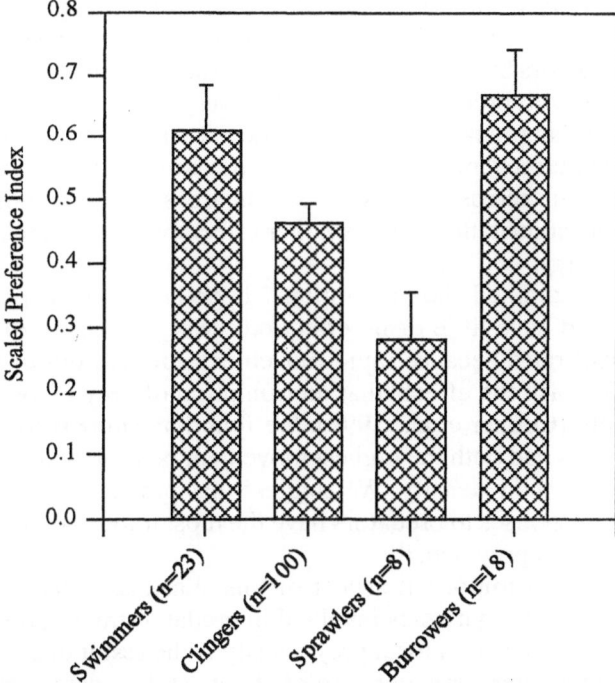

Figure 7. Predator preference for prey belonging to the four movement groups. The preference index has been scaled such that within each study, the most preferred prey is given a value of one and the least preferred prey is given a value of zero. Bars indicate the mean preference value and vertical lines represent one standard error. Sample sizes are given next to the name of each group.

suggest that the relationship is more complex than Figure 3 suggests and that it varies between taxa with different movement rates. Our first hypothesis, the "attack rate" hypothesis, predicted that predators will have large impacts on the density of taxa with high movement rates because these prey will have high encounter rates with predators. However, we found that while predator preference appears to play a very important role in determining predator impact on prey density (note the similarity in the patterns of Figs. 5 and 7), it does not appear to be related to prey movement rate in the manner we had envisioned because the prey groups with the highest (swimmers) and lowest (burrowers) movement rates had similar and high predator preference values (Fig. 7).

Our second hypothesis involved movement responses of prey to predators, the "activity response" hypothesis. We predicted that predators would have large effects on the density of prey with high movement rates because those prey would respond to the presence of predators by increasing their movement rates. While predators had the largest impact on the density of swimmers, they also had a strong effect on the density of burrowers, the group with the lowest movement rate (Fig. 5). However, we found that activity responses might play an important role in the impact that predators have on the density of swimmers and clingers. Swimmers respond to predators by increasing movement rates, presumably in an effort to escape areas with predators (Fig. 6). This response should contribute to the large effect that predators have on the density of swimmers. In contrast, clingers decrease movement rates in response to predators, which should have the effect of increasing prey density in areas with predators (Fig. 6). Potentially, this contributes to the smaller effect that predators have on the density of clingers.

Finally, we examined the "swamping" hypothesis which predicts that predators should have little or no effect on the local density of prey with high movement rates because rapid movement into and out of areas with predators will mask the effects that consumption of prey by predators has on prey density (Cooper et al., 1990). We found no support for this hypothesis, swimmers had both the highest movement rats and predators had the largest effect on their density. We suggest that predator consumption of prey and prey responses to predators have the most important influences on predator impact on prey density.

Perhaps the most important aspect of this analysis is that it illustrates the complexity of the dynamics involved in predator impacts on local prey density. Predator impact on local prey density is the result of an interaction between prey movement responses to predators and predator consumption of prey. However, this interaction varies between types of prey. We found that predators have large effects on the density of swimmers, and that this is due to both a high predator preference for swimmers and an increase in movement rate in response to the presence of predators. Predators have only a moderate impact on the density of clingers. This could be the result of an interaction between predator consumption of clingers (which should drive local clinger density down) and clingers decreasing their movement rates in response to predators (which should increase their local density in areas with predators; see above). Burrowers showed no change in movement rate in response to predator presence; the large impact that predators have on their density should be the result of predator consumption alone.

Can prey belonging to different movement groups be found within the same study designed to examine predator impact on prey density? For the analysis on predator impact on prey density we collected data from 19 published studies that examined predator impact on prey density; four of the studies examined prey taxa within the same enclosures that belonged to all four movement groups. Thus, it may be fairly common for a study examining predator impact on prey density to have prey from different movement groups in the same enclosures, resulting in very different predator-prey dynamics occurring between different prey within the same study. This highlights the importance of examining movement responses of prey into and out of enclosures when examining predator impact on local prey density in streams.

In the next section we consider the role that spatial scale plays in influencing the impact that predators have on prey density, and how the relative roles that prey emigration responses to predators and predator consumption of prey change as spatial scale changes.

Effects of scale on prey movements and predator impact

Ecological patterns often depend on our scale of observation because the relative influence of different pattern-forming processes change with scale (Allen and Hoeckstra, 1991; Levin 1992). The spatial scale of predation experiments in streams spans a relatively large range (0.001 to > 100 m²). Do the results of experiments performed at different scales reflect different underlying processes? Englund (in press) addressed this issue by using simulation models that are extensions of the models presented earlier (Cooper et al., 1990; Sih and Wooster, 1994; see section *Prey exchange and predator impacts on prey: Two models*). The most general result derived from his models is that the results of small-scale experiments should primarily reflect prey movements while larger-scale experiments should reflect predation mortality. This conclusion follows from the assumptions that predation rate is scale-independent, whereas *per capita* migration rate (from patches) decreases with increasing spatial scale. At small scales, the perimeter-to-area ratio is large and any random movement has a high probability of crossing the perimeter. In contrast, at larger scales the perimeter-to-area ratio is very small and most movements will occur within the experimental arena. Thus, we may expect predator impact on prey density in small-scale experiments to be influenced by prey movements as described earlier. However, in very large-scale experiments, the *per capita* migration rates out of and into experimental enclosures will approach zero and predator impact will be determined by predation rate.

Prey movements also influence predation rates. For example, it is reasonable to assume that escape behaviors reduce individual predation risk, and thus predation rates. Also, background movements, not induced by predators, can increase encounter rates and thus predation rates (see Sih et al., 1992). To account for the effect of movements on predation rates, Englund (in press) and Englund and Olsson (1996) modeled predator impact for experiments performed at different scales in a habitat with refuge and predator patches arranged in a checker board pattern (Fig. 8). This model differs from those presented earlier (Cooper et al., 1990; Sih and Wooster, 1994) in that experimental units do not necessarily correspond to natural predator and refuge patches. Depending on the scale of the study an experimental unit can contain a variable number of patches. This corresponds to the common situation where it is not possible to identify natural spatial units that either have or lack predators. In such a situation cages typically include both predator-accessible and refuge areas. In the model, prey

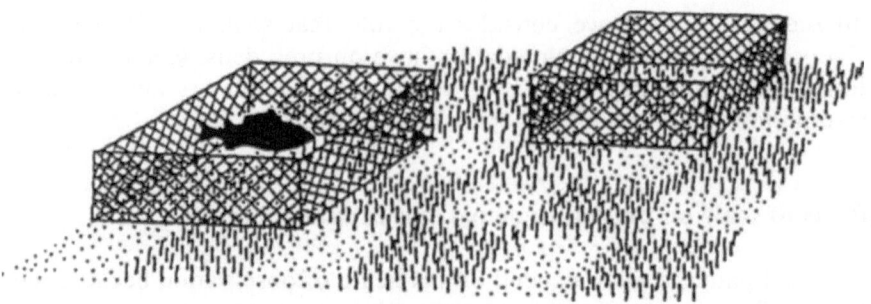

Figure 8. A model habitat used by Englund and Olsson (1996) to study the effects of spatial scale on the perceived impact of predators on prey density. It was assumed that predators could feed only in non-vegetated patches (stippled areas) and that prey move between the two types of patches. The effect of a local manipulation was found by comparing the densities in cages that contain predators with cages lacking predators.

movement behavior is specified by the per capita movement rates out of predator and refuge patches (m_p and m_r, respectively). These migration rates should not be confused with per capita migration rates out from cages. In a small-scale study that includes few patches, a large proportion of the patches will border the cage perimeter and per capita migration rates will be high; in a large-scale study the same prey behavior (specified by m_p and m_r) will produce much smaller per capita migration rates. Here we give a non-technical presentation of model results and refer the reader to Englund and Olsson (1996) for details of the model.

An intriguing result from these models is that prey migration can have contrasting influences on predator impact in large- and small-scale experiments. Consider first a situation where movement rates from predator and refuge patches are equal ($m_p = m_r$). In a small-scale experiment where each experimental unit includes only a few patches, the effect of high movement rates will be to swamp predation rates as discussed by Cooper et al. (1990). Predator impact will thus decrease with increasing movement rates (Fig. 9 (A)). However, in a large-scale experiment, predator impact will only be influenced by predation rates. Because prey movements between refuge and predator patches increase the total number of encounters between predators and prey, there will be a positive relationship between movement rates and predator impact. Next, consider a situation where predators cause prey to increase movement rates out of predator patches ($m_p > m_r$). In a small-scale experiment, this type of behavior increases predator impact because prey leave predator cages (Fig. 9 (B)). However, in a large-scale experiment, most movements will occur within the cage and the effect of this escape behavior will be to decrease prey exposure (i.e., prey will accumulate in refuges) and, as a result, predator impact on prey density. Finally, if prey freeze ($m_p < m_r$), this will decrease predator impact in small-scale experiments as discussed earlier. Given the assumption that

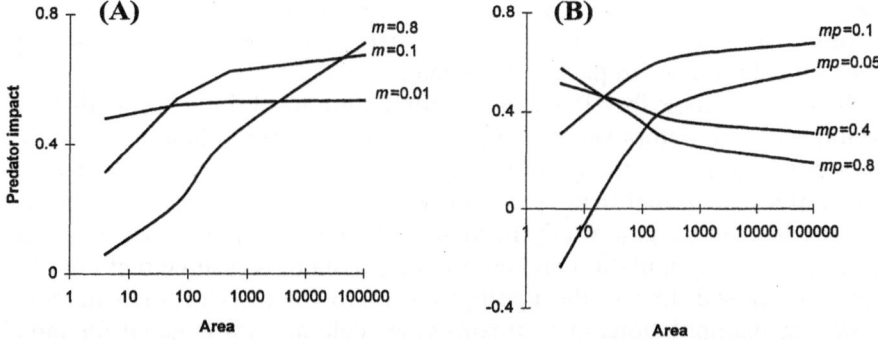

Figure 9. Predicted relationship between the size of a manipulated area and predator impact for different types of movement in the model habitat depicted in Fig. 8. The two figures illustrate (A) a situation where prey do not alter their movement rates in response to predators (mp = mr), and (B) where predators either increase or decrease their movement rates out from predator patches. The movement rate from refuge patches (m_r) is 0.1, and movement rate out from predator patches (m_p) varies between 0.05 and 0.8. Here predator impact is calculated as $-\log(N_p/N_c)$, where N_p is prey density in the predator patch and N_c is prey density in the refuge patch (after Cooper et al., 1990).

such a behavior will occur only if it reduces predation rates, it will presumably also reduce predator impact in large-scale experiments (Fig. 9 (B)).

An important implication of these findings is that the results from small-scale experiments that allow prey exchange cannot be extrapolated to larger scales. It is quite conceivable that a small-scale manipulation will strongly affect one species but leave another unaffected, whereas a larger experiment would show the opposite result (see Figs. 9 (A), (B)). For example, Forrester (1994) examined the effects of fish predation and prey emigration rates on the density of several mayfly species in enclosures (35 meters long), and found significant effects on *Baetis* but no effect on *Eurylophella*. However, because *Eurylophella* experienced a much higher predation rate, we would expect it to be more strongly affected in a larger-scale experiment.

These results also emphasize the importance of choosing the appropriate experimental scale with respect to the hypothesis being tested. We can broadly distinguish two types of predator effects. First, the influence of a patchy distribution of predators on prey distribution can be examined with small scale open (i.e., no enclosure) experiments. For example, by contrasting stream pools with and without fish it is possible to better understand local prey distribution patterns (Cooper et al., 1990, Sih et al., 1992). Predator impact at this scale is influenced both by predation and prey movement rates. However, because movement rates are scale-dependent, it is important that the scale of the experiment is similar to that of natural predator and refuge patches. The use of screened cages may be problematic because it is likely that screens impede migration rates (Cooper et al.,

1990). If migration rates in and out of screened cages are lower than they would be in a natural patch of the same size, then cages do not accurately assess local predator impact at that scale.

We can also look for effects at the population level, i.e., effects that we would see if predators were manipulated in the entire habitat used by the prey population. For many lotic prey this corresponds to a stream-wide scale. At such a global scale, prey populations are in effect closed systems, and predator impacts are only influenced by demographic rates. Because most models of population and community dynamics assume that populations are closed, this is the appropriate scale for tests of such models. However, manipulations at a stream-wide scale are not feasible for most lotic predator-prey systems. A critical question therefore is how large does an experimental arena have to be for the response of the enclosed population to approximate the behavior of a closed population? Englund (in press) used a simple model to estimate a threshold migration rate from arenas that can be accepted for an experiment to yield a local predator impact that is within $\pm 10\%$ of the predator impact on the prey population. His calculations suggest that for realistic predation rates, threshold per capita migration rates need to be $1-5\%$ per day if migration is not to be influenced by the predator treatment. However, if prey show escape or freezing behaviors, the threshold value could be much lower. Englund (in press) concluded that for many prey taxa, very large cages are needed to estimate population predator impact. Even in the experiment using the largest experimental arenas published so far (Forrester, 1994), the results seem to primarily reflect prey movements.

Suggestions for future study

Our review reveals the importance that prey dispersal, prey emigration responses to predators, and spatial scale have in determining predator impacts on prey density. Specifically, our overview has shown that variation in prey emigration responses to different predators, as well as variation among different types of prey, can have profound influences on predator impact. Unfortunately, very few studies have examined the impact that different predators within the same system have on prey behavior and density (Soluk and Collins, 1988 a, b; Soluk, 1993; Scrimgeour and Culp, 1994). While there have been a fairly large number of studies that have examined the impact that one type of predator has on the density of multiple species of prey within the same stream, very little work has been conducted to determine sources of variation in predator impact among prey taxa (e.g., differences in inherent movement rates of prey, emigration responses to predators, and attack rates of predators). In addition, it will be enlightening to examine variation in the responses of multiple types of prey with different inherent movement rates to a single type of predator. These types of studies should be important in revealing sources of variation in predator impact. The patchiness and availability of refuges, and prey movement rates into and out of refuges, should also play important roles in determining predator impacts. Future studies that use field enclosures to manipulate predator density should also consider manipulating refuge availability (e.g., Dudgeon, 1993). Finally, our review suggests that the spatial scale at which predator-prey interactions are examined can have profound influences on conclusions reached regarding predator impact and the contribution that prey emigration responses make in determining the level of predator impact. Experimental studies that manipulate predator presence at several spatial scales within the same stream should prove rewarding. We also note that published experimental studies probably tell us very little about the population effects of predation. Attempts to measure such effects with very large-scale experiments will greatly increase our understanding of the role of predation in streams.

References

Allan, J.D. (1975) The distributional ecology and diversity of benthic insects in Cement Creek, Colorado. *Ecology* 56:1040–1053.

Allan, J.D. (1978) Trout predation and the size composition of stream drift. *Limnol. Oceanogr.* 23:1231–1237.

Allan, J.D. (1981) Determinants of diet of brook trout (*Salvelinus fontinalis*) in a mountain stream. *Can. J. Fish. Aquat. Sci.* 38:184–192.

Allan, J.D. (1982) The effects of reduction in trout density on the invertebrate community of a mountain stream. *Ecology* 63:1444–1455.

Allen, T.H.F. and Hoeckstra, T.W. (1991) Role of heterogeneity in scaling of ecological systems under analysis. *In:* J. Kolasa and S.T.A. Pickett (eds): *Ecological Heterogeneity.* Springer-Verlag, New York, pp 47–68.

Angradi, T.R. and Griffith, J.S. (1990) Diel feeding chronology and diet selection of rainbow trout (*Oncorhynchus mykiss*) in the Henry's Fork of the Snake River, Idaho. *Can. J. Fish. Aquat. Sci.* 47:199–209.

Arnqvist, G. and Wooster, D. (1995) Meta-analysis: Synthesizing research findings in ecology and evolution. *Trends Ecol. Evol.* 10:236–240.

Bechara, J.A., Moreau, G. and Planas, D. (1992) Top-down effects of brook trout (*Salvelinus fontinalis*) in a boreal forest stream. *Can. J. Fish. Aquat. Sci.* 49:2093–2103.

Bechara, J.A., Moreau, G. and Hare, L. (1993) The impact of brook trout (*Salvelinus fontinalis*) on an experimental stream benthic community: The role of spatial and size refugia. *J. Anim. Ecol.* 62:451–464.

Beers, C.E. and Culp, J.M. (1990) Plasticity in foraging behaviour of a lotic minnow (*Rhinichthys cataractae*) in response to different light intensities. *Can. J. Zool.* 68:101–105.

Benton, M.J. and Pritchard, G. (1990) Mayfly locomotory responses to endoparasitic infection and predator presence: The effects on predator encounter rate. *Freshw. Biol.* 23:363–371.

Brusven, M.A. and Rose, T.S. (1981) Influence of substrate composition and suspended sediment on insect predation by torrent sculpin, *Cottus rhothenus*. *Can. J. Fish. Aquat. Sci.* 38:1444–1448.

Chesson, J. (1983) The estimation and analysis of preference and its relationship to foraging models. *Ecology* 64:1297–1304.

Ciborowski, J.J.H. and Clifford, H.F. (1984) Short-term colonization patterns of lotic macroinvertebrates. *Can. J. Fish. Aquat. Sci.* 41:1626–1633.

Clifford, H.F., Casey, R.J. and Saffran, K.A. (1992) Short-term colonization of rough and smooth tiles by benthic macroinvertebrates and algae (chlorophyll *a*) in two streams. *J. N. Amer. Benthol. Soc.* 11:304–315.

Cooper, H. and Hedges, L.V. (1994) *The Handbook of Research Synthesis*. Russell Sage Foundation, New York.

Cooper, S.D. (1984) Prey preferences and interactions of predators from stream pools. *Verhandl. Intern. Verein. theor. angew. Limnol.* 22:1853–1857.

Cooper, S.D., Walde, S.J. and Peckarsky, B.L. (1990) Prey exchange rates and the impact of predators on prey populations in streams. *Ecology* 71:1503–1514.

Culp, J.M. (1986) Experimental evidence that stream macroinvertebrate community structure is unaffected by different densities of coho salmon fry. *J. N. Amer. Benthol. Soc.* 5:140–149.

Culp, J.M., Glozier, N.E. and Scrimgeour, G.J. (1991) Reduction of predation risk under the cover of darkness: Avoidance responses of mayfly larvae to a benthic fish. *Oecologia* 86:163–169.

Dudgeon, D. (1993) The effects of spate-induced disturbance, predation and environmental complexity on macroinvertebrates in a tropical stream. *Freshw. Biol.* 30:189–197.

Englund, G. (1997) Importance of spatial scale and prey movements in predator caging experiments. *Ecology; in press.*

Englund, G. and Olsson, T.I. (1996) Treatment effects in a predator caging experiment: Influence of predation rate and prey movements. *Oikos* 72:519–528.

Feltmate, B.W. (1987) Predator-prey interactions in streams: A combined field and laboratory study of *Paragnetina media* (Plecoptera) and *Hydropsyche sparna*. *Can. J. Zool.* 65:448–451.

Feltmate, B.W. and Williams, D.D. (1989) Influence of rainbow trout (*Oncorhynchus mykiss*) on density and feeding behavior of a perlid stonefly. *Can. J. Fish. Aquat. Sci.* 46:1575–1580.

Flecker, A.S. (1984) The effect of predation and detritus on the structure of a stream insect community: A field test. *Oecologia* 64:300–305.

Flecker, A.S. and Allan, J.D. (1984) The importance of predation, substrate and spatial refugia in determining lotic insect distributions. *Oecologia* 64:306–313.

Forrester, G.E. (1994) Influence of predatory fish on the drift dispersal and local density of stream insects. *Ecology* 75:1208–1218.

Fuller, R.L. and Hynes, H.B.N. (1987) Feeding ecology of three predacious aquatic insects and two fish in a riffle of the Speed River, Ontario. *Hydrobiologia* 150:243–255.

Fuller, R.L. and Rand, P.S. (1990) Influence of substrate type on vulnerability of prey to predacious aquatic insects. *J. N. Amer. Benthol. Soc.* 9:1–8.

Gilliam, J.F., Fraser, D.F. and Sabat, A.M. (1989) Strong effects of foraging minnows on a stream benthic invertebrate community. *Ecology* 70:445–452.

Gurevitch, J. and Hedges, L.V. (1993) Meta-analysis: Combining the results of independent experiments. *In:* S.M. Scheiner and J. Gurevitch (eds): *Design and Analysis of Ecological Experiments.* Chapman and Hall, London, pp 378–398.

Gurevitch, J., Morrow, L.L., Wallace, A. and Walsh, J.S. (1992) A meta-analysis of competition field experiments. *Amer. Nat.* 140:539–572.

Hedges, L.V. and Olkin, I. (1985) *Statistical Methods for Meta-Analysis.* Academic Press, Orlando, Florida.

Hershey, A.E. (1987) Tubes and foraging behavior in larval Chironomidae: Implications for predator avoidance. *Oecologia* 73:236–241.

Holomuzki, J.R. and Stevenson, R.J. (1992) Role of predatory fish in community dynamics of an ephemeral stream. *Can. J. Fish. Aquat. Sci.* 46:2322–2330.

Johnson, J.H. (1983) Diel food habits of two setipalpian stoneflies (Plecoptera) in tributaries of the Clearwater River, Idaho. *Freshw. Biol.* 13:105–111.

Kohler, S.L. and McPeek, M.A. (1989) Predation risk and the foraging behavior of competing stream insects. *Ecology* 70:1811–1825.

Kovalak, W.P. (1978) On the feeding habits of *Phasganophora capitata* (Plecoptera: Perlidae). *Great Lakes Entomol.* 11:45–49.

Lancaster, J. (1990) Predation and drift of lotic macroinvertebrates during colonization. *Oecologia* 85:48–56.

Lancaster, J., Hildrew, A.G. and Townsend, C.R. (1991) Invertebrate predation on patchy and mobile prey in streams. *J. Anim. Ecol.* 60:625–641.

Levin, S.A. (1992) The problem of scale in ecology. *Ecology* 73:1943–1967.

Lima, S.L. and Dill, L.M. (1990) Behavioral decisions made under the risk of predation: A review and prospectus. *Can. J. Zool.* 68:619–640.

Malmqvist, B. and Sjöström, P. (1987) Stream drift as a consequence of disturbance by invertebrate predators: Field and laboratory experiments. *Oecologia* 74:396–403.

Mann, C. (1990) Meta-analysis in the breech. *Science* 249:476–480.

McIntosh, A.R. and Townsend, C.R. (1994) Interpopulation variation in mayfly antipredator tactics: Differential effects contrasting predatory fish. *Ecology* 75:2078–2090.

Menge, B.A., Berlow, E.L., Blanchette, C.A., Navarrete, S.A. and Yamada, S.B. (1994) The keystone species concept: Variation in interaction strength in a rocky intertidial habitat. *Ecol. Monogr.* 64:249–286.

Merritt, R.W. and Cummins, K.W. (1984) *An Introduction to the Aquatic Insects of North America,* Second Edition. Kendall/Hunt Publishing Co., Dubuque, Iowa.

Michael, D.I. and Culver, D.A. (1986) Influence of plecopteran and megalopteran predators on *Hydropsyche* (Trichoptera: Hydropsychidae) microdistribution and behavior. *J. N. Amer. Benthol. Soc.* 6:46–55.

Molles, M.C. and Pietruszka, R.D. (1983) Mechanisms of prey selection by predaceous stoneflies: Role of prey morphology, behavior and predator hunger. *Oecologia* 57:25–31.

Murdoch, W.W. (1994) Population regulation in theory and practice. *Ecology* 75:271–287.

Novak, J.K. and Estes, R.D. (1974) Summer food habits of the black sculpin, *Cottus baileyi,* in the Upper South Fork Holston River drainage. *Trans. Amer. Fish. Soc.* 2:270–276.

Ode, P.R. and Wissinger, S.A. (1993) Interaction between chemical and tactile cues in mayfly detection of stoneflies. *Freshw. Biol.* 30:351–357.

Peckarsky, B.L. (1985) Do predaceous stoneflies and siltation affect the structure of stream insect communities colonizing enclosures? *Can. J. Zool.* 63:1519–1530.

Peckarsky, B.L. (1986) Colonization of natural substrates by stream benthos. *Can. J. Fish. Aquat. Sci.* 43:700–709.

Peckarsky, B.L., Horn, S.C. and Statzner, B. (1990) Stonefly predation along a hydraulic gradient: A field test of the harsh-benign hypothesis. *Freshw. Biol.* 24:181–191.

Peckarsky, B.L. and Wilcox, R.S. (1989) Stonefly nymphs use hydrodynamic cues to discriminate prey. *Oecologia* 79:265–270.

Power, M.E. (1992) Habitat heterogeneity and the functional significance of fish in river food webs. *Ecology* 73:1675–1688.

Reice, S.R. and Edwards, R.L. (1986) The effect of vertebrate predation on lotic macroinvertebrate communities in Quebec, Canada. *Can. J. Zool.* 64:1930–1936.

Rosenthal, R. (1984) *Meta-Analytic Procedures for Social Research.* Applied Social Research Methods Series, Vol. 6. Sage Publications, Newbury Park, California.

Schlosser, I.J. (1992) Effects of life-history attributes and stream discharge on filter-feeder colonization. *J. N. Amer. Benthol. Soc.* 11:366–376.

Schlosser, I.J. and Ebel, K.K. (1989) Effects of flow regime and cyprinid predation on a headwater stream. *Ecol. Monogr.* 59:41–57.

Scrimgeour, G.J. and Culp, J.M. (1994) Foraging and evading predators: The effect of predator species on a behavioural trade-off by a lotic mayfly. *Oikos* 69:71–79.

Shapiro, D.A. (1985) Recent applications of meta-analysis in clinical research. *Clin. Psychol. Rev.* 5:13–34.

Shaw, D.W. and Minshall, G.W. (1980) Colonization of an introduced substrate by stream macroinvertebrates. *Oikos* 34:259–271.

Siegfried, C.A. and Knight, A.W. (1976) Prey selection by a setipalpian stonefly nymph, *Acroneuria (Calineuria) californica* Banks (Plecoptera: Perlidae). *Ecology* 57:603–608.

Sih, A. (1987) Predators and prey lifestyles: An evolutionary and ecological overview. *In:* W.C. Kerfoot and A. Sih (eds): *Predation: Direct and Indirect Impacts on Aquatic Communities.* University Press of New England, Hanover, NH, pp 203–224.

Sih, A. and Moore, R.D. (1990) Interacting effects of predator and prey behavior in determining diets. *In:* R.N. Hughes (ed.): *Behavioral Mechanisms of Food Selection.* NATO ASI Series, Springer-Verlag, New York, pp 771–796.

Sih, A. and Wooster, D.E. (1994) Prey behavior, prey dispersal and predator impacts on stream prey. *Ecology* 75:1199–1207.

Sih, A., Crowley, P.H., McPeek, M.A., Petranka, J.W. and Strohmeier, K. (1985) Predation, competition and prey communities: A review of field experiments. *Annu. Rev. Ecol. Syst.* 16:269–311.

Sih, A., Kats, L.B. and Moore, R.D. (1992) Effects of predatory sunfish on the density, drift and refuge use of stream salamander larvae. *Ecology* 73:1418–1430.

Sjöström, P. (1985) Hunting behaviour of the perlid stonefly nymph *Dinocras cephalotes* (Plecoptera) under different light conditions. *Anim. Behav.* 33:534–540.

Soluk, D.A. (1993) Multiple predator effects: Predicting combined functional response of stream fish and invertebrate predators. *Ecology* 74:219–225.

Soluk, D.A. and Collins, N.C. (1988a) Synergistic interactions between fish and stoneflies: Facilitation and interference among stream predators. *Oikos* 52:94–100.

Soluk, D.A. and Collins, N.C. (1988b) Balancing risks? Responses and non-responses of mayfly larvae to fish and stonefly predators. *Oecologia* 77:370–374.

Taylor, A.D. (1990) Metapopulations, dispersal, and predator-prey dynamics: An overview. *Ecology* 71:429–436.

Tikkanen, P., Muotka, T. and Huhta, A. (1994) Predator detection and avoidance by lotic mayfly nymphs of different size. *Oecologia* 99:252–259.

Tippets, W.E. and Moyle, P.B. (1978) Epibenthic feeding by rainbow trout (*Salmo gairdneri*) in the McCloud River, California. *J. Anim. Ecol.* 47:549–559.

Townsend, C.R. and Hildrew, A.G. (1976) Field experiments on the drifting, colonization, and continuous redistribution of stream benthos. *J. Anim. Ecol.* 45:759–772.

Walde, S.J. and Davies, R.W. (1984) Invertebrate predation and lotic prey communities: Evaluation of *in situ* enclosure/exclosure experiments. *Ecology* 65:1206–1213.

Walde, S.J. and Davies, R.W. (1987) Spatial and temporal variation in the diet of a predaceous stonefly (Plecoptera: Perlodidae). *Freshw. Biol.* 17:109–115.

Walton, O.E., Jr. (1980) Invertebrate drift from predator-prey associations. *Ecology* 61:1486–1497.

Williams, D.D. (1986) Factors influencing the microdistribution of two sympatric species of Plecoptera: An experimental study. *Can. J. Fish. Aquat. Sci.* 43:1005–1009.

Williams, D.D. (1987) A laboratory study of predator-prey interactions of stoneflies and mayflies. *Freshw. Biol.* 17:471–490.

Williams, D.D. and Moore, K.A. (1982) The effect of environmental factors on the activity of *Gammarus pseudolimnaeus* (Amphipoda). *Hydrobiologia* 96:137–147.

Williams, D.D. and Moore, K.A. (1985) The role of semiochemicals in benthic community relationships of the lotic amphipod *Gammarus pseudolimnaeus*: A laboratory analysis. *Oikos* 44:280–286.

Wooster, D. (1994) Predator impacts on stream benthic prey. *Oecologia* 99:7–15.

Wooster, D. and Sih, A. (1995) A review of the drift and activity responses of stream prey to predator presence. *Oikos* 73:3–8.

Aspects of life-history evolution

Evolutionary Ecology of Freshwater Animals
ed. by B. Streit, T. Städler and C.M. Lively
© 1997 Birkhäuser Verlag Basel/Switzerland

Rotifer life history strategies and evolution in freshwater plankton communities

N. Walz

Institut für Gewässerökologie und Binnenfischerei, Müggelseedamm 260, D-12587 Berlin, Germany

Summary. Rotifers play an important role in many freshwater plankton communities. The populations are controlled from "bottom-up" depending on different food quantities and qualities. This regulation, however, is also depending on characteristics of rotifer life histories. Rotifers are controlled from "top-down" by predators, especially by copepods, by instars of *Chaoborus* and by predatory rotifers. Mechanical interference by *Daphnia* is considered here as a special case of predation as the rotifers are mostly killed by this action. Different defense mechanisms are discussed. Simultaneously rotifer densities are controlled from "bottom-up." Many species are generalists and feed in the same range of algae as other zooplankton. Specialists, however, feed on larger forms, e.g., dinoflagellates. As threshold food levels for rotifers are higher than those for cladocerans they are often outcompeted when food concentrations are lowered by the clearance activity of cladocerans. At the cost of higher food concentrations (high K_s-food levels) rotifers may exhibit high maximum growth rates (r_{max}) and short times for their population development. These abilities increase with rotifer body size and render rotifers susceptible to environmental fluctuations.

Introduction

Rotifers are generally not considered to have high impact on their environment; that is, they are not regarded as being keystone species. This title distinguishes their close competitors, the cladocerans, especially the genus *Daphnia* (De Bernardi and Peters, 1987). Rotifers are, however, important in ecological studies, but this role often has been limited to ecotoxicology and aquaculture (Watanabe et al., 1983). Nevertheless, because they are often exceedingly abundant and some of them are easily cultured, rotifers are very suitable as experimental organisms in studies of population ecology (Walz, 1993). In fact, because their population size and productivity can exceed that of other micrometazoa (Yan and Geiling, 1985), their importance should not be underestimated.

A variety of reviews have emphasized these and related issues (Starkweather, 1987; Wallace and Snell, 1991; Nogrady et al., 1993; Walz, 1993c), but in spite of several decades of intensive research, we have yet to reach a comprehensive understanding of rotifers in their habitat. In fact, the questions that remain are fundamental, and they are varied. Chief among these issues are the following: How are rotifers involved in the aquatic food web? What sorts of impact do they have on the food available? How are they controlled by predation? What species are their main competitors? Do rotifers

and cladocerans divide the habitat spatially and/or temporally? Does efficiency of movement as calculated by Reynolds number influence rotifer
biology (Wieser, 1986) and if cladocerans are more effective in movement,
how can rotifers coexist with them?

Answers to these and related questions must lie in the life history of the
rotifer *Bauplan*. Here I will examine bottom-up (food limitation) and top-
down (predation) control mechanisms as they relate to rotifer biology.

Top-down control of rotifer communities

Predation by fishes is generally not considered to be a significant aspect of
rotifer populations (see below). However, predation by invertebrates does
appear to be significant nearly all of the time (Dodson, 1974). This chapter
reviews the impact of predation on rotifers. In general, there are two strategies that they adopt to avoid the threat posed by predation. The first is to
evolve a high growth rate (r_{max}); birth rates are higher than death rates due
to predation. The second is to evolve defence mechanisms. Both strategies
are energy consuming and may compete within the organisms for scarce
energy reserves.

Invertebrate predation of rotifers

Interference between invertebrate predators and rotifers is relatively well
known. Many species, especially the crustacean plankton are predators.
The copepod *Cyclops vicinus* (older copepodite stages > stage 4 and
adults) prefers *Polyarthra* spp., *Keratella quadrata, K. cochlearis, Synchaeta* spp. and *Asplanchna* spp. with the highest coefficient of selection
for *Asplanchna* (Brandl and Fernando, 1978). On the other hand, Zankai
(1984) showed that *C. vicinus* did not select rotifers, but nevertheless
consumed them in large quantities. Recently, it has become clear that adults
of *C. vicinus* are omnivorous and feed at least 50% on algae (Tóth et al.,
1987). They reproduce even if fed on the alga diet alone (Santer and Van
den Bosch, 1994).

In Lake Constance, Germany, *Cyclops vicinus* represents the key factor
controlling rotifer development at the end of the spring phytoplankton
bloom (Walz et al., 1987). The abundance of rotifers begins to increase only
after *C. vicinus* has gone into dormancy and disappeared. Then populations of *Polyarthra vulgaris/dolichoptera, Asplanchna priodonta, Keratella
cochlearis* and *K. quadrata* begin a dramatic population increase. This increase happens after *C. vicinus* disappears, even though when the copepod
is present rotifer birth rates are high. At such times *Synchaeta* populations
are totally depressed. Similar observations were reported between *Cyclops
bicuspidatus* and *Synchaeta* sp. in Lake Michigan (USA) (Stemberger and

Evans, 1984). This copepod prefers soft-bodied rotifers like *Synchaeta* spp. and *Polyarthra major*. However, *Polyarthra vulgaris* (perhaps because of its escape behaviour) and loricate rotifers like *Keratella cochlearis* and *Kellicottia longispina* are avoided. *Mesocyclops leuckarti* feeds on *Asplanchna* sp. and *Synchaeta pectinata* (Gophen, 1977), but apparently prefers *Polyarthra dolichoptera* and *Synchaeta oblonga* and avoids *Keratella cochlearis* (Karabin, 1978). In Lake Constance just the opposite pattern of rotifer (Walz et al., 1987) and of *Mesocyclops leuckarti* (Stich, 1985) numbers led to the hypothesis that this copepod controlled the population of *Polyarthra vulgaris/dolichoptera* in autumn. *Mesocyclops edax* preys upon *Asplanchna girodi* and *Polyarthra vulgaris*, but not on *K. cochlearis* (Gilbert and Williamson, 1978).

Cyclopoid copepods, formerly considered to be obligate predators, are actually omnivorous (see above) and calanoid copepods, formally believed to be filter feeders, are raptors. For example, *Diaptomus pallidus* has been reported to prey upon rotifers, and actually improves its survival and reproduction. This calanoid copepod is supported by *Synchaeta oblonga*, which is defenceless, less by *Polyarthra ramata*, which is able to escape by jumping, and not at all by the loricate species *Keratella cochlearis* (Williamson and Butler, 1986).

The predatory rotifer *Asplanchna priodonta* preys upon *Polyarthra* spp. and *Keratella* spp., *Synchaeta* spp. and also on large algae (Salt et al., 1978). This species prefers *K. cochlearis* and *K. quadrata* (Ejsmont-Karabin, 1974; Guiset, 1977). According to Gilbert and Williamson (1978) *A. priodonta* selects *Keratella cochlearis*, but avoids *Polyarthra vulgaris*. Further experiments with *Asplanchna girodi* demonstrate the easy capture of individuals of *Synchaeta pectinata, K. cochlearis,* and *Conochilus* spp. (Gilbert, 1980). *A. girodi* rarely captures *Kellicottia bostonensis* and intact *Conochilus* colonies and rejects the peritrich ciliate *Rhabdostylis* sp. and the dinoflagellate *Peridinium*.

In Lake Constance after the reduction of *Cyclops vicinus* in late spring, a second predator appears, *Asplanchna priodonta* (Walz et al; 1987). This rotifer exerts a strong predatory pressure on *Keratella cochlearis*, which develops only after the complete disappearance of the predator. However, *K. quadrata* is not affected as much and coexists with *Asplanchna*. An analogous situation on a spatial scale rather than time is observed in a small, oligo-mesotrophic pre-Alpine lake, where *Asplanchna* and *K. quadrata/hiemalis* live at different depths (Höffgen, 1987). In the Fasaneriesee, a eutrophic gravel pit near Munich (Germany), *K. cochlearis* also evades the predator rotifers *Asplanchna girodi* and *A. priodonta* by moving to deeper (Huber, 1982). Apparently, *Asplanchna* is prevented to following because of its higher requirements for oxygen contents (Mikschi, 1989). In a Dutch man-made lake, high death rates of *K. cochlearis* are reported in the presence of high *Asplanchna* densities (Van der Bosch and Ringelberg, 1985). In a shallow Japanese lake, *B. angularis, B. forficularis,* and

Keratella sp. decrease or remain at low densities when the predator *Asplanchna brightwelli* develops (Urabe, 1992). This happens even when the egg ratios of the prey indicate favourable birth rates. These three prey species are encountered frequently in the stomach of *A. brightwelli*. Other rotifer species, *Polyarthra vulgaris, B. calyciflorus*, and *Hexarthra intermedia*, which are not eaten by *A. brightwelli*, become abundant in the plankton.

Synchaeta species possess a virgate mastax and several species are carnivorous (Naumann, 1923; Nauwerck, 1963). Similar conditions are found in the *Trichocerca* species. Many species suck out the contents of large algae, however, the same species are carnivorous on other rotifers (Naumann, 1923, Nauwerck, 1963; Williamson, 1983). *Trichocerca capucina* is known to suck out the contents of eggs of other rotifers (Pourriot, 1970), especially those of *Keratella* spp. (Naumann, 1923; Nauwerck, 1963).

The instars of *Chaoborus* larvae are predators on a variety of zooplankton species preferring smaller prey with rotifers being the main prey of early instars (Moore and Gilbert, 1987). However, *Chaoborus* does not prefer any particular species. According to Williamson (1983). *Chaoborus* preys on *Asplanchna*. The first two instars of *C. trivittatus* feeds on *Keratella cochlearis* and *Kellicottia longispina* if they are present in high numbers after the elimination of *Daphnia rosea* (Neill, 1985). Because *Chaoborus* remains near the bottom during the day and is an active predator only at night, its influence on rotifer population dynamics might have been often neglected.

Vertebrate predation of rotifers

Most adult fishes do not take rotifers as food as they are too small to be seen. Large *Asplanchna* and colonies of *Conochilus unicornis* have been found in stomachs of some pelagial fish (Stenson, 1982). Adult fish appear to have a more indirect impact on rotifers by feeding on their microcrustacean competitors (Gliwicz and Pijanowska, 1989; Neill, 1984). On the other hand, the first prey of larval fishes is often in the size range of rotifers, and carp, perch, and roach are known to consume rotifers (Filatov, 1972; Guma'a, 1978; Hammer, 1985). The significance of this predation may be of little ecological importance to lakes with large pelagial regions as young fishes live in the littoral. In some unusual cases, rotifers appear to be unpalatable to small zooplanktivorous fish (Felix et al., 1995). Nevertheless, aquaculture relies on the fact that rotifers, especially *B. calyciflorus* (freshwater) and *B. plicatilis* (marine), are excellent foods for a variety of commercially important organisms (Emmerson, 1984; Lubzenz et al., 1989; Widigdo, 1988).

Mechanical interference by cladocerans acting as predation

The alternate appearance of cladocerans and rotifers known in the plankton of many lakes was reported even by Dieffenbach and Sachse (1911) for *Daphnia* and *Keratella cochlearis*. The convential explanation is that suppression of rotifers takes place when the high grazing rates of *Daphnia* lower phytoplankton densities below a threshold level required by the rotifers (bottom-up control, see below). This is known as exploitative competition (Vanni, 1986). However, *Daphnia* also sweep rotifers into their grazing chambers and kill them. This was called mechanical interference competition (Gilbert, 1985). Rotifers, especially *Keratella cochlearis* and *B. calyciflorus*, are damaged when swept into the filtering apparatus of *Daphnia* (Burns and Gilbert, 1986). Rotifers have even been found in the guts of *Daphnia*.

Does this mechanism, first observed in laboratory experiments, play an important role in natural communities? In semi-natural enclosures, *K. cochlearis* was suppressed only by *Daphnia* through resource exploitation (Threlkeld and Choinski, 1987). However, field studies have shown interference competition. Even at low *Daphnia* densities (one to five individuals per liter), mechanical interference has been reported to play a role in natural communities (Gilbert, 1988a). In the case of Loch Leven, Scotland. MacIsaac and Gilbert (1990) argued that correlations between *Kerattella cochlearis* birth rates and chlorophyll-*a* showed that rotifers were not limited by low food concentrations.

The classical inverse relationship between rotifers and *Daphnia* was demonstrated by Lampert and Rothhaupt (1991) for a shallow, North German lake. In this study, the authors showed that *Daphnia* reduced seston concentration and that because rotifer densities were positively correlated with seston concentration they were negatively correlated with *Daphnia* densities. This is sufficiently explained by exploitative competition, but this view neglects, as the authors argued, that the rotifer abundance was an order of magnitude higher before *Daphnia* was present. This fact, however, may also be explained by the lower food level caused by *Daphnia*-grazing. A similar observation of interference competition is seen in another shallow lake, this time in Japan. According to Urabe's (1992) study, in early July *Daphnia similis* (a large cladoceran) was present but the population density of *Keratella valga* was low even though food was sufficiently plentiful to provide the rotifer with enough energy to have a high egg ratio. Other potential predators were absent at this time.

Because they are small (*Keratella cochlearis, Synchaeta oblonga*) or delicate (*Ascomorpha ecaudis*) many rotifers are especially vulnerable to *Daphnia* damage. Others, equipped with escape behaviour (*Polyarthra*) or with larger size (colonies of *Conochilus unicornis*) are better protected (Gilbert, 1988b). In contrast to large *Daphnia*, smaller cladocerans (< 1.2 mm) do not seem to inhibit rotifers. They often co-occur at high densities and do not mechanically interfere with the rotifers (Gilbert,

1988a). Ciliates are much more depressed by *Daphnia*-interference than rotifers (Wickham and Gilbert, 1991). Mortality rates of heterotrophic flagellates, ciliates, and rotifers are positively related to the mean body size of *Daphnia* (Pace and Vaqué, 1994).

Interference and exploitative competition can operate simultaneously at low food concentrations (MacIsaac and Gilbert, 1991), but at high food concentrations the daphnids may eliminate rotifers by mechanical interference alone. Therefore, mechanical interference is a very rapid method for cladocerans to extirpate rotifers; this elimination can be done at a higher rate than by resource competition. Of course, ciliates also are suppressed by these mechanisms (Jack and Gilbert, 1994).

Defense mechanisms

Rotifers have evolved many defence mechanisms against invertebrate predators. These include rapid escape responses (e.g., *Polyarthra, Filinia*), hard lorica (e.g., *Keratella*), spines (e.g., *Filinia, Kellicottia, Brachionus*), coloniality and mucus sheets (Stemberger and Gilbert, 1987). In a Spanish reservoir, *Asplanchna girodi* selects only *Keratella cochlearis* with spine length less than 15 µm while rejecting for individuals with longer spines. Whatever gains are achieved by producing spines, there are the costs of production. At high food concentrations maximum population growth rates of spined *K. testudo* (0.15 d^{-1}) are less than half of those of the unspined forms (0.39 d^{-1}). At lower food concentrations, spined *K. testudo* have a lower survivorship and fecundity (Conde-Porcuna et al., 1993).

Bottom-up control of life histories

Impact of food size – Rotifer generalists

According to the competitive exclusion principle (Hardin, 1960), species competing for the same limiting food resources cannot coexist indefinitely under equilibrium conditions. Therefore, a prerequisite for coexistence is selectivity for different food types. Food can be classified by size, energetic value and biochemical content (e.g., fatty acids). Because zooplankton species differ greatly in size, food size-selectivity was examined first. It was argued that smaller species, especially rotifers, should be able to ingest smaller particles, while larger sized zooplankton will ingest larger foods (Gliwicz, 1969). For that reason – and because the diet of protozoa was unknown – rotifers were considered to be the main bacterivors in the plankton. A linear relationship was demonstrated between the length of the consumer (in the case of cladocerans) and food size (Burns, 1968).

Larger crustaceans, however, may not fit this pattern. The size-efficiency hypothesis assumed a competition between all herbivores for the fine particles, larger zooplankton with higher efficiency (Brooks and Dodson, 1965). According to analyses of the distance between their filter-setae and their ingestion- and growth rates, many cladocerans filter unselectively over a wide size range from bacteria to nanoplankton of about 30 µm. To this group belong *Diaphanosoma, Chydorus, Ceriodaphnia* and many species of the genus *Daphnia* (e.g., *D. pulex*). Another cladoceran group, however, filters bacteria and picoplankton (<3 µm) very inefficiently, if at all: These include *Bosmina, Simocephalus, Sida,* and other *Daphnia*-species, e.g., *D. pulicaria* (Bogdan and Gilbert, 1987; DeMott, 1985; Pace and Orcutt, 1981). In the group capable of ingesting bacteria, capacity to ingest bacteria drops with increasing distance between the setae of the feeding appendages. Setae distance also is correlated with individual body size (Brendelberger and Geller, 1985, Porter et al., 1983a). Consequently, cladocerans, especially daphnids, feed unselectively in a broad range of nanoplankton from 3 to about 20 µm (Sterner, 1989). However, this is roughly the same range in which rotifers feed (Pourriot, 1977). That consumer size is not directly coupled with food size is evident in the small rotifer *Anuraeopsis fissa* (body weight = 0.002 µg). This species has been cultured using *Scenedesmus obliquus* (10–13 µm) (Dumont et al., 1995), the same type of food as used in many studies for cladocerans. Growth with coenobia of *Scenedesmus* sp., however, was not possible (Esparcia, 1994).

The effect of cladocerans on bacteria remains controversial. Cladoceran feeding on bacteria did not have an important grazing effect on them (Børsheim and Andersen, 1987), and the nutritional value for these zooplankton is very low (Platzek, 1986). However, other cladocerans may be significantly involved in bacterial grazing (Güde, 1989). It appears that the heterotrophic nano-flagellates (HNF) are the most important bacteria grazers in limnic biotopes (Arndt, 1993).

As previously assumed based on their size, rotifers are not the dominant bacteria and picophytoplankton feeders (Bogdan and Gilbert, 1987; Ooms-Wilms et al., 1993; Pace et al., 1990; Rothhaupt, 1990a; Vadstein and Olsen, 1993). Rotifers ingest bacteria very inefficiently so that a very high density of particles is required for rotifers to survive (Seaman et al., 1986; Starkweather et al., 1979). Only *Keratella cochlearis, Kellicottia longispina,* and *Conochilus unicornis* were found to consume bacteria and small aflagellate algae efficiently (Bogdan and Gilbert, 1984; Bogdan and Gilbert, 1987; Gilbert and Bogdan, 1981; Ross and Munawar, 1981). *Filinia* spp. are often present in bacterial horizons of eutrophic stratified lakes with a chemocline (Guerrero et al., 1978). They feed on bacteria and minute detritus particles (Nauwerck, 1963) and are very rare in larger lakes which probably have a low bacterial density. Sanders et al. (1989) found *Gastropus* sp., *Hexarthra* sp., and *Filinia longiseta* feeding on bacteria-sized microspheres, but reported lower ingestion rates for *Anuraeopsis fissa, Kellicottia bostonensis* and *Keratella* spp. on these particles. *Conochilus* sp. has been labeled bacteriophagous, but not *Polyarthra* sp. or *Trichocerca* sp. The relative bacterial grazing impact of rotifers in eutrophic Lake Oglethorpe (Georgia, USA) accounted for a maximum of 13%, compared with 55–99% for heterotrophic flagellates, 2–45% for phagotrophic phytoflagellates, and 14–80% for ciliates. Ooms-Wilms et al. (1995) added *Lepadella* sp. to the list of bacteria feeding rotifers. In a review of rotifer impact in the microbial web, Arndt (1993) drew the conclusion that rotifers were not able to control bacterial production.

Those rotifers that feed on bacteria also feed non-selectively on algae. Thus they may be characterized as generalists (Gilbert and Bogdan, 1981). *Keratella quadrata* is a polyphagous species capable of feeding on particles smaller than 10 μm. These include Chlorococcales, Volvocales, Chrysomonadales, *Euglena, Rhodomonas*, centric diatoms, and detritus (Nauwerck, 1963; Pourriot, 1977). Similar feeding preferences are typical for *Keratella hiemalis* (Zimmermann, 1974). Rothhaupt (1990b) found that the optimal size for three species of *Brachionus* was < 12 μm. Thus, genus *Brachionus*, and probably the entire family Brachionidae appears to be generalists feeders (Edmondson, 1965; Gilbert and Starkweather, 1977; Gilbert, and Jack, 1993).

Impact of food size – rotifer specialists

Some rotifer species may be considered as specialists (Gilbert and Bogdan, 1984). *Polyarthra* species apparently prefer larger foods including the flagellate *Cryptomonas* and *Euglena* (up to 45 μm). However, they have a lower ingestion efficiency when feeding on small aflagellate algae (Bogdan and Gilbert, 1987; Gilbert and Bogdan, 1981). When *Polyarthra vulgaris* was the dominant rotifer a significant decrease in large cryptomonads was observed by Scheda and Cowell (1988). Some *Synchaeta* species (especially *S. pectinata*) are pronounced specialists for larger particles. *Synchaeta pectinata* sucks out large *Cryptomonas*. *Trichocerca* does the same thing, but it feeds on filamentous algae (Pourriot, 1965). *Ascomorpha ovalis* and *A. saltans* are specialists on dinoflagellates (Koste, 1978; Pourriot, 1977) while *Ascomorpha ecaudis* captures green algae and Chrysomonades (Pourriot, 1965; Ruttner-Kolisko, 1972). In laboratory conditions *A. ecaudis* captures *Cryptomonas erosa* by using its gelatinous envelope as a sticky surface on which the particles adhere (Stemberger, 1987). Other *Synchaeta, Trichocerca,* and *Asplanchna* are carnivorous. These genera feed on large algae and rotifers (Ejsmont-Karabin, 1974; Guiset, 1977). Cladocerans scarcely graze on particles larger than 20 μm (Gliwicz and Sidlar, 1980; Sterner, 1989). Thus these rotifers feeding on larger particles ("inedible algae") may compete with *Eudiaptomus gracilis* and *Cyclops vicinus* (Horn, 1985) and generally with other copepods that prefer larger algal cells. However, copepods show a highly selective feeding behaviour, grasping individual algal cells. Physical and chemical characteristics, e.g., taste (DeMott, 1986), of the particles play a dominant role. Indeed, in this range of particles, the netplankton, different specialist consumers may find specific niches.

Effects of rotifer and cladoceran grazing on phytoplankton are very different. In short-term enclosure experiments macrozooplankton (*Daphnia galeata*) suppressed phytoplankton, while microzooplankton (*K. cochlearis*) enhanced phytoplankton populations (Havens, 1993). In similar treatments, Bergquist et al. (1985) demonstrated that small zooplankton (especially rotifers) enhanced smaller phytoplankton growth while suppressed that of larger cells. At the same time larger zooplankters dominated by *Daphnia* caused declines of smaller algae, while larger phytoplankton increased.

Impact of cyanobacteria on food availability

Cyanobacteria, especially filamentous forms, have different effects on zooplankton communities. Larger cladocerans seem to be more inhibited than smaller cladocerans, copepods, or rotifers (Dumont, 1977; Orcutt and Pace, 1984; Richman and Dodson, 1983). The inhibition may be the result of mechanical interference, poor food quality, or direct toxicity of the cyanobacteria. Rotifers, including generalists like *Brachionus* species, are notably uninhibited by cyanobacteria and may profit in such situations. For example, *B. calyciflorus* feeds on filamentous *Aphanizomenon* sp. by nibbling the ends of the filaments (Dumont, 1977). In Lake Nakuru, an East-African soda lake, *B. plicatilis* and *B. dimidiatus* feed exclusively on the blue-green *Spirulina platensis* (Varesci and Jacobs, 1984). By an analysis of stomach contents Infante (1978) found that 35 % of *B. calyciflorus* individuals had eaten cyanobacteria. Although some strains of cyanobacteria are toxic to rotifers (e.g., *Anabaena flos-aquae* on *Asplanchna girodi*, Snell, 1980), most cladoceran-toxic strains do not affect rotifers. In contrast to cladocerans, *B. calyciflorus* grows in the presence of toxic strains of *Anabaena flos-aquae* (Starkweather and Kellar, 1983) and *Microcystis aeruginosa* (Fulton and Pearls, 1987). In competition experiments however, blooms of non-toxic *M. aeruginosa* do not improve competitive ability of *B. calyciflorus* as the rotifers were outcompeted by *Daphnia ambigua* at low concentrations of algae (Fulton and Pearl, 1988). The advantage of rotifers seems to lie in their resistance to toxicity. *Microcystis aeruginosa* is ingested by *Brachionus rubens*, although it produces endotoxis. Those individuals feeding on *M. aeruginosa* die faster than individuals no fed this cyanobacterium. At high concentrations ingestion is stopped. The tough filaments of *Cylindrospermopis* are not ingested (Rothhaupt, 1991). Inhibition of *Daphnia pulex* and *Keratella cochlearis* growth is dependent on *Microcystis aeruginosa* concentration, but *Daphnia* appear to be more susceptible than *Keratella* (Smith and Gilbert, 1995).

In conditions without cyanobacteria large *Daphnia*, (e.g., *D. pulicaria*) exhibit the lowest threshold food concentration, while the smallest (*D. cucullata*) exhibit the greatest. This pattern is reversed in presence of cyanobacteria. *Cylindrospermopsis raciborski* filaments increase the threshold food concentration for body growth and reproduction (Gliwicz and Lampert, 1990). Thus, rotifers may gain competitive superiority over large *Daphnia* under conditions when those crustaceans are adversely affected by cyanobacteria. In those situations the impact of the interference competition by the *Daphnia* are negated.

Gilbert (1990) studied the effect of *Anabaena affinis* filaments on growth rate of cladocerans and rotifers. He demonstrated that *Anabaena* was toxic only when ingested (i.e., it functions as an endotoxin). Under these experimental conditions, growth rate of *Daphnia*, especially of larger species, was reduced. However, in rotifers such reduction was not observed. In competition experiments, *Daphnia* excluded the rotifer *Synchaeta pectinata*, but when *Anabaena* was present *Daphnia* disappeared and the *Synchaeta* population increased.

B. plicatilis grows better with a mixed diet of *Chlorella* and the blue-green *Schizothrix calcicola* than on either of these algae separately (Snell et al., 1983). *Anabaena flos-aquae*, together with *Monoraphidium*, is a better food for *B. rubens* than *Monoraphidium* alone. *Anabaena* alone apparently had no food value (Rothhaupt, 1991). *Planktothrix agardhii*, however, when used alone is found to be a poor food resource for *B. calyciflorus*. Likewise, additions of this cyanobacterium increased growth rates, compared with *Monoraphidium* alone (Weithoff and Walz, 1995). On the other hand, low concentrations of non-toxic *Anabaena flos-aquae* filaments mechanically interfere with the ability of *D. pulex* and *D. galeata* to feed on *Chlamydomonas* (Gilbert and Durand, 1990). Clearly the food quality of *A. flos-aquae* can compensate for the lower ingestion rates. In contrast, *K. cochlearis* is inhibited only at high filament concentrations, but these are not eaten efficiently and have no effect on growth rates. In a plankton community, mechanical interference of daphnids by cyanobacteria may reduce the danger for interference on rotifers.

Impact of clay on food quality

Suspended clay and silt particles inhibit cladocerans but not rotifers and may reverse the competitive advantage of the former (Kirk and Gilbert, 1990). In their study, Kirk and Gilbert found that the presence of suspended clay decreased ingestion rates for five cladoceran species but not for three rotifer species. Rotifers appear to be more selective for phytoplankton when algae is suspended with clay. On the other hand, cladocerans ingest more suspended clay particles than rotifers do, but the influence of clay on cladocerans is size specific. Smaller cladocerans are more inhibited by clay than larger cladocerans because they are less selective for phytoplankton at high clay concentrations. At concentrations of suspended clay above 50 mg/l the frequencies of postabdominal rejections done by *Daphnia ambigia* increased while the frequency beat of thoracic appendages decreased (Kirk, 1991). From enclosure experiments with mineral turbidity it was concluded (Cuker and Hudson, 1992) that copepods and rotifers were favored over cladocerans, except the clay-tolerant genus *Diaphanosoma*. According to Cuker (1993), however, there was no bottom-up effect of clay by changing the competition between rotifers and cladocerans. However, a top-down effect was noticed, as clay diminished predation of larger *Chaoborus* instars by fish, which in turn increased the effect on crustaceans. Throughout these experiments the rotifer, *Keratella cochlearis*, was unaffected.

Impact of biochemical content on food quality

Growth is not only dependent on food particle size but a basic prerequisite is also food quality. For example, the marine rotifer *Synchaeta cecilia* may grow on 13 algal species of different sizes and taxa but they grow best for the long-term with *Cryptomonas ovoidea* (Egloff, 1988). Twenty-four algal species did not support rotifer growth although some of them were ingested. When six of the most nutritious species were offered as food in mixtures of two species reproduction of *Synchaeta* was greatly enhanced by some of the combinations, at least in short-term experiments. Similar examples have been shown for freshwater rotifers (Pourriot, 1965). *Keratella taurocephala* is capable of selective and flexible feeding behaviour in Little Rock Lake (Wisconsin, USA) but the algae selected are not correlated to the ability of this algae to support rotifer growth (Sierszen, 1990).

In a two-stage chemostat, the growth-ingestion efficiency (K_1) of *B. plicatis* depends on the growth rate of the food alga *Brachimonas submarina* (Scott, 1980). The proportion of carbohydrats decreases whereas lipid and protein contents of the alga increases with higher algal growth rates. In rotifer aquaculture it has long been known that yeasts are inadequate as food and higher population growth rates are obtained with mixed diets of several algae or algae and yeast. This is confirmed by more recent studies on life history characteristics of *B. plicatilis* (Korstad et al., 1989b).

Additions of cobalamin (Vitamin B_{12}) to the culture medium enhance *B. plicatilis* growth with *Chlorella*, which alone produces suppressed growth (Hirayama et al., 1989) On the other hand, rotifers grown on baker's yeast are deficient in total lipids. In rotifers the fatty acid structure is determined by the diet (Lubzens et al., 1985). Rotifer growth especially depends on the content of highly unsaturated fatty acids (HUFA) in the food (Lubzens,

1987). Müller-Navarra (1995a) was able to explain 92 % of the variation of *Daphnia galeata* growth rate by the seston content of eicosapentaenoic acid (20:5ω3).

Daphnia-species have higher phosphorus to carbon (P:C) values than their food (about 30 µg P/mgC) (Hessen, 1992). Unfortunately, only two values have been found for the rotifers with 26 µgP/mgC for *Euchlanis dilatata* (Gulati et al., 1989) and 23.5 µgP/mgC for *B. rubens* (Rothhaupt, 1995). These values are not very much lower than for the daphnids. This shows a high P demand of both groups and, therefore, a high tendency to be P-limited. This stands in contrast to copepods which have P values of about 10 µgP/mgC. Hessen (1992) concluded that zooplankton suffer from direct P-limitation. Severe P-limited algae were nutritionally poor as food for *B. rubens* (Rothhaupt, 1993). However, it is not clear if reduced P-content or another biochemical constituent is limiting. Growth rate of *Daphnia galeata* is higher in P-limited *Cyclotella* at higher eicosapentaenoic acid content (20:5ω3) than in P-limited *Scenedesmus* which have a much lower fatty acid content (Müller-Navarra, 1995b). Both P-limited food algae, however, are of low food value for *B. rubens* (Rothhaupt, 1995). In contrast, the amino acid profiles of rotifers are not different when the animals were fed different algae. Therefore, this research fails to explain the better nutritive value of certain algae (Lubzens et al., 1989). Rothhaupt (1995) reports that feeding of N-limited *Scenedesmus* in concentrations >ILL results in low C-yield for *Brachionus*. However, N-yield is not different from unlimited algae; therefore, direct N-limitation of *Brachionus* seems possible.

Impact of protozoans as food for rotifers

Many rotifers are able to feed on protozoans. Arndt (1993) presented a list of rotifers feeding on heterotrophic and mixotrophic flagellates, among them many planktonic genera (*Synchaeta, Asplanchana, Brachionus, Keratella, Kellicottia, Polyarthra*). Autotrophic flagellates are well known as a good food for rotifers (Pourriot, 1977). The feeding of predatory rotifers, such as *Asplanchna* on ciliates is not surprising (Naumann, 1923). *Paramecium* is known a good culture feed (Robertson and Salt, 1981). Gilbert and Jack (1993) observed *B. calyciflorus,* and particularly *S. pectinata* to feed on ciliates. *S. pectinata* seem to feed on some ciliates even more efficiently than large cryptomonads that are presumed to be its preferred food. The clearance rates on *Tetrahymena* are twice as high as those on *Strobilidium gyrans* probably because of *Strobilidium's* saltatory behavior. *B. calyciflorus* have high clearance rates on *Tetrahymena*, higher than those on yeast and algae at low densities. Gilbert and Jack (1993) concluded that large rotifers are effective predators on ciliates in the 45–60 µm range. Except for special cases, however, rotifers are not be able to control the abundance of planktonic protozoans (Arndt, 1993) which have high reproductive rates.

Cladocerans, on the other side, because of their higher clearance capacity may exert a top-down control of nanoplanktonic flagellates and small ciliates (Jürgens, 1994). This, however, depends on the cladoceran species and other conditions. In the presence of cyanobacteria filaments rotifers were responsible for 68 and 93% of the total metazoan grazing on nanoflagellates (Sanders et al. 1994). A question remains for the consumer: is the ingestion of protozoans a subsistence for the growth of the consumer? At least for *Paramecium*-fed *Asplanchna* (see above) and for *Daphnia pulex* it seems so (Wickham et al., 1993). Cladoceran growth rates increase up to 50% when ciliates were added to the diet (*Cryptomonas* sp.) but no more than 10% when rotifers were added.

Impact of food quantity on growth rate

It is obvious that there are not as many food particle classes as consumer species. In contrast, in the plankton there seems to be a broad generalist group, ranging from rotifers to cladocerans, all of which compete for the same food sizes, but they do not do so with the same success rate. Several different boundary conditions (e.g., clay, toxins and filament length) have been mentioned, but these mechanisms are confined only to special cases. A further factor to be considered for all zooplankters is the capacity to which they can adapt to different concentrations of food. This is closely associated with the life history strategies of the species.

The ability of zooplankton to capture algae and convert that food into their own biomass is critical to comprehending how zooplankton communities develop. In aquatic herbivors this relationship may be expressed by using K_s as defined by a modification of the Monod-equation:

$$r = r_{max} * C/K_s + C$$

where r = specific growth rate, C = food concentration, and K_s = food concentration for half-maximum growth rate.

Species with low K_s values and high k values, where $k = r_{max}/K_s$ = slope at low food concentrations (Walz, 1993a), have higher affinities for food. At low food concentrations their rates of ingestion and growth are higher than those of species with high r_{max} exhibiting lower affinities to food. For example, *Keratella chochlearis* is adapted to lower food concentrations. On the other hand, *Brachionus angularis* is a superior competitor at higher concentrations (Walz, 1993a). Food affinity of *B. rubens* was higher (higher k-values, lower K_s-values) than for *B. calyciflorus* (Rothhaupt, 1990a). This concept is in agreement with the lowest resource requirement, R* (Tilman, 1982).

These relationships differ for rotifer species with different life strategies. *K. cochlearis* reaches higher maximum growth rates at relatively much

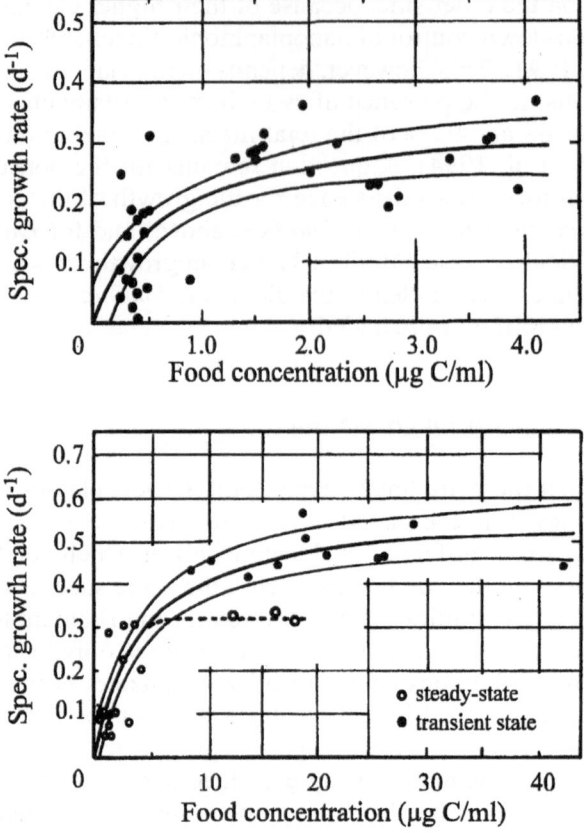

Figure 1. Specific growth rates of rotifers as a function of food concentration. Upper panel: *Keratella cochlearis,* lower panel: *Brachionus angularis.* (After Walz, 1993a).

lower food concentrations than *B. angularis* which in turn attains growth at much higher food concentrations (Fig. 1). In the estuarine Potomac River (USA), *Synchaeta* sp., shows similar growth kinetics (Heinbokel et al., 1988). The maximum growth rate ($r_{max} = 0.36$ d^{-1}) was found only at very high chlorophyll concentrations (expressed as carbon: >1.0 µg C ml^{-1}).

The concept of K_s in the Monod model (Walz, 1993a) has not been widely applied to cladocerans. Nevertheless, other indications for the use of different food concentrations by the cladocerans and the rotifers exist. From experimental studies the incipient limiting levels (ILL) can be derived. ILL is the food concentration at which maximal ingestion rates are reached. These ILL seem to correspond to K_s. Comparisons of ILL show that cladocerans do not depend as much on high food resources as rotifers. Most studies on daphnids indicate that these animals reach maximum ingestion rates at a food concentration of about 0.3 µg C ml^{-1} (Geller, 1975;

Lampert and Muck, 1985; Porter et al., 1982). Within cladocerans, the ILL and ingestion rates seem to be higher for larger species, at least when comparing a smaller species (*Bosmina longirostris*) to a larger one *(Daphnia rosea)* (DeMott, 1982).

In contrast the ILL for rotifers is generally higher. For example, using *Stichococcus bacillaris* and *Coccomyxa* sp. as food Walz (1993a) determined the ILL for *Keratella cochlearis* to be >1.5 µg C ml^{-1} and 2 µg C ml^{-1} for *Brachionus angularis*. A smaller form of *B. angularis* had an ILL of 1.3 µg C ml^{-1} with food particle sizes comparable to these algae (Duncan and Gulati, 1983). The ILL of rotifers seems to depend on particle size. It is relatively lower for optimum food sizes, but increases for smaller particles by a nonlinear function (Rothhaupt, 1990a). Yúfera and Pascual (1985) reported ILLs for *B. plicatilis* for different algae that were above 5 µg C ml^{-1} (if C = 50% dry weight). Korstad et al. (1989a), however, found ILLs of 1.46 and 1.90 µg C ml^{-1} for this species. The ILL for *B. calyciflorus* was 2.5 µg C ml^{-1} with the green alga *Euglena gracilis* as food (Starkweather and Gilbert, 1977) and 1.5 µg C ml^{-1} with the cyanobacterium, *Anabaena flos-aquae* (Starkweather, 1981).

Threshold food concentrations (TFC) also demonstrate that rotifers depend on high food concentrations to survive. TFC are food concentrations below which no positive growth is possible. Such studies must take into account that there are different thresholds for the metabolism of individuals and populations (Walz, 1993b). The population-based threshold includes an effort to offset losses due to mortality (i.e., r = 0). In general TFC is higher for rotifers than for cladocerans (Duncan, 1989; Gliwicz, 1990; Gliwicz and Lampert, 1990; Stemberger and Gilbert, 1985) and those of cladocerans tend to surpass those of calanoid copepods (Lampert and Muck, 1985). Ranges between the groups, however, overlap and depend on the specific species and on resource fluctuations (MacIsaac and Gilbert, 1991). Schiemer (1985) arranged aquatic pelagic animals according to increasing threshold concentrations, with calanoids having thresholds of 0.007–0.11 µg C ml^{-1}, cladocerans with 0.04–0.12 µg C ml^{-1}, and ciliates with 0.6–4.0 µg C ml^{-1}. An overall trend based on different taxonomical groups indicates an inverse relationship with body size (Duncan, 1989).

From combined experiments of several authors it is evident that TFC increases with body size in rotifers (Fig. 2) (Hartmann, 1987; Rothhaupt, 1990a; Stemberger and Gilbert, 1985; Walz and Rothbucher, 1987). A similar relationship to body weight also is obvious in the genus *Asplanchna* (Stemberger and Gilbert, 1984).

This positive correlation between TFC and body weight in rotifers, however, was not found in other zooplankton groups. Thresholds in cladocerans decrease with increasing body size (Gliwicz, 1990; Gliwicz and Lampert, 1990). TFC for the large *Daphnia magna* are about 0.015 µg C ml^{-1}, and, for the smallest tested cladoceran, *Ceriodaphnia reticulata*, about

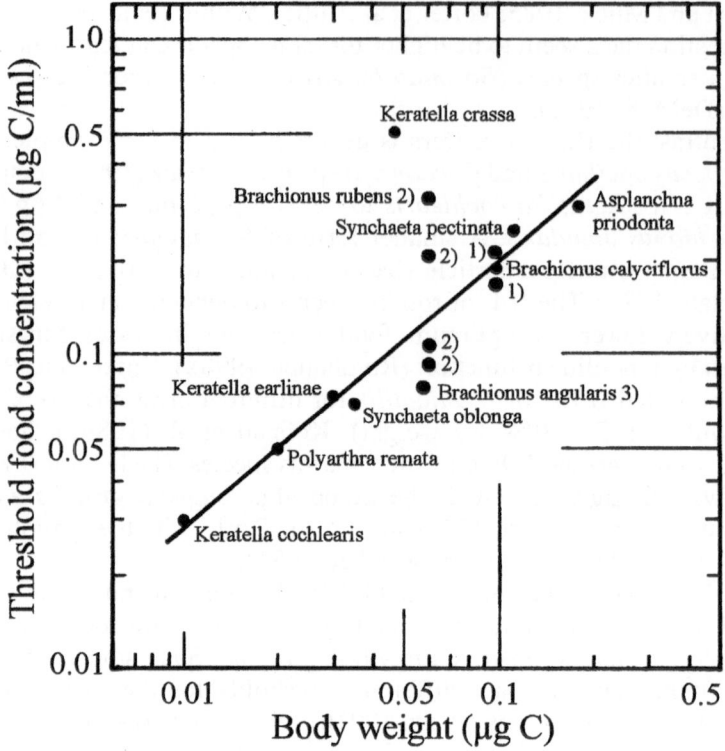

Figure 2. Threshold food concentrations as a function of body weight. 1,2, *B. calyciflorus* and *B. rubens* according to Rothhaupt (1990a); 3, *B. angularis* according to Hartmann (1987). (After Stemberger and Gilbert, 1985).

0.04 µg C ml^{-1}. According to Frost (1985) the thresholds of copepods, however, are about 0.05 µg C ml^{-1} for the larger *Calanus finnmarchicus* and 0.01 µg C ml^{-1} for the smaller *Pseudocalanus* sp.

At food concentrations below the threshold levels, individuals consume stored food reserves, starve and eventually die. At low food concentrations Cladocera are superior in two respects: first, by having lower thresholds and second by possessing higher energy reserves. Resistance to starvation in cladocerans depends on animal age and total lipid reserves (Goulden and Henry, 1987; Tessier et al., 1983). *Daphnia* can withstand significantly longer periods of food shortage than *K. cochlearis* (MacIsaac and Gilbert, 1991). This rotifer had little ability to survive scarce food conditions. The LT$_{50}$ (lethal times 50%, i.e., time when 50% of the individuals are dead) is between 2 days (Gilbert, 1985) and 4 days (MacIsaac and Gilbert, 1991), about ¼ to ½ as short as for *Daphnia ambigua*.

Impact of food quantity on other life history parameters

With increasing food concentration the duration of juvenile phase (D_p, pre-reproductive period between hatching and laying of the first egg) diminishes. In Figure 3 the rate of juvenile development ($1/D_p$) is shown to increase with food concentration in *B. angularis* (Walz and Rothbucher, 1987). This shortening of the pre-reproductive phase with higher food concentrations has been established for many rotifers (King, 1967, Pilarska, 1977) and planktonic crustaceans (Weglenska, 1971). In *B. plicatilis*, however, the periods were shortest at intermediate food concentrations, whereas at lower and higher concentrations these periods are prolonged (Schmid-Araya, 1991). The juvenile period did not differ with food concentration in *Encentrum linnhei* (Schmid-Araya, 1991).

Apparently there is no consistent pattern of embryonic development (D_c) when food concentration is altered. For example, no relationship was found in *B. plicatilis* (Schmid-Araya, 1991), *Euchlanis dilatata* (King, 1967), *B. calyciflorus* (Pourriot and Deluzarches, 1971). On the other hand, Schmid-Araya (1991) reported shorter D_e in *Encentrum linnhei* with higher food levels. For *B. angularis*, D_e are shortest and the rates of development ($1/D_e$) are highest in medium range food concentrations where relative egg size is smallest (Walz and Rothbucher, 1987). But increasing juvenile development rate compensates for the prolongation of the time of embryonic development (D_e). The rate of generation development ($1/D_g = 1/D_e + 1/D_p$), as an analogue to the population growth rate, therefore, reaches a plateau

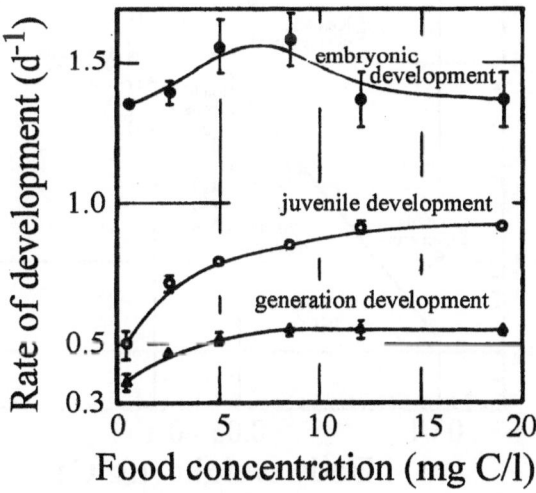

Figure 3. Relationship between food concentration and rate of embryonic development ($1/D_e$), rate of juvenile development ($1/D_p$), and rate of generation development ($1/D_g$) of *B. angularis*. (After Walz and Rothbucher, 1987).

according to a saturation function of the Monod model. The kinetics of rates of development resemble those of the more general Monod-saturation functions of population growth rates (Fig. 1).

Body size and life histories of rotifers

These findings discussed above correlate well to an interesting relationship between K_s-food concentration and body size in rotifers. The smaller species have a higher food affinity with lower K_s-values. Stemberger and Gilbert (1985) demonstrated this for eight rotifer species to which one may add an additional four (Fig. 4) from the works of Rothhaupt (1990a) and Walz (1993a). This generalization also holds within the genus *Asplanchna* (Stemberger and Gilbert, 1984).

A general positive interspecific relationship exists also between maximum growth rate (r_{max}) and body size. Thus, smaller rotifers are not able to

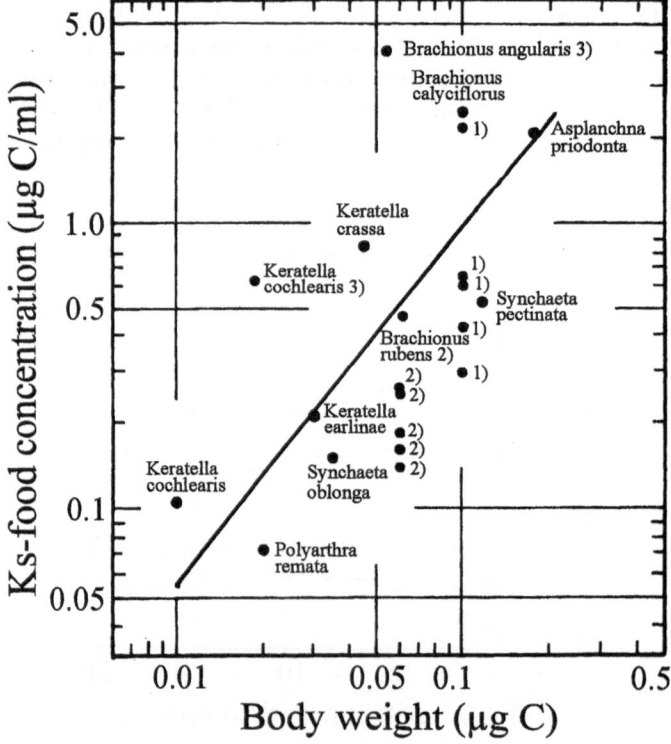

Figure 4. K_s-food concentrations as a function of rotifer body weight. 1,2, *B. calyciflorus* and *B. rubens* according to Rothhaupt (1990a); 3, *B. angularis* and *K. cochlearis* according to Walz (1993a). (After Stemberger and Gilbert, 1985).

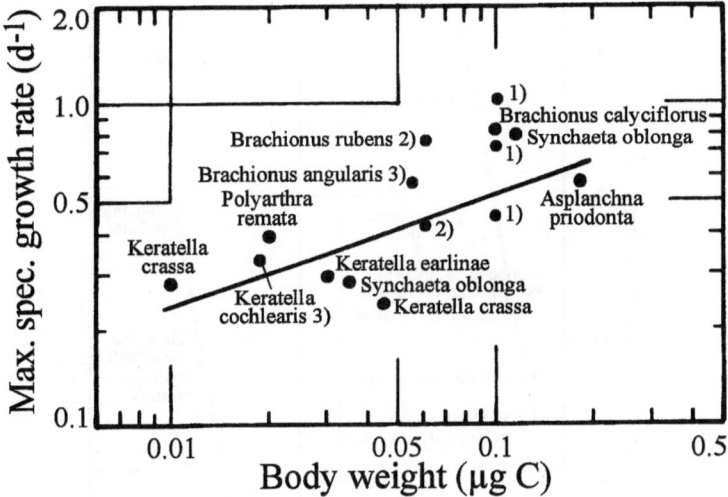

Figure 5. Maximum specific growth rates of rotifers in relation to body weight, 1,2, *B. calyciflorus* and *B. rubens* according to Rothhaupt (1990a); 3, *B. angularis* and *K. cochlearis* according to Walz (1993a). (After Stemberger and Gilbert, 1985).

grow as fast as larger species (Fig. 5). The values depicted in this figure were obtained from laboratory cultures of several authors (Rothhaupt, 1990a; Stemberger and Gilbert, 1985; Walz, 1993b). A further positive relationship is demonstrated between r_{max} and body weight for the rotifers and planktonic crustaceans of Lake Constance (Germany) (Walz, 1993c) (Fig. 6). Both rotifers and crustaceans have higher maximum growth rates with increasing body size. It is interesting to see that rotifers with the highest r_{max} are viviparous (*Asplanchna*) or do not carry their eggs (*Synchaeta*), whereas all K_s-strategists bear their eggs until hatching.

Combining both relationships, K_s and r_{max} to body weight it is evident that large species (e.g., *Asplanchna*) are able to grow faster but with a higher requirement for food (energetic limitation). Smaller species (e.g., *Keratella*) are rather limited by kinetic constraints. Figure 7 shows an allometric relationship between body volume and relative egg size in 43 rotifers species (Walz et al., 1995). The exponent 0.6 (at $p < 0.001$ significantly different from 1) means larger rotifers had relatively smaller eggs. This corresponds to a functional relationship (Fig. 8): species with relatively smaller eggs had higher rates of embryonic development, shorter generation times and higher maximum birth rates (r_{max}-strategy). Similar trade-offs between relative egg size and body weight, maximum birth rate and rate of juvenile growth as is seen in rotifers, also hold for cladocerans (Romanovsky, 1985). As in rotifers large cladocerans have, relatively smaller eggs (Lynch, 1980).

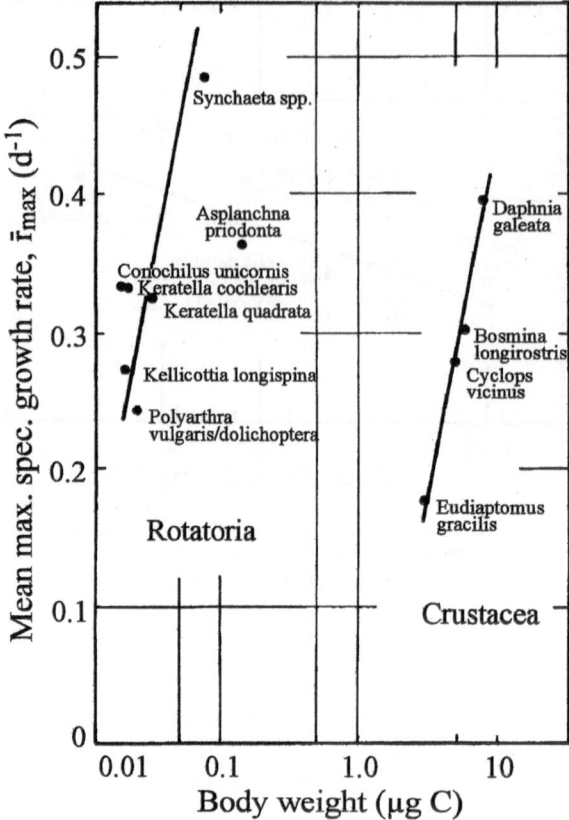

Figure 6. Maximum specific growth rates of rotifers and crustaceans in Lake Constance 1977/1978 in relation to body weight. To calculate a representative value for r_{max} for every population, the five highest growth rates between two sampling days in the growing season were averaged. (After Walz, 1993c).

Egg volumes seem to have an influence on the duration of embryonic development (D_e). In *B. calyciflorus* and *B. caudatus* embryonic development was prolonged with increase in egg size (Duncan and Gulati, 1983; Pourriot, 1973). In contrast, Bennett and Boraas (1989) found that a 70% reduction in egg volume had little effect on egg development time in *B. calyciflorus*. In *B. angularis*, no continuous relationship between absolute egg volume and D_e was found, but D_e shortened with higher relative egg size (Walz and Rothbucher, 1987), i.e., with a smaller body size of the mother.

Relatively larger eggs may produce a higher individual fitness of the offspring, which should be responsible for a higher survival rate for the younger age classes. Survival of young *K. cochlearis* was better than that of *B. angularis* (Walz, 1987). *K. cochlearis* had relatively larger eggs. A higher survival ratio of individual larger offspring was shown in *B. calyciflorus* (Galindo et al., 1993).

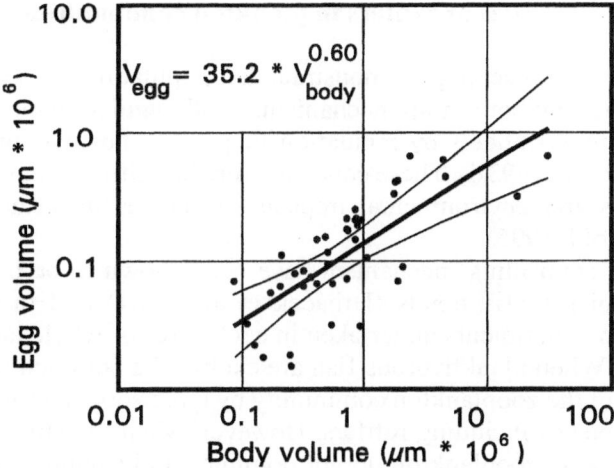

Figure 7. Relationship between body volume and egg volume in rotifer species. (After Walz et al., 1995).

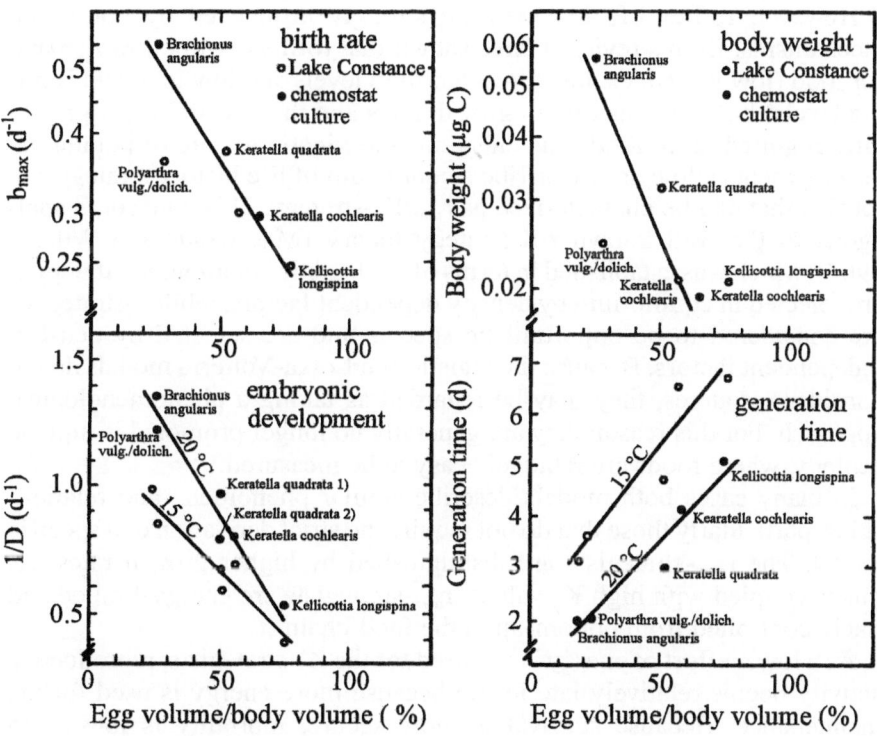

Figure 8. Relationships of different rotifer population dynamic parameters to the relative egg size. (After Walz, 1993 c).

Life history strategies of rotifers in plankton communities

Studies of plankton ecology demonstrate that populations are regulated by both top-down and bottom-up mechanisms. Both regulation modes are tied up in community-models by regulation loops with positive and negative feedbacks (Walz, 1993 d). These concepts were identified as environmental requirements and environmental impacts of species in terms of niche-theory (Leibold, 1995).

Top-down controlling mechanism have been shown to be important in many examples in fishponds (Hrbácek et al., 1961) and in many bio-manipulation experiments undertaken in enclosures or whole lakes (Benndorf, 1990). When planktivorous fish are excluded a common result is the domination of the zooplankton community by large zooplankton at the cost of smaller species including rotifers. However, when vertebrate predators are present, large zooplankton do not dominate and smaller zooplankton, including rotifers, prosper. Therefore, top-down effects by vertebrates are, in a real sense, beneficial to rotifers when their main competitors (large cladocerans) are reduced in numbers (Neill, 1984). As a result, the life histories of rotifers (e.g., population age distribution and reproduction) are not directly affected by vertebrate predation.

However, rotifer life histories are strongly influenced by bottom-up mechanisms. In this review I have shown that rotifers with a low demand for food (low K_s-values, low threshold food levels and low incipient limiting levels) have low maximum growth rates and those with higher growth rates required more food. The energetic and kinetic feature of population developments allow us to describe a continuum of life history strategies in rotifers that can be characterized as r_{max}/K_s-strategy. This concept is analogous to the well known r/K-strategy theory (MacArthur and Wilson, 1967), which was established in terms of the logistic equation. K-strategists are selected in equilibrium by density dependent factors, while r-strategists are considered to be opportunistic species and are selected by density-independent factors. Because the logistic and Lotka-Volterra models do not consider resources, they may be regarded as taking a phenomenological approach. For this reason they are generally no longer promoted in aquatic ecology, where foods are relatively easy to be measured.

In many cases both models describe similar phenomena and relationships, particularly those that do not require the strict derivation of a specific model. The r_{max}-strategists are distinguished by higher growth rates. As this is coupled with high K_s-values, r_{max}-strategists are energy-limited and likely controlled from bottom-up in the food-chain.

As a lower effort of energy is required for the K_s-strategists, reproductive activity begins relatively late in life because more energy is used for the maintenance. Because survival is more secure, mortality is lower and reproduction may be distributed over a longer period, i.e., at a lower rate. This strategy does not demand a high food concentration. As populations of K_s-strategists never attain high r_{max}-values, they are likely to suffer from predator pressure and become controlled from the top-down.

The r_{max}-/K_s-hypothesis also gives a model for the different strategies of rotifers and their main competitors, the cladocerans. As described above, rotifers are r_{max}-strategists when they are compared with cladocerans requiring higher food concentrations. This shows that rotifers depend on pioneer conditions when the cladocerans with their greater filtering efficiency are not present. In such situations rotifers use rapid development thereby achieving larger population levels in a few days only. Cladocerans react more slowly, attaining higher population densities only after a delay of between 8 and 14 days (Jacobs, 1978, Porter et al., 1983). Because their population development is slower, cladocerans are more likely to be controlled by top-down processes.

Obviously, one of the preconditions for success in the plankton is an adequate food supply. In the long run, species with higher demands on food resources are eliminated by those that can graze down the food supply at least under stable conditions. This scenario assumes that population densities are in equilibrium with their resources. That is, the zooplankton are food-limited and therefore population densities will fluctuate around some mean value as long as the resource levels do not change (Wiens, 1984). Under these circumstances the principle of competitive exclusion principle must be in operation. However, this view does not consider any temporal aspect of the interactions between species. Nevertheless, in most cases one assumes that a simultaneous occurrence of similar species cannot be explained without invoking a temporal or spatial variability of resources (Schulze et al., 1996).

On the other hand, non-equilibrium models presume that changing conditions act as stabilizing elements permitting the coexistence of species that would otherwise be serious competitors (DeAngelis and Waterhouse, 1987). Environmental fluctuations, interactions with predators, competition, among other processes, may be responsible for species succession (DeMott, 1989; Rothhaupt, 1990c). In variable environments it is the speed with which species are able to respond to environmental changes which possess an advantage over competitors. For this purpose a species should maximize the kinetic features of its growth. However, this is coupled with large energy expenditure. Therefore, this strategy depends on the amplitude and frequency of the food supply. The intermediate disturbance hypothesis (Connell, 1978) argues that the frequencies and intensities of perturbations are deciding factors in increasing species diversity (Reynolds et al., 1993). In semicontinuous multispecies-cultures of marine phytoplankton this hypothesis was supported within disturbance intervals of from 1 to 14 days (Sommer, 1995). This argument may be brought into agreement with the kinetic considerations of the different species, r_{max}-strategists should have an advantage under fluctuating conditions, while K_s-strategists may have an advantage in more constant environments. For that reason environmental fluctuations should benefit many rotifers in plankton communities.

Fluctuations may be introduced from the outside mostly by variation of the food resources and by meteorological events: i.e., by factors operating from bottom-up. Physical factors set fluctuations by climatic and meteorological changes over the season. These external fluctuations then influence internal metabolic cycles, e.g., by time delays within organisms (metabolic times between food uptake and reproduction or delays between populations on different trophic levels) (Walz, 1993 e). Rotifers are good candidates for studies of the influence of environmental fluctuations and for experimental tests of the intermediate disturbance hypothesis.

In contrast, most vertebrate predators (the main top-down actors) have life-times of several years. Therefore, they have more of the character of steering factors. Nevertheless, their predatory behaviour, which is active only during the daylight hours, does have an immediate response: diel vertical migration of many zooplankton species (Stich and Lampert, 1981). However, this behaviour is highly predictive for the prey. Thus such diurnal fluctuations have too high frequencies – in terms of intermediate disturbance hypothesis – to be considered as perturbations. Likewise, differences in the strength of year-class of fishes has too low a frequency to be considered as perturbation.

Acknowledgments
I would like to thank R.L. Wallace and K.O. Rothhaupt, who contributed constructive criticism to improve the paper.

References

Arndt, H. (1993) Rotifers as predators on components of the microbial web (bacteria, heterotrophic flagellates, ciliates) – a review. *Hydrobiologia* 255/256:231–246.
Benndorf, J. (1990) Conditions for effective biomanipulation: conclusions derived from whole-lake experiments in Europe. *Hydrobiologia* 200/201:187–203.
Bennett, W.N. and Boraas, M.E. (1989) A demographic profile of the fastest growing metazoan: a strain of *Brachionus calyciflorus* (Rotifera). *Oikos* 55:365–369.
Bergquist, A.M., Carpenter, S.R. and Latino, J.C. (1985) Shifts in phytoplankton size structure and community composition during grazing by contrasting zooplankton assemblages. *Limnol. Oceanogr.* 30:1037–1045.
Bogdan, K.G. and Gilbert, J.J. (1987) Quantitative comparison of food niches in some freshwater zooplankton. A multi-tracer-cell approach. *Oecologia* 72:331–340.
Børsheim, K.Y. and Andersen, S. (1983) Grazing and food size selection by crustacean zooplankton compared to production of bacteria and phytoplankton in a shallow Norwegian mountain lake. *J. Plankton Res.* 9:367–379.
Brandl, Z. and Fernando, C.H. (1978) Prey selection by the cyclopoid copepods *Mesocyclops edax* and *Cyclops vicinus. Verh. Internat. Verein. Limnol.* 20:2505–2510.
Brendelberger, H. (1985) Filter mesh-size and retention efficiency for small particles: comparative studies with Cladocera. *Arch. Hydrobiol. Beih. (Ergebn. Limnol.)* 21:135–146.
Brendelberger, H. and Geller, W. (1985) Variability of filter structures in eight *Daphnia* species: mesh sizes and filtering areas. *J. Plankton Res.* 7:473–486.
Brooks, J.L. and Dodson, S.I. (1965) Predation, body size, and composition of plankton. *Science* 150:28–35.

Burns, C.W. (1968) The relationship between body size of filter feeding Cladocera and the maximum size of particles ingested. *Limnol. Oceanogr.* 13:675–678.

Burns, C.W. and Gilbert, J.J. (1986) Effects of daphnid size and density on interference between *Daphnia* and *Keratella cochlearis. Limnol. Oceanogr.* 31:848–858.

Conde-Porcuna, J.M., Morales-Baquero, R. and Cruz-Pizaro, L. (1993) Effects of *Daphnia longispina* on rotifer populations in a natural environment: relative importance of food limitation and interference competition. *J. Plankton Res.* 16:691–706.

Connell, J.H. (1978) Diversity in tropical rainforest and coral reefs. *Science* 109:1304–1310.

Cuker, B.E. (1993) Suspended clay alter trophic interactions in the plankton. *Ecology* 74: 944–953.

Cuker, B.E. and Hudson, Jr., L. (1992) Type of suspended clay influences zooplankton response to phosphorus loading. *Limnol. Oceanogr.* 37:566–576.

DeAngelis, D.L. and Waterhouse, J.C. (1987) Equilibrium and nonequilibrium concepts in ecological models. *Ecol. Monogr.* 57:1–21.

De Bernardi, R. and Peters, R.H. (1987) Why Daphnia? *In:* R. De Bernardi and R.H. Peters (eds): *Daphnia. Mem. 1st. Ital. Idrobiol.* 45:1–9.

De Mott, W.R. (1982) Feeding selectivities and relative ingestion rats of *Daphnia* and *Bosmina. Limnol. Oceanogr.* 27:518–527.

DeMott, W.R. (1985) Relations between filter mesh-size, feeding mode, and capture efficiency for cladoceran feeding on ultrafine particles. *Arch. Hydrobiol. Beih. (Ergebn. Limnol.)* 21: 125–134.

DeMott, W.R. (1986) The role of taste in food selection by freshwater zooplankton. *Oecologia* 69:334–340.

Dieffenbach, H. and Sachse, R. (1991) Biologische Untersuchungen an Rädertierchen in Teichgewässern. *Int. Revue ges. Hydrobiol. Suppl.* 3:1–93.

Dodson, S.I. (1974) Zooplankton competition and predation: an experimental test of the size-efficiency hypothesis. *Ecology* 55:605–613.

Dumont, H.J. (1977) Biotic factors in the population dynamics of rotifers. *Arch. Hydrobiol. Beih. (Ergebn. Limnol.)* 8:98–112.

Dumont, H.J., Sarma, S.S.S. and Jawahar Ali, A. (1995) Laboratory studies on the population dynamics of *Anuraeopsis fissa* (Rotifer) in relation to food density. *Fresh. Biol.* 33:39–46.

Duncan, A. (1989) Food limitation and body size in the life cycle of planktonic rotifers and cladocerans. *Hydrobiologia* 186/187:11–28.

Duncan, A. and Gulati, R.D. (1983) Feeding studies with natural food particles on tropical species of planktonic rotifers. *In:* F. Schiemer (ed.): *Limnology of Parakrama Samudra, Sri Lanka.* W. Junk Publishers, The Hague, pp 117–125.

Egloff, D.A. (1988) Food and growth relations of the marine microzooplankter, *Synchaeta cecilia* (Rotifera). *Hydrobiologia* 157:129–141.

Ejsmont-Karabin, J. (1974) Studies on the feeding of planktonic polyphae *Asplanchna priodonta* Gosse (Rotatoria). *Ekologia Polska Ser. A* 22:311–317.

Emmerson, W.D. (1984) Predation and energetics of *Penaeus indicus* (Decapoda, Penaeidae) larvae feeding on *Brachionus plicatilis* and *Artemia* nauplii. *Aquaculture* 38:201–209.

Esparcia, A. (1994) Feeding in the rotifer *Anuraopsis fissa. Verh. Int. Verein. Limnol.* 25: 2324–2326.

Felix, A., Stevens, M.E. and Wallace, R.L. (1995) Unpalatability of a colonial rotifer, *Sinantherina socialis*, to small zooplanktivorous fishes. *Invertebr. Biology* 114(2):139–144.

Filatov, V.P. (1972) The efficiency of natural food by carp larvae (*Cyprinus carpio*). *Vopr. Ikhtiol. (Moskva)* 12:886–892 (in Russian).

Frost, B.W. (1985) Food limitation of the planktonic marine copepods *Calanus pacificus* and *Pseudocalanus sp.* in a temperate fjord. *Arch. Hydrobiol. Beih. (Ergebn. Limnol.)* 21:1–13.

Fulton, R.S.III. and Pearl, H.W. (1987) Toxic and inhibitory effects of the blue-green alga *Microcystis aeruginosa* on herbivorous zooplankton. *J. Plankton Res.* 9:837–855.

Fulton, R.S.III. and Pearl, H.W. (1988) Effects of the blue-green alga *Microcystis aeruginosa* on zooplankton competitive relations. *Oecologia* 76:383–389.

Galindo, M.D., Guisande, C. and Toja, J. (1993) Reproductive investment of several rotifer species. *Hydrobiologia* 255/256:317–324.

Geller, W. (1975) Die Nahrungsaufnahme von *Daphnia pulex* in Abhängigkeit von der Futterkonzentration, der Temperatur, der Körpergröße und dem Hungerzustand der Tiere. *Arch. Hydrobiol./Suppl.* 48:47–107.

Gilbert, J.J. (1980) Observation of the susceptibility of some protists and rotifera to predation by *Asplanchna girodi. Hydrobiologia* 73:87–91.

Gilbert, J.J. (1985) Competition between rotifers and *Daphnia. Ecology* 66:1943–1950.

Gilbert, J.J. (1988a) Suppression of rotifer populations by *Daphnia*. A review of the evidence, the mechanisms, and the effects on zooplankton community structure. *Limnol. Oceanogr.* 33:1286–1303.

Gilbert, J.J. (1988b) Susceptibilities of ten rotifer species to interference from *Daphnia pulex. Ecology* 69:1826–1838.

Gilbert, J.J. (1990) Differential effects of *Anabaena affinis* on cladocerans and rotifers: Mechanisms and implications. *Ecology* 71:1727–1740.

Gilbert, J.J. and Bogdan, K.G. (1981) Selectivity of *Polyarthra* and *Keratella* for flagellate and aflagellate cells. *Verh. Internat. Verein. Limnol.* 21:1515–1521.

Gilbert, J.J. and Durand, M.W. (1990) Effect of *Anabaena flos-aquae* on the abilities of *Daphnia* and *Keratella* to feed and reproduce on unicellular algae. *Freshw. Biol.* 24:577–596.

Gilbert, J.J. and Jack, J.D. (1993) Rotifers as predators on small ciliates. *Hydrobiologia* 255/256:247–253.

Gilbert, J.J. and Starkweather, P.L. (1977) Feeding in the rotifer *Brachionus calyciflorus*. I. Regulatory mechanisms. *Oecologia* 28:125–131.

Gilbert, J.J. and Williamson C.E. (1978) Predator prey behavior and its effect on rotifer survival in associations of *Mesocyclops edax. Asplanchna girodi, Polyarthra vulgaris*, and *Keratella cochlearis. Oecologia* 37:13–22.

Gliwicz, Z.M. (1969) Studies on the feeding of pelagic zooplankton in lakes with varying trophy. *Ekol. Pol. Ser. A.* 17:663–708.

Gliwicz, Z.M. (1990) Food thresholds and body size in cladocerans. *Nature* 343:638–640.

Gliwicz, Z.M. and Lampert, W. (1990) Food thresholds in *Daphnia* species in the absence and presence of blue-green filaments. *Ecology* 7:691–702.

Gliwicz, Z.M. and Siedlar, E. (1980) Food size limitation and algae interfering with food collection in *Daphnia. Arch. Hydrobiol.* 88:155–177.

Gliwicz, Z.M., and Pijanowska, J. (1989) The role of predation in zooplankton succession. *In*: U. Sommer (ed.): *Plankton Ecology. Succession in Plankton Communities*. Springer-Verlag, Berlin, etc. Brock/Springer Series in Contemporary Bioscience. pp 253–298.

Gophen, M. (1977) Food and feeding habits of *Mesocyclops leuckarti* (Claus) in Lake Kinneret (Israel). *Freshw. Biol.* 7:513–518.

Goulden, C.E. and Henry, L.B.D. (1987) Egg size, postembryonic yolk, and survival ability. *Oecologia* 72:28–31.

Güde, H. (1989) The role of grazing on bacteria in plankton succession. *In:* U. Sommer (ed.): *Plankton Ecology. Succession in Plankton Communities*. Springer-Verlag, Berlin, etc. Brock/Springer Series in Contemporary Bioscience. pp 337–358.

Guerrero, R., Abella, C. and Miracle, M.R. (1978) Spatial and temporal distribution of bacteria in a meromictic karstic lake basin relationships with physicochemical parameters and zooplankton. *Verh. Internat. Verein. Limnol.* 20:2264–2271.

Guiset, A. (1977) Stomach contents in *Asplanchna* and *Ploesoma. Arch. Hydrobiol. Beih. (Ergebn. Limnol.)* 8:126–129.

Gulati, R.D., Ejsmont-Karabin, J., Rooth, J. and Siewertsen, K. (1989) A laboratory study of phosphorus and nitrogen excretion of *Euchlanis dilatata lucksiana. Hydrobiologia* 186/187:347–354.

Guma'a, S.A. (1978) The food and feeding habits of young perch *Perca fluviatilis* in Windermere. *Freshw. Biol.* 8:177–187.

Hammer, C. (1985) Feeding behavior of roach (*Rutilus rutilus*) larvae and the fry of perch (*Perca fluviatilis*) in Lake Lankau. *Arch. Hydrobiol.* 103:61–74.

Hardin, G. (1960) The competitive exclusion theory. *Science* 131:1292–1297.

Hartmann, U. (1987) Die Populationsdynamik der pelagischen Rotatorien *Brachionus angularis* und *Notholca caudata* in Abhängigkeit von der Futterkonzentration und der Temperatur. Diploma thesis, Fac. Biology, University of Munich, 91 pp.

Havens, K.E. (1993) An experimental analysis of macrozooplankton, microzooplankton and phytoplankton interactions in a temperate eutrophic lake. *Arch. Hydrobiol.* 127:9–20.

Heinbokel, J.F., Coats, D.W., Henderson, K.W., and Tyler, M.A. (1988) Reproduction rates and secondary production of three species of the rotifer genus *Synchaeta* in the estuarine Potomac River. *J. Plankton Res.* 10:659–674.

Hessen, D.O. (1992) Nutrient element limitation of zooplankton production. *Amer. Nat.* 140: 799–814.

Hirayama, K., Maruyama, I. and Maeda, T. (1989) Nutritional effect of freshwater Chlorella on growth of the rotifer Branchionus plicatilis. *Hydrobiologia* 186/187:39–42.

Höffgen, B. (1987) Abundanz- und Populationsdynamik des Mikrozooplanktons im Brunnsee. Diploma thesis, Fac. Biology, University of Munich, 179 pp.

Horn, W. (1985) Investigation into the food selectivity of the plantic crustaceans *Daphnia hyalina, Eudiaptomus gracilis* and *Cyclops vicinus. Int. Rev. ges. Hydrobiol.* 70:603–612.

Hrbácek, J., Dvoráková, M., Korínek, V. and Procházkóva, L. (1961) Demonstration of the effect of the fish stock on the species composition of zooplankton and the intensity of the whole plankton assemblage. *Verh. Internat. Verein. Limnol.* 14:192–195.

Huber, M. (1982) Populationsdynamik der Rädertiere des Fasanerie-Sees im Sommer 1981. Diploma thesis, Fac. Biology, University of Munich, 99 pp.

Infante, A. (1978) Natural food of herbivorous zooplankton of Lake Valencia (Venezuela). *Arch. Hydrobiol.* 82:347–358.

Jack, J.D., Gilbert, J.J. (1994) Effects of *Daphnia* on microzooplankton communities. *J. Plankton Res.* 16:1499–1512.

Jacobs, J. (1978) Coexistence in similar zooplankton species by differential adaptation to reproduction and escape in an environment with fluctuating food and enemy densities. III. Laboratory experiments. *Oecologia* 35:35–54.

Jürgens, K. (1994) Impact of *Daphnia* on planktonic microbial food webs – a review. *Mar. Microb. Food Webs* 8:295–324.

Karabin, A. (1978) The pressure of pelagic predators of the genus *Mesocyclops* (Copepoda, Crustacean) on small zooplankton. *Ekol. Polska* 26:241–257.

King, C.E. (1967) Food, age, and the dynamics of laboratory population of rotifers. *Ecology* 48:111–128.

Kirk, K.L. (1991) Inorganic particles alter competition in grazing plankton: The role of selective feeding. *Ecology* 72:915–923.

Kirk, K.L. and Gilbert, J.J. (1990) Suspended clay and the population dynamics of planktonic rotifers and cladocerans. *Ecology* 71:1741–1755.

Korstad, J., Vadstein, O. and Olsen, Y. (1989a) Feeding kinetics of *Brachionus plicatilis* fed *Isochrysis galbana. Hydrobiologia* 186/187:51–57.

Korstad, J., Olsen, Y. and Vadstein, O. (1989b) Life history characteristics of *Brachionus plicatilis* (Rotifera) fed different algae. *Hydrobiologia* 186/187:43–50.

Koste, W. (1978) *Synchaeta grandis*, ein in Mitteleuropa vom Aussterben bedrohtes Rädertier. *Mikrokosmos* 1978:331–336.

Lampert, W. and Muck, P. (1985) Multiple aspects of food limitation in zooplankton communities the *Daphnia – Eudiaptomus* example. *Arch. Hydrobiol. Beih. (Ergebn. Limnol.)* 21: 311–322.

Lampert, W. and Rothhaupt, K.O. (1991) Alternating dynamics of rotifers and *Daphnia magna* in a shallow lake. *Arch. Hydrobiol.* 120:447–456.

Leibold, M.A. (1995) The niche concept revisited: mechanistic models and community context. *Ecology* 76:1371–1382.

Lubzens, E. (1987) Raising rotifers for use in aquaculture (a review). *Hydrobiologia* 147: 245–255.

Lubzens, E., Marko, A. and Tietz, A. (1985) *De novo* synthesis of fatty acids in the rotifer *Brachionus plicatilis. Aquaculture* 47:27–37.

Lubzens, E., Tandler, A. and Minkoff, G. (1989) Rotifers as food in aquaculture. *Hydrobiologia* 186/187:387–400.

Lynch, M. (1980) The evolution of cladoceran life histories. *Quart. Rev. Biol.* 55:23–41.

MacArthur, R.H. and Wilson, E.O. (1967) *The theory of Island Biogeography.* Princeton University Press, Princeton NJ, 216 pp.

MacIsaac, H.J. and Gilbert, J.J. (1990) Does exploitative competition from *Daphnia* limit the abundance of *Keratella* in Loch Leven? A reassessment of May and Jones (1989). *J. Plankton Res.* 12:1315–1322.

MacIsaac, H.J. and Gilbert, J.J. (1991) Competition between *Keratella cochlearis* and *Daphnia ambigua* effects of temporal patterns of food supply. *Freshw. Biol.* 25:189–198.

Mikschi, E. (1989) Rotifer distribution in relation to temperature and oxygen content. *Hydrobiologia* 186/187:209–214.

Moore, M.V., and Gilbert, J.J. (1987) Age-specific *Chaoborus* predation on rotifer prey. *Freshw. Biol.* 17(1987) 223–236.

Müller-Navarra, D. (1995a) Evidence that a highly unsaturated fatty acid limits *Daphnia* growth in nature. *Arch. Hydrobiol.* 132:297–307.

Müller-Navarra, D. (1995b) Biochemical versus mineral nutrition in *Daphnia. Limnol. Oceanogr.* 40:1209–1214.

Naumann, E. (1923) Spezielle Untersuchungen über die Ernährungsbiologie des tierischen Limnoplanktons II. (Copepoden, Rotatoria). *Lunds Univ. Arskr. NF.* 19:3–17.

Nauwerck, A. (1963) Die Beziehungen zwischen Zooplankton und Phytoplankton im See Erken. *Smyp. Bot. Uppsala* 17:1–163.

Neill, W.E. (1984) Regulation of rotifer densities by crustacean zooplankton in an oligotrophic montane lake in British Columbia. *Oecologia* 61:175–181.

Neill, W.E. (1985) The effects of herbivore competition upon the dynamics of *Chaoborus* predation. *Arch. Hydrobiol. Beih. (Ergebn. Limnol.)* 21:483–491.

Nogrady, T., Wallace, R.I. and Snell, T.W. (1993) *Rotifera* Vol. 1. *Biology, Ecology and Systematics.* SBP Academic Publishers bv, The Hague, 142 pp.

Ooms-Wilms, A.L., Postema, G. and Gulati, R.D. (1993) Clearance rates of bacteria by the rotifer *Filinia longiseta* (Ehrb.) measured using three tracers. *Hydrobiologia* 255/256: 255–260.

Ooms-Wilms, A.L., Postema, G. and Gulati, R.D. (1995) Evaluation of Rotifera based on measurements of *in situ* ingestion of fluorescent particles, including some comparisons with Cladocera. *J. Plankton Res.* 17:1057–1077.

Orcutt, J.D.J. and Pace, M.L. (1984) Seasonal dynamics of rotifer and crustacean zooplankton populations in a eutrophic, monomictic lake with a note on rotifer sampling techniques. *Hydrobiologia* 119:73–80.

Pace, M.L. and Orcutt, J.D. (1981) The relative importance of protozoans, rotifers, and crustaceans in a freshwater zooplankton community. *Limnol. Oceanogr.* 26:822–830.

Pace, M.L. and Vaqué, D. (1994) The importance of *Daphnia* in determining mortality rates of protozoans and rotifers in lakes. *Limnol. Oceanogr.* 39:985–996.

Pace, M.L., McManus, G.B., and Findlay, S.E.G. (1990) Planktonic community structure determines the fate of bacterial production in a temperate lake. *Limnol. Oceanogr.* 35: 795–808.

Platzek, J. (1986) *Die Nahrungsaufnahme planktischer Crustaceen. Grazing – Experimente in einem eutrophen Baggersee.* Diploma thesis, Fac. Biology, University of Munich, 217 pp.

Porter, K.G., Gerritsen, J. and Orcutt, J.D.J. (1982) The effect of food concentration on swimming patterns, feeding behavior, ingestion, assimilation, and respiration by *Daphnia. Limnol. Oceanogr.* 27:935–949.

Porter, K.G., Orcutt, J.D. and Gerritsen, J. (1983) Functional response and fitness in a generalist filter feeder, *Daphnia magna* (Cladocera, Crustacea). *Ecology* 64:735–742.

Pourriot, R. (1965) Quelques *Trichocera* (Rotifères) et leurs regimes alimentaires. *Ann. Hydrobiol.* 1:155–171.

Pourriot, R. (1970) Recherches sur l'ecologie des rotifères. *Vie et milieu* 21 5–181.126

Pourriot, R. (1973) Rapports entre la température, la taille des adultes, la longuer des oufs et le taux de développement embryonnaire chez *Brachionus calyciflorus* Pallas (Rotifère). *Ann. Hydrobiol.* 4:103–115.

Pourriot, R. (1977) Food and feeding habits of the Rotifera. *Arch. Hydrobiol. Beih. (Ergebn. Limnol).* 8:243–260.

Pourriot, R., Deluzarches, M. (1971) Recherches sur la biologie des Rotiferès. II. Influence de la température sur la duree du développement embryonnaire et post-embryonnaire. *Ann. Limnol.* 7:25–52.

Reynolds, C.S., Padisák, J. and Sommer, U. (1993) Intermediate disturbance in the ecology of phytoplankton and the maintenance of species diversity: a synthesis. *Hydrobiologia* 249: 183–188.

Richman, S. and Dodson, S.I. (1983) The effect of food quality on feeding and respiration by *Daphnia* and *Diaptomus. Limnol. Oceanogr.* 28:948–956.

Robertson, J.R. and Salt, G.W. (1981) Responses in growth, mortality, and reproduction to variable food levels by the rotifer, *Asplanchna girodi. Ecology* 62:1585–1596.

Romanovsky, Y.E. (1985) Food limitation and life-history strategies in cladoceran crustaceans. *Arch. Hydrobiol. (Ergebn. Limnol.)* 21:363–372.

Rothhaupt, K.O. (1990a) Population growth rates of two closely related rotifer species: effects of food quantity, particle size, and nutritional quality. *Freshw. Biol.* 23:561–570.

Rothhaupt, K.O. (1990b) Differences in particle-size dependent feeding efficiencies of closely related rotifer species. *Limnol. Oceanogr.* 35:16–23.

Rothhaupt, K.O. (1990c) Resource competition of herbivorous zooplankton. A review of approaches and perspectives. *Arch. Hydrobiol.* 118:1–29.

Rothhaupt, K.O. (1991) The influence of toxic and filamentous blue-green algae on feeding and population growth of the rotifer *Brachionus rubens. Int. Revue ges. Hydrobiol.* 76: 67–72.

Rothhaupt, K.O. (1993) Rotifers and continuous culture techniques: model systems for testing mechanistic concepts of consumer-resource interactions. *In:* N. Walz (ed.): *Plankton Regulation Dynamics. Experiments and Models in Rotifer Continuous Cultures.* Springer-Verlag, Berlin, Ecological Studies Vol. 98: pp 178–192.

Rothhaupt, K.O. (1995) Algal nutrient limitation affects rotifer growth but not ingestion rate. *Limnol. Oceanogr.* 40:1201–1208.

Salt, G.W., Sabbadini, G.F. and Commins, M.L. (1978) Trophi morphology relative to food habits in six species of rotifers (Asplanchnidae). *Trans. Amer. Micros. Soc.* 97:496–485.

Sanders, R.W., Porter, K.G., Bennett, S.J. and Debiase, A.E. (1989) Seasonal pattern of bacterivory by flagellates, ciliates, rotifer and cladocerans in a freshwater planktonic community. *Limnol. Oceanogr.* 34:673–687.

Sanders, R.W., Leeper, D.A., King, C.H., Porter K.G. (1994) Grazing by rotifers and crustacean zooplankton on nanoplantonic protists. *Hydrobiologia* 288:167–181.

Santer, B. and Van den Bosch, F. (1994) Herbivorous nutrition of *Cyclops vicinus.* The effect of a pure algal diet on feeding, development, reproduction and life cycle. *J. Plankton Res.* 16:171–195.

Scheda, S.M. and Cowell, B.C. (1988) Rotifer grazers and phytoplankton seasonal experiments on natural communities. *Arch. Hydrobiol.* 114:31–44.

Schiemer, F. (1985) Bioenergetic niche differentiation of aquatic invertebrates. *Verh. Internat. Verein. Limnol.* 22:3014–3018.

Schmid-Araya, J.M. (1991) The effect of food concentration on the life histories of *Brachionus plicatilis* (O.F.M.) and *Encentrum linnhei* Scott. *Arch. Hydrobiol.* 121:87–102.

Schulze P.C., Zagarese, H.E. and Williamson, C.E. (1996) Competition between crustacean zooplankton in continuous cultures. *Limnol. Oceanogr.* 40:33–45.

Scott, J.M. (1980) Effect of growth rate of the food alga on the growth/ingestion efficiency of a marine herbivore. *J. Mar. Biol. Ass. U.K.* 60:681–702.

Seaman, M.T., Gophen, M., Cavari, B.Z. and Azoulay, B. (1986) *Brachionus calyciflorus* Pallas as an agent for removal of *E. coli* in sewage ponds. *Hydrobiologia* 135:55–56.

Sierszen, M.E. (1990) Variable selectivity and the role of nutritional quality in food selection by a planktonic rotifer. *Oikos* 59:241–247.

Smith, A.D. and Gilbert, J.J. (1995) Relative susceptibilities of rotifers and cladocerans to *Microcystis aeruginosa. Arch. Hydrobiol.* 132:309–336.

Snell, T.W. and King, C.E. (1977) Lifespan and fecundity pattern in rotifers the cost of reproduction. *Evolution* 31:882–890.

Snell, T.W. (1980) Blue-green algae and selection in rotifer populations. *Oecologia* 46: 343–346.

Snell, T.W., Biebrich, C.J. and Fuerst, R. (1983) The effects of green and blue-green algal diets on the reproductive rate of the rotifer *Brachionus plicatilis. Aquaculture* 3:21–30.

Sommer, U. (1995) An experimental test of the intermediate disturbance hypothesis using cultures of marine phytoplankton. *Limnol. Oceanogr.* 40:1271–1277.

Starkweather, P.L. (1981) Trophic relationships between the rotifer *Brachionus calyciflorus* and the blue-green alga *Anabaena flos-aquae. Verh. Internat. Verein. Limnol.* 21:1507–1514.

Starkweather, P.L. (1987) Rotifera. *In:* T.S. Pandian and J.S. Vernberg (eds): *Animal Energetics.* Academic Press, New York, pp 159–183.

Starkweather, P.L. and Gilbert, J.J. (1977) Feeding in the rotifer *Brachionus calyciflorus.* 2. Effect of food density on feeding rates using *Euglena gracilis* and *Rhodoturula glutinis. Oecologia* 28:133–139.

Starkweather, P.L. and Kellar, P.E. (1983) Utilization of cyanobacteria by *Brachionus calyciflorus: Anabaena flos-aquae* (NCR-44-l) as a sole or complementary food source. *Hydrobiology* 104:373–377.

Starkweather, P.L., Gilbert, J.J. and Frost, T.M. (1979) Bacterial feeding by the rotifer *Brachionus calyciflorus*. Clearance and ingestion rate, behavior and population dynamics. *Oecologia* 44:26–30.

Stemberger, R.S. (1987) The potential for population growth of *Ascomorpha ecaudis*. *Hydrobiologia* 147:297–301.

Stemberger, R.S. and Evans, M.S. (1984) Rotifer seasonal succession and copepod predation in Lake Michigan. *J. Great Lakes Res.* 10:417–428.

Stemberger, R.S. and Gilbert, J.J. (1984) Body size, ration level, and population growth in *Asplanchna*. *Oecologia* 64:355–359.

Stemberger, R.S. and Gilbert, J.J. (1985) Body size, food concentration, and population growth in plantonic rotifers. *Ecology* 66:1151–1159.

Stemberger, R.S. and Gilbert, J.J. (1987) Defenses of planktonic rotifers against predators. *In:* W.C. Kerfoot and A. Sih (eds): Predation: *Direct and Indirect Impacts on Aquatic Communities*. University Press of New England, Hannover, London, pp 227–239.

Stenson, J.A.E. (1982) Fish impact on rotifer community structure. *Hydrobiologia* 87:57–64.

Sterner, R.W. (1989) The role of grazers in phytoplankton succession. *In:* U. Sommer (ed.): *Plankton Ecology. Succession in Plankton Communities*. Springer-Verlag, Berlin, Brock/Springer Series in Contemporary Bioscience. pp 107–170.

Stich, H.B. (1985) *Untersuchungen zur tagesperiodischen Vertikalwanderung planktischer Crustaceen im Bodensee*. Diss. Fac. Biology, University of Freiburg, 282 pp.

Stich, H.B. and Lampert, W. (1981) Predator evasion as an explanation of diurnal vertical migration by zooplankton. *Nature* 293:396–398.

Tessier, A.J., Henry, L.L., Goulden, C.E. and Durnad, M.W. (1983) Starvation in *Daphnia*: energy reserves and reproductive allocation. *Limnol. Oceanogr.* 28:667–676.

Threlkeld, S.T. and Choinski, E. (1987) Rotifers, cladocerans and planktivorous fish: What are the major interactions? *Hydrobiologia* 147:239–243.

Tilman, D. (1982) *Resource Competition and Community Structure*. Princeton University Press, Princeton N.J., 296 pp.

Tóth, L.G., Zankai, N.P. and Messner, O.M. (1987) Alga consumption of four dominant planktonic crustaceans in Lake Balaton (Hungary). *Hydrobiologia* 145:323–332.

Urabe, J. (1992) Midsummer succession of rotifer plankton in a shallow eutrophic pond. *J. Plankton Res.* 14:851–866.

Vadstein, O., Olsen, Y., Reinertsen, H. and Jemsen, A. (1993) The role of planktonic bacteria in phosphorus cycling in lakes – sink and link. *Limnol. Oceanogr.* 38:1539–1544.

Van der Bosch, F. and Ringelberg, J. (1985) Seasonal succession and population dynamics of *Keratella cochlearis* (Ehrb.) and *Kellicottia longispina* (Kellicott) in Lake Maarsseven I (Netherlands). *Arch. Hydrobiol.* 103:273–290.

Vanni, M. (1986) Competition in zooplankton communities: suppression of small species by *Daphnia pulex*. *Limnol. Oceanogr.* 31:1039–1056.

Vareschi, E. and Jacobs, J. (1984) The ecology of Lake Nakuru (Kenya). V. Production and consumption of consumer organisms. *Oecologia* 61:83–98.

Wallace, R.L. and Snell, T.W. (1991) Rotifera. *In:* J.H. Thorp and A.P. Covich (eds): *Ecology and Classification of North American Freshwater Invertebrates*. Academic Press, San Diego, New York, pp 187–248.

Walz, N. (1987) Comparative population dynamics of the rotifers *Brachionus angularis* and *Keratella cochlearis*. *Hydrobiologia* 147:209–213.

Walz, N. (1993a) Carbon metabolism and population dynamics of *Brachionus angularis* and *Keratella cochlearis*. *In:* N. Walz (ed.) *Plankton Regulation Dynamics. Experiments and Models in Rotifer Continuous Cultures*. Springer-Verlag, Berlin, Ecological Studies, Vol. 98, pp 89–105.

Walz, N. (1993b) Elements of energy balance of *Brachionus angularis*. *In:* N. Walz (ed.): *Plankton Regulation Dynamics. Experiments and Models in Rotifer Continuous Cultures*. Springer-Verlag, Berlin, Ecological Studies Vol. 98, pp 106–122.

Walz, N. (1993c) Life history strategies of rotifers. *In:* N. Walz (ed.): *Plankton Regulation Dynamics. Experiments and Models in Rotifer Continuous Cultures*. Springer-Verlag, Berlin, Ecological Studies, Vol. 98, pp 193–214.

Walz, N. (1993d) Chemostat regulation principles in natural plankton communities. *In:* N. Walz (ed.): *Plankton Regulation Dynamics. Experiments and Models in Rotifer Continuous Cultures*. Springer-Verlag, Berlin, Ecological Studies, Vol. 98, pp 226–242.

Walz, N. (1993 e) Model simulations of continuous cultures. *Hydrobiologia* 255/256: 165–170.

Walz, N. and Rothbucher, F. (1987) Effect of food concentration on body size, egg size, and population dynamics of *Brachionus angularis* (Rotatoria). *Verh. Internat. Verein. Limnol.* 24: 2750–2753.

Walz, N., Elster, H.-J. and Mezger, M. (1987) The development of the rotifer community structure in Lake Constance during its eutrophication. *Arch. Hydrobiol. Suppl. (Monogr. Beitr.)* 74: 452–487.

Walz, N., Sarma, S.S.S. and Benker, U. (1995) Egg size in relation to body size in rotifers. An indication of reproductive strategy? *Hydrobiologia* 313/314: 165–170.

Watanabe, T., Kitajima, C. and Fujita, S. (1983) Nutritional value of live organisms used in Japan for mass culture of fish: A review. *Aquaculture* 34: 115–143.

Weglenska, T. (1971) The influence of various concentrations of natural food on the development fecundity and production of plantonic filtrators. *Ekol. Pol. Ser. A* 19: 427–473.

Weithoff, G. and Walz, N. (1995) The influence of the filamentous cyanobacteria *Planktothrix agardhii* on population growth and reproductive pattern of the rotifer *Brachionus calyciflorus*. *Hydrobiologia* 313/314: 381–386.

Wickham, S.A. and Gilbert, J.J. (1991) Relative vulnerability of natural rotifer and ciliate communities to cladocerans laboratory and field experiments. *Freshw. Biol.* 26: 77–86.

Wickham, S.A., Gilbert, J.J. and Berninger, U.G. (1993) Effects of rotifers and climates on the growth and survival of *Daphnia*. *J. Plankton Res.* 15: 317–334.

Widigdo, B. (1987) *Experimentelle Untersuchungen zur Eignung des Rotators Brachionus calyciflorus als Erstfutter für Karpfenlarven* (Cyprinus carpio L.). Diss. Fac. Biology, University of Munich, 119 pp.

Wiens, J.A. (1984) Resource systems, populations, and communities. *In:* P.W. Price, C.N. Slobodchikoff and W.S. Gaud (eds): *A New Ecology Novel Approaches to Interactive Systems*. John Wiley and Sons, New York, pp 397–436.

Wieser, W. (1986) Bioenergetik. Energietransformation bei Organismen. G. Thieme Verlag, Stuttgart, 245 pp.

Williamson, C.E. (1983) Invertebrate predation on planktonic rotifers. *Hydrobiologia* 104: 385–396.

Williamson, C.E. and Butler, N.M. (1986) Predation on rotifers by the suspension-feeding calanoid copepod *Diaptomus pallidus*. *Limnol. Oceanogr* 3: 393–402.

Yan, N.D. and Geiling, W. (1985) Elevated planktonic rotifer biomass in acidified metal-contaminated lakes near Sudbury, Ontario. *Hydrobiologia* 120: 199–205.

Yúfera, M. and Pascual, E. (1985) Effects of algal concentration on feeding and ingestion rate of *Brachionus plicatilis* in mass culture. *Hydrobiologia* 122: 181–187.

Zankai, N.P. (1984) Predation of *Cyclops vicinus* (Copepoda: Cylopoida) on small zooplankton animals in Lake Balaton (Hungary). *Arch. Hydrobiol.* 99: 360–378.

Zimmermann, C. (1974) Die pelagischen Rotatorien des Sempachersees, mit spezieller Berücksichtigung der Brachioniden und der Ernährungsfrage. *Schweiz. Z. Hydrol.* 36: 205–300.

Evolutionary Ecology of Freshwater Animals
ed. by B. Streit, T. Städler and C. M. Lively
© 1997 Birkhäuser Verlag Basel/Switzerland

The evolution and genetics of maturation in *Daphnia*

D. Ebert

*NERC Centre for Population Biology, Imperial College at Silwood Park, Ascot, SL5 7PY, UK
and Institut für Zoologie, Universität Basel, Rheinsprung 9, CH-4051 Basel, Switzerland*

Summary. Maturation is a crucial stage in the life of every organism, and theoretical and empirical studies have provided good evidence that evolution maximizes fitness at both age and size at maturity. Although the ultimate (evolutionary) determinants of maturation are well established, the proximate processes which initiate maturation are described for very few species. In the planktonic crustacean *Daphnia magna* it has been suggested that a "threshold size" must be reached before maturation is initiated. This type of maturation mechanism has important implications for the evolution of age and size at maturity. The threshold size means that smaller juveniles have to grow through more instars than larger juveniles to reach maturity. Similarly, newborn of equal size, but raised under different growth conditions, reach the threshold at different times, depending on the growth rate. I suggest that the threshold mechanism serves to canalize size at maturity, and that this has consequences in form of higher variances for instar number and age at maturity. As the maturation phenotype is decoupled from variation in size at birth and variation in juvenile growth, variation in size at birth can evolve and be maintained independently from the evolution of size at maturity.

Introduction

Body size contributes to fitness through its effects on survival, reproduction, competitive ability, and sexual attractiveness (Bergmann, 1847; Darwin, 1859; Lack, 1947; Hutchinson and MacArthur, 1959; Stearns, 1976). Therefore, among the key traits of any organism are their size at birth and the size at which maturation is initiated. Life-history theory predicts optimal body sizes for these traits (Roff, 1992; Stearns, 1992). Important assumption in life-history theory are the functional relationships between life-history traits considered; i.e., the signs and contours of trade-offs between life-history components such as offspring size, growth rates, age and size at maturity. Most life-history models use a "plausible" trade-off structure, derived from the experience of the investigator or derived from empirical data. Very little work incorporates developmental processes to assess the correlation matrix between life-history traits. Here I review the evidence for a maturation threshold mechanism in *Daphnia*, which integrates development, physiology and life-history evolution. This threshold model allows us to predict life-history covariances, and provides new insights into the evolution of life-history in *Daphnia*.

Since Hrbácek and Hrbácková-Esslová (1960) first described a relationship between the fish stock of ponds and the size and growth rate of zooplankton, many studies have documented the strong impact of size-dependent selection on zooplankton (e.g., Lynch, 1977; Pace et al., 1984). Factors known to influence the size structure of cladoceran populations include size-selective predation (Confer and Blades, 1975; De Bernardi and Giusani, 1975; Lynch 1980; Murtaugh, 1981; Scott and Murdoch, 1983; Kerfoot and Sih, 1987), size-dependent metabolic efficiency (Brooks and Dodson 1965; Hall et al., 1976; Gliwicz, 1990), size-dependent response to food shortage (Neill, 1975; Threlkeld, 1976; Tessier et al., 1983; Goulden et al., 1987; Tessier and Consolatti, 1989), and size-dependent infection rates with microparasites (Vidtmann, 1993; Stirnadel and Ebert, 1997). Although it is widely accepted that size-dependent fecundity and mortality are the key determinants in the evolution of cladoceran life-history, it was only in the last years that the importance of body size at birth for the entire life-history of daphnids has been put forward (Ebert,1991, 1993; Glazier, 1992; Gliwicz and Guisande, 1992; Boersma, 1995).

In this paper, I describe how age and size at maturity depend on size at birth and juvenile growth. The basic mechanism that underlies this function is the maturation process, which is initiated after a size threshold is passed (Ebert, 1992). I will illustrate this model and the consequences of variation in length at birth and juvenile growth for the maturation phenotype using data from *Daphnia magna*. Combining the knowledge on the natural variation of length at birth and juvenile growth with the mechanics of the maturation process will allow me to explore variances and covariances of length, age and instar number at maturity. The model provides a tool for explaining previously unexplained variation in *Daphnia* life-history traits.

The threshold model

The described model provides a mechanistic framework of how different life-history traits relate to each other. The model is based on a small set of well know observations from *Daphnia* life-history and is purely deterministic. The purpose of the model is to understand the mechanics of *Daphnia* maturation and to explain some phenomena which have previously puzzled experimentalists. The main assumption of the model is that maturation is initiated after a daphnid has reached a certain size (the threshold size). I will show that other maturation criteria which have been proposed (e.g., an age threshold) are not supported by data.

Definitions

Maturity in *Daphnia* is defined as the instar in which the first eggs are deposited into the brood chamber (= "primiparous instar"). The instar before the primiparous instar is the "adolescent instar". Females that mature in the same number of instars represent an "instar-group." "Increment" is a measure of growth and is defined as the ratio of body-length after a molt to the body-length before the molt. The increment gives an estimate of length increase which is independent of time and is therefore not a growth rate, although it is strongly correlated with growth rate. Typically, the increment is in the range of 1.10 to 1.35, i.e., 10–35% increase in body length during one molt. Throughout this paper I use the mean juvenile increment, i.e., the arithmetic mean of all juvenile instars until the adolescent instar. This can be calculated from the data as

$$\text{mean juvenile increment} = \left(\frac{\text{length at adolescence}}{\text{length at birth}}\right)^{(1/(\text{instar number at maturity} -2))}$$

Since the juvenile increment strongly depends on environmental quality (e.g., larger increments are found under better conditions), juvenile increment can be used as an indicator for environmental quality.

The model

Increase in body-length in *Daphnia* is stepwise. Figure 1 A shows this for 12 hypothetical female *Daphnia* ranked by their body-length at birth. For simplicity, the increment was assumed to be constant in all instars and for all females. This is clearly an oversimplification, because the last increment before maturation is usually reduced due to reproduction, but Figure 1 serves to demonstrate the mechanism of maturation only in a qualitative way. Maturation is initiated after maturation size threshold is passed. The first eggs are laid two instars after the initiation of egg production, which occurs in the instar after the threshold is first passed. This assumption is based on the observation that egg production takes two instars in *Daphnia* (Makrushin, 1981; Zaffagnini, 1987). The model in Figure 1 (A) and 1 (B) make the following predictions:

1) Daphnids that are smaller at birth need more instars to reach the maturation threshold.
2) There is a positive correlation between length at birth and length at maturity among females maturing in the same number of instars. Thus, within an instargroup, the larger a female is at birth, the larger it will be at maturity (Fig. 1 (A), 1 (B)).
3) Two daphnids with nearly the same length at birth but different numbers of juvenile instars differ markedly in their length at maturity, as compared to two daphnids with similar length at birth and maturation in the same instar (Fig. 1 (A), 1 (B)).

Figure 1 (B) shows the expected relationship between length at birth and length at maturity assuming the stepwise model as shown in Figure 1 (A). Within instar-groups, size at birth is positively correlated with size at maturity.

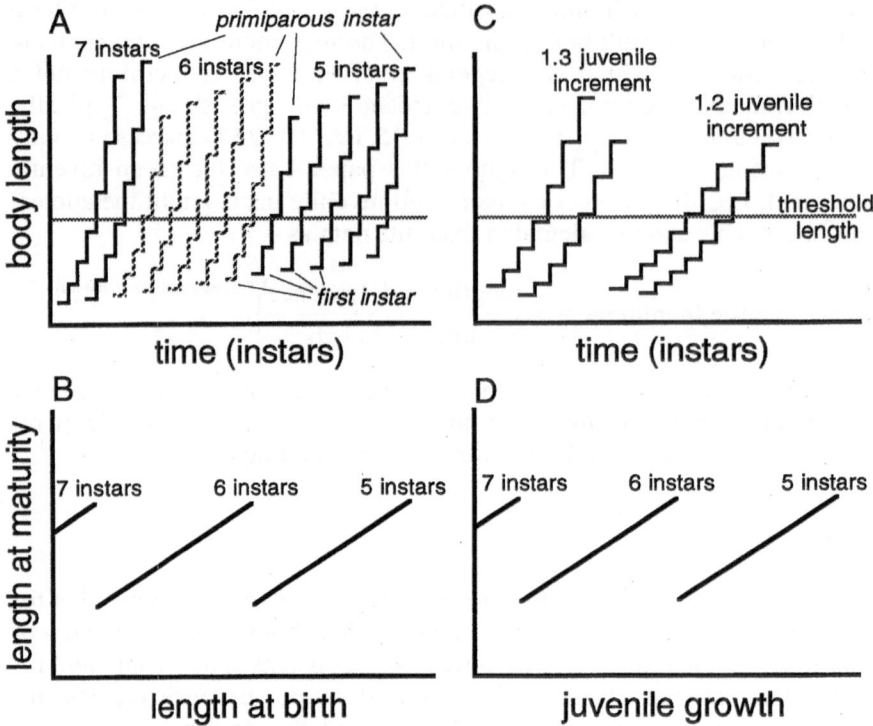

Figure 1. A simplified threshold model. (A) Calculated increase of body-length for 12 hypothetical *Daphnia* from the first instar until the primiparous instar (first egg laying instar). Each
step represents one instar. The next instar length is 1.3 times the length of the previous instar,
i.e., 30% increase per instar. After the threshold length is reached maturation is initiated. Since
egg production takes two instars, the first eggs are laid at the beginning of the third instar after
the threshold is passed. Daphnids smaller at birth need more instars to reach the threshold than
larger born animals. To avoid an accumulation of lines the beginning of individual growth
curves is shifted by two instar to the left. (B) The predicted relationship between length at birth
and length at maturity. It is assumed that no genetic or environmental variation for growth is
present. Note that within instar groups, a positive correlation between length at birth and at
maturity exists, but not across instar groups. (C) As A, but only four females are shown, two
under poor growth conditions (juvenile increment = 1.2) and two under good growth conditions
(juvenile increment = 1.3). Lower increments result in more instars until the threshold is
reached and on average a smaller size at maturity. (D) The predicted relationship between
juvenile growth (juvenile increment) and the length at maturity. It is assumed that no genetic or
environmental variation for length at birth exists.

Introducing different growth conditions

The threshold model in Figure 1 (A) can be adapted to variation in growth
conditions. The effect of reduced increment (e.g., due to poor feeding
conditions) is similar to the effect of being smaller at birth (Fig. 1 (C)). The
smaller the mean increment of a female, the more instars are needed to
reach the threshold. Thus we can add two more predictions:

4) For equal sizes at birth, the average number of instars at maturity will increase as the mean juvenile increment is reduced.

5) Within instar-groups, size at maturity will increase with more favorable growth conditions (Fig. 1(D)).

Age at maturity

The predictions for age at maturity are similar to those for length at maturity. For many crustaceans it has been shown that the time period between two successive molts is positively correlated with body-length (Hartnoll, 1982, 1985; Botsford, 1985), i.e., larger animals have longer intermolt intervals. For *Daphnia* these relationships were first described by Banta (1939) and Hrbáčková-Esslová (1962). These studies show that variation in the number of instars to reach maturity has a stronger impact on the age at maturity than variation in instar duration. These relationships lead to two more predictions:

6) The more instars, the greater the age at maturity. Strictly, this prediction is only valid of growth conditions and lengths at birth do not vary, but, as I show below, the prediction largely holds even when some variance for length at birth or juvenile increment is present.

7) Since larger newborn undergo molting later than smaller newborn (and the same pattern is prevalent in all later instars), one can predict a positive correlation between length at birth and age at maturity within instar groups. This prediction is much more sensitive to variation in juvenile increment than prediction (6).

Bringing it all together

Using very simple mathematics, one can now calculate the growth of hypothetical *Daphnia* females. By multiplying the body length at birth by the mean juvenile increment one obtains the length of the second instar. Multiplying this again with the increment, one gets the length of the third instar, and so on. After the hypothetical female becomes larger, and the assumed maturation threshold size is reached (or surpassed), egg production is initiated and two more instars follow before the eggs are deposited into the brood pouch.

Using parameters derived from empirical studies on *Daphnia* one can calculate length at maturity and the number of instars at maturity for any combination of length birth and juvenile increment. Using a functional relationship between body length and at inter-molt period one can further calculate age at maturity. The detailed method for this has been described elsewhere (Ebert, 1994). Figure 2 shows the result of such a simulation. Figure 2 (top) can be seen as a combination of the two effects seen in Figures 1(B) and 1(D). The empirical parameters were obtained from the detailed data given by Banta (1939) for *Daphnia longispina*. For these

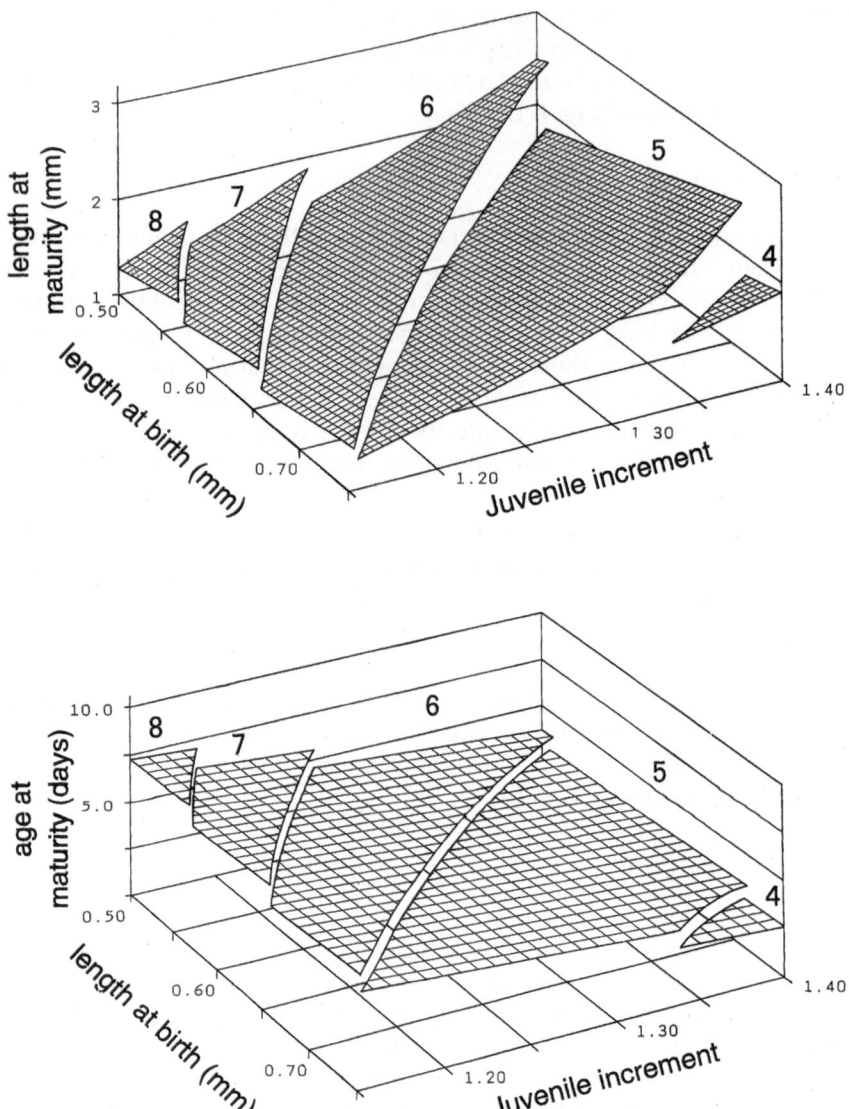

Figure 2. Simulation of the dependence of length at maturity (top) and age at maturity (bottom) on the length at birth and the juvenile increment. Each layer in the graphs represents all females with the same number of instars at maturity. The numbers in the graphs indicate the number of instars. The parameter ranges used in this simulation are for *D. longispina*, taken from Banta (1939).

illustrations I chose ranges of juvenile increment and length at birth that are approximately the extremes measured for this species. Further, one has to keep in mind that under normal conditions (laboratory and field) not all possible combinations of length at birth and juvenile increment are realized, but rather that environmental conditions and the length of newborn are correlated. For example, newborn are generally larger at birth (and less numerous) under poor environmental conditions (Ebert, 1993).

Introducing genetic variation

Figure 3 introduces genetic variance into the model. Figure 3(A) relates body-length at birth to body-length at maturity for three clones that differ in their rigid maturation threshold length (everything else being equal). For each clone, body-length at birth and at maturity are positively correlated within instar-groups, but the relation changes abruptly when the number of instars at maturity changes. A genetic difference in the maturation threshold size leads to, on average, reduced length at maturity, because the minimum and the maximum possible lengths at maturity decrease. Such genetic difference also shift the relation between body-length at birth and at maturity. Within clones there is a well defined length at birth above which all females belong to one instar group and below which all females belong to another instar group, which has one instar more. When the maturation threshold differs among clones, one can find females from different clones with the same length at birth but different numbers of instars at maturity. Females from different clones with equal length at birth, can, but do not necessarily mature in different instars.

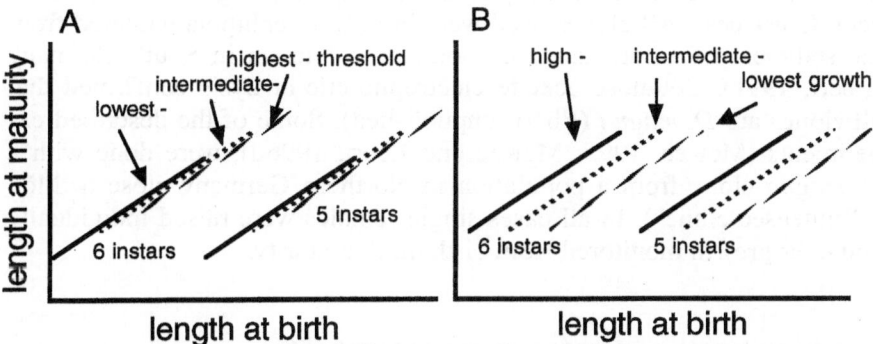

Figure 3. Schematized relation between length at birth and length at maturity for 3 clones. (A) The three clones vary in their maturation threshold length. Note that depending on the threshold length, newborn of equal length might grow through different numbers of instars before they mature and therefore differ markedly in their size at maturity. It is assumed that juvenile growth is equal among clones. (B) The three clones differ in their clone-specific growth (juvenile increment).

Figure 3(B) shows three clones with the same threshold size, but with different clone-specific growth rates. For most sizes at birth, higher clone-specific growth rates result in larger sizes at maturity. However, for some birth lengths the faster growing clones may need one instar less to mature, resulting in a smaller length at maturity compared to females which were of equal length at birth but have slower growth. As was the case for genetic variance in threshold size, genetic variance in growth also causes an overlap of instar groups with respect to the birth-size-axis, i.e., for some birth sizes females from different clones might differ in their number of instars at maturity.

Figure 3(A) shows a positive correlation between birth and maturation length within instar-groups regardless of threshold size variation. However, it is not possible to estimate the correlation when the number of instars at maturity varies among females.

Variation in the clone-specific juvenile increment however, could change this drastically (Fig. 3(B)). Within instar groups, the total phenotypic correlation between length at birth and at maturity becomes less strong when clonal variation for growth is high. However, the environmental correlation (after correcting for the clone effects), should still be positive within instar groups.

The maturation threshold model as outlined here is clearly an oversimplification. Nevertheless, its qualitative predictions are testable so that a failure of the predictions would allow rejection of the model (On the other hand, a consistency between the model and the data does not necessarily verify the model).

Before presenting experimental support for the threshold model I summarize some general aspects for the experiments. More details can be found in Ebert (1991, 1992, 1994; McKee and Ebert, 1996a). All experiments were done with laboratory clones of *Daphnia magna* Straus (Cladocera, Crustacea). All clones used were bred from ephippia gathered from the sediment of a carp stocking pond near Munich in South Germany (Ebert, 1991). Cellulose acetate electrophoretic analysis confirmed that all clones are *D. magna* (Ebert, unpublished). Some of the described experiments (McKee, 1995; McKee and Ebert, 1996a), were done with a *D. magna* clone from a population in Northern Germany close to Plön ("Binnensee clone"). In all cases single females were raised individually and their growth monitored from birth until maturity.

Age, body-length and instar number at maturity

From a large pool of newborn daphnids (< 12 h old), 15 females had been selected to represent a large range of lengths at birth. These 15 newborn *D. magna* were all kept under the same conditions and their lengths were measured in every instar until the first eggs were laid (Ebert, 1991). The

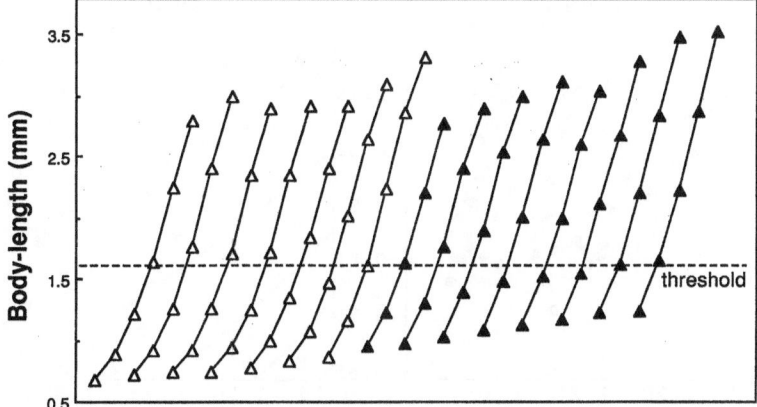

Figure 4. Instar wise increase of body length in successive instars of 15 female *D. magna* reared under constant food conditions. Each triangle represents the length of one instar, beginning with length at birth until the primiparous instar (see text). The females are sorted from left to right in increasing body-length at birth. To avoid an accumulation of lines, the beginnings of individual growth curves are shifted by one instar to the left. The white symbols indicate females which needed six instars to reach maturity, the black symbols show females with five instars at maturity. The stippled line indicates the assumed maturation threshold. After passing the threshold size it takes two more instars to reach the primiparous instar (data from Ebert, 1991). The length data in this graph differ slightly from those for length at maturity in Figure 1 a in Ebert (1991). Here measurements are used of total body-length excluding the spina base. This is necessary when plotting length increase across instars of juvenile daphnids, because very young daphnids do not have a spina base.

results are consistent with predictions 1–3: The eight females that were largest at birth matured in five instars, while the seven smallest females took six instars (Fig. 4). Within each instar group, the correlation between length at birth and length at maturity was positive ($r = 0.95$, $n = 8$, $p = 0.0002$ in the six-instar-group; $r = 0.89$, $n = 7$, $p = 0.007$ in the five-instar-group). Two females of nearly the same length at birth but different numbers of instars differed markedly in their length at maturity (the seventh and the eighth female in Fig. 4).

All lengths at maturity fall in a well defined size range (about 2.8–3.6 mm). Altogether smaller at birth, the females with six instars at maturity matured on average at similar lengths as the females that matured in five instars ($t = 1.23$, $n = 15$, $p = 0.23$). Thus, with respect to the length at maturity, growth through additional instars compensates for a smaller length at birth.

The model predicts that variation in juvenile growth should also influence the maturation process. Slower growing females are predicted to take more instars to mature. To test this effect I did an experiment similar to the one described previously (Ebert, 1994). I measured numerous females from a large pool of newborn *D. magna* (born from mothers of different age, but reared in the same environmental conditions), and sorted

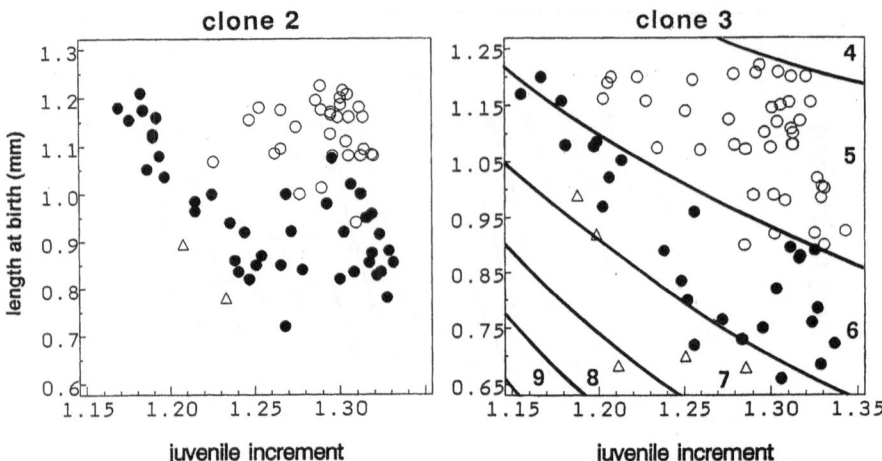

Figure 5. The number of instars at maturity in relation to the length at birth and mean juvenile increment. The white circles indicate females with five instars at maturity, black circles females with six instars at maturity and the white triangles females with seven instars at maturity. The numbers inside the right graph are the predicted number of instars at maturity for all parameters combinations which fall in the areas in between two curved lines. Note that the predicted and observed numbers of instars at maturity agree very well with each other. From Ebert (1994), with permission.

them individually according to their length. To vary juvenile growth I assigned these newborn to 13 food treatments (very low to very high food concentration), such that within each food treatment I had six females with a large range of length at birth. I did this with two clones, resulting in $6 \times 13 \times 2 = 156$ replicates. Replicates were checked twice daily and length was measured after every molt until maturity (Ebert, 1994). Figure 5 shows the result of this experiment with respect to the number of instars a female had grown through at maturity. The results clearly confirm predictions 1 and 4 of the model; larger length at birth and higher juvenile increments lead to maturity in fewer instars than smaller newborn growing with a smaller increment.

The graph for clone 3 (Fig. 5) has the predicted number of instars at maturity superimposed on the data (Ebert, 1994). I obtained these predictions as described in the model section (see also Ebert, 1994). It is important to note that to calculate the predicted number of instars, all that was needed was the threshold size, which has been estimated to be 1.58 mm for clone 3 (Ebert, 1994). Thus it was possible to calculate for each hypothetical length at birth and juvenile increment the predicted number of instars at maturity. No additional parameter estimates were used. The predicted and the observed numbers of instars at maturity correspond very well. The same is true for clone 2 (not shown here).

Figure 5 shows also that two other threshold types can be excluded as maturation criteria for *Daphnia*. If *Daphnia* would mature always in a cer-

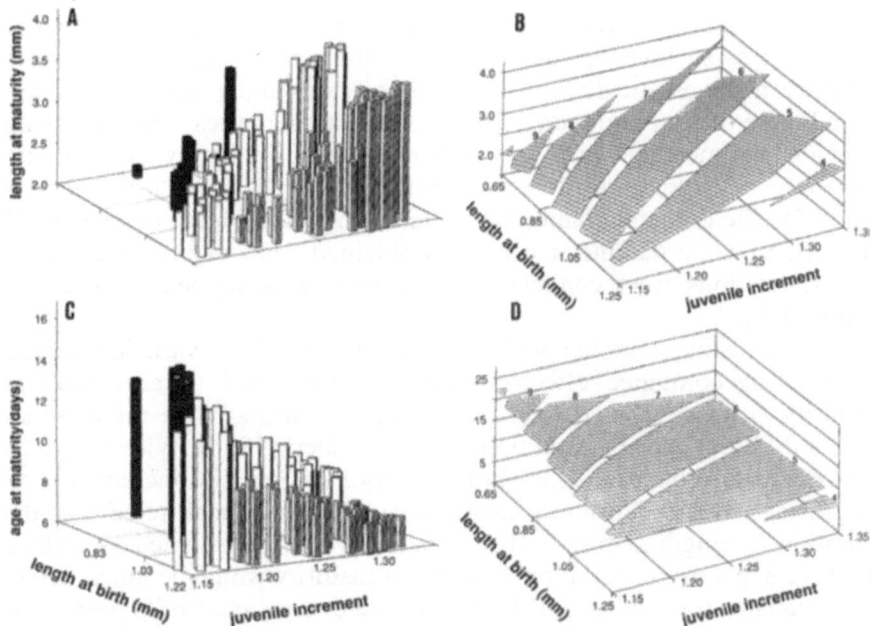

Figure 6. The dependence of length at maturity (top) and age at maturity (bottom) on the length at birth and the juvenile increment. The data from two clones are pooled in this figure. The left side shows the result of an experiment, while the graphs on the right shows a simulation based on the parameter estimates obtained from the experiment. The numbers of instars at maturity (instar groups) are indicated with different shades on the left graphs (black squared columns: seven instars, white round columns: six instars, hatched squared columns: five instars), but with numbers on the right graphs (further details of the graphs on the right are given in Fig. 2). Note the agreement between the experimental results and the simulated results. From Ebert (1994), with permission.

tain instar (the fifth instar was proposed by Banta (1939)), no systematic variation in the number of instars at maturity would be found. Figure 5 shows that this is clearly not the case. Instar number at maturity varies greatly within clones. If daphnids would mature after a fixed time period (an age threshold, Paloheimo et al., 1982), the distribution of instar groups in Figure 5 would look very different. Rapidly growing daphnids (large increment) would grow through more instars until maturity than slowly growing daphnids. Thus, for a given length at birth, the smaller the juvenile increment the lower the number of instars at maturity would be. This is clearly the opposite of what the data in Figure 5 suggest. Therefore I reject the age threshold hypothesis as a growth model for *Daphnia*.

Figure 6 shows the effect length at birth and juvenile increment have on age and length at maturity (Ebert, 1994). Within instar groups, length at maturity increases with increasing length at birth and with increasing juvenile increment, as predicted (predictions 1, 2, 4 and 5, Fig. 1). However, the lengths at maturity of two females with nearly the same length at birth or nearly the same juvenile increment, but different numbers of instars at maturity, differ drastically, causing the large steps in the landscape depicted in Figure 6(A) (prediction 3). Age at maturity increases with decreasing juvenile increment and with each additional instar, but hardly changes with variation in length at birth within instar groups (Fig. 6(C), predictions 6 and 7). The computer-simulated patterns shown in Figure 6(B) and (D) are based on the parameter estimates obtained from the empirical data. The simulations were conducted in the manner described for Figure 2 (Ebert, 1994).

One interesting point to note here is that, despite all the variation created through the combined effects of length-at-birth and instar-number at maturity, the variation of length at maturity appears to be rather small across different growth conditions (Fig. 7) (Ebert, 1994). Within a given environmental condition (i.e., little variation in juvenile increment), all lengths at maturity fall into a small range of values (small relative to the variation in length at birth). If maturation would be initiated in a fixed instar or a given age, variation in length at maturity would be much larger. This "canalization" of length at maturity is a consequence of higher variation in both the number of instars at maturity and the age at maturity. I return to this point later.

Genetic and environmental variation of the threshold

The threshold itself might vary as a consequence of environmental and/or genetic factors. These factors can be separated experimentally (Ebert, 1991, 1994). Genetic variation has been detected for almost every trait which has been studied in *Daphnia* and it is therefore not surprising to find genetic variation for the threshold length as well (Ebert, 1991, 1994).

The experiment with these two clones revealed indeed a significant difference in the threshold estimate (Ebert, 1994). Using a PROBIT analysis the threshold of the two clones were estimated to be 1.58 mm (95% confidence interval: 1.54–1.62) for clone 3 and 1.71 mm (95% c.i.: 1.67–1.74) for clone 2.

This two-clone experiment was done over a range of feeding conditions, with the aim of detecting a possible increase or decrease of the threshold size with food quantity (Ebert, 1992, 1994). No such influence could be detected; it seems as if the threshold was independent of the food level in both clones.

In a similar experiment, Enserink et al. (1995) largely confirmed the food independence of the threshold size, but noted that at very low food levels, reproduction was in some cases postponed beyond the threshold. This does not necessarily invalidate the threshold model, since under these extreme food conditions it might often be impossible for *D. magna* to

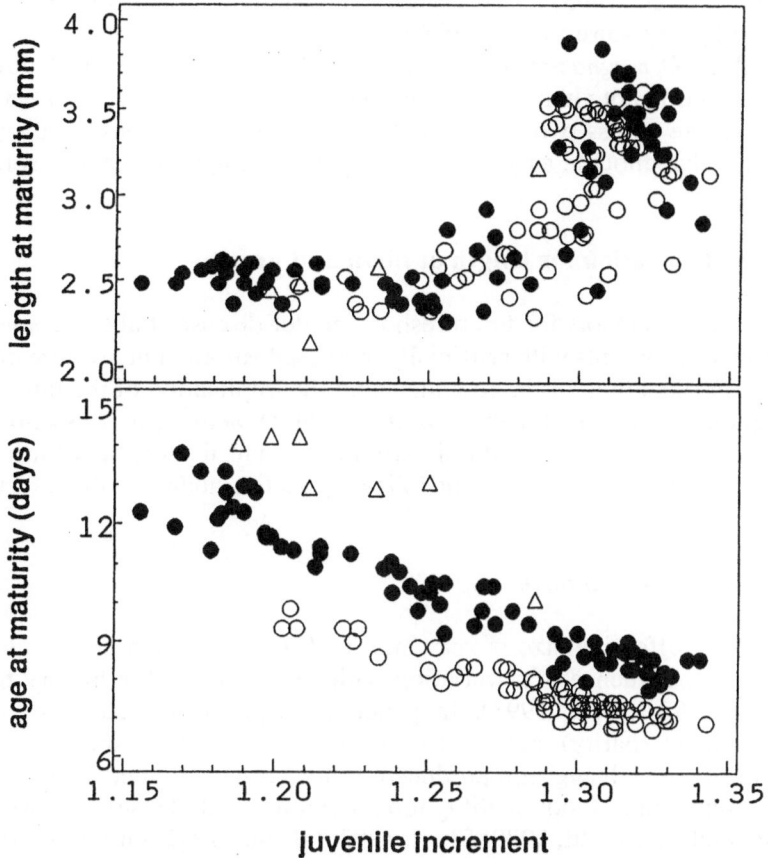

Figure 7. Length at maturity (top) and age at maturity (bottom) in relation to the mean juvenile increment. The white circles indicate females with five instars at maturity, black circles females with six instars at maturity and the white triangles are females with seven instars at maturity. From Ebert (1994), with permission.

produce even a single egg. In their experiment, the few clutches which were produced under very low food condition contained only one egg each. Under such conditions the presence of eggs in the brood pouch is not a reliable indicator for maturation. A female might mature physiologically, but still be unable to reproduce due to lack of resources. In such cases it might be necessary to dissect the females and look at the state of the ovaries to confirm maturation.

McKee (1995) and McKee and Ebert (1996a) tested the sensitivity of the threshold size across different temperatures and fool levels in a two-way design. There was again no indication of a food effect, but the threshold estimate decreased across the three temperatures from about 1.75 mm at 12°C to about 1.56 mm at 22°C.

Enserink et al. (1995) tested whether the maturation threshold size is influenced by exposure to lead (PbCl$_2$ in the water). They found that lead stress affects *D. magna* in a manner similar to food stress, and that the threshold size itself is not altered. However, they found a difference in the mean threshold size across two experiments (but no statistical tests were used), indicating that another, as yet unidentified factor might also play a role.

Sources of variation for length at birth and growth

The empirical support for the threshold model discussed above comes in part from experiments with artificially increased variation in juvenile increment and/or length at birth. To understand the implication of the threshold maturation mechanism for the life-history of *D. magna*, it is essential to understand the amount of natural variation present for length at birth and juvenile growth in this species. The following section addresses these factors.

Variation in length at birth

Variation in offspring size is very high in *Daphnia*, which appears to be largely a consequence of covariation with various other life-history traits (Glazier, 1992; Ebert, 1993). In general, offspring size increases with clutch number (parity), maternal size, maternal age, and duration of the instar during which eggs are produced. It decreases with increasing clutch size, temperature, food quality, food quantity and density (crowding) (Tessier and Consolatti, 1989; Ebert, 1991; Glazier, 1992; Guisande, 1993; Boersma, 1995; McKee and Ebert, 1996b). Comparatively little work has been done on the genetics of offspring size (Ebert et al., 1993a, b), but both broad-sense and narrow-sense heritabilities have been shown to be approximately around 30% across the first six clutches of *D. magna* (Ebert et al., 1993a). An interesting question that arises at this point is whether variation in offspring size within environments is adaptive or constrained (Glazier, 1992; Ebert, 1993; Boersma, 1995).

How a large range of offspring sizes can be obtained from a single clone of *Daphnia* has already been indicated in Figure 5. A better picture of individual variation in newborn size can be obtained by considering the experiment done with 88 clones from one population (carp pond population, Ebert, 1991). All females were raised under the same conditions and two offspring of each of their first four clutches were measured. The individual variation in offspring length was large when the first four clutches were considered together. The largest minus the smallest offspring born by one mother (kept under constant conditions) within the first four clutches provides an estimate of the total offspring body-length variation for individual mothers. The mean of these differences was 0.26 mm (range 0.02 to

0.52 mm, $n = 273$), which means that the largest offspring was on average about double as heavy as the smallest offspring. This is an unusual large range for newborn size for an animal (but not compared to plant seeds, e.g., Giles (1990)). Broad-sense heritability of variation in offspring length was moderate. For example, heritability of the range across the first four clutches was 11.4%, heritability of the variance was 14%. Although small, these estimates indicate the possibility of an evolutionary change in offspring size variation if selection acted on variance.

The most prominent source of covariance in offspring size appears to be clutch number. It might be maternal age or maternal length which underlies this co-variation, but for simplicity I use clutch number here, because it can be estimated without error. To test if newborn size and its relation to clutch number has a genetic component, I analyzed the newborn length – clutch number relationship in more detail for the data set containing the 88 clones of one *D. magna* population (Ebert, 1991). I calculated linear regressions of mean newborn length on clutch number (I raised length to the fourth power for linearization of the relation; the third power would be more appropriate because it scales better with body weight, but statistically this is irrelevant). These regressions were done separately for each female. Four clutches were used per individual female. The slopes were significantly ($p < 0.05$) positive for 269 of the 273 individual females tested (average $r^2 = 0.80$), indicating that variation in length at birth between clutches is not random but increases systematically with clutch number. The slope estimates obtained from each female can be used in a standard quantitative genetic analysis. Broad-sense heritability of the slopes was 27% ($p < 0.01$), indicating genetic variance for between-clutch newborn length variation. Broad-sense heritability for the intercept was 13% ($p < 0.05$); a higher intercept for a given slope indicates that the offspring length of all clutches is larger.

Variation in size at birth is not a unique feature of *D. magna*. Many *Daphnia* species show pronounced variation in body-length at birth. The ranges listed in Table 1 reflect differences in biomass of greater than factor 2 between the smallest and the largest length.

In summary, size at birth varies strongly in *Daphnia*. This variation is due partially to maternal effects and environmental factors, but genetic factors also play a role. Remarkably, variation is found among the offspring of individual females and this variance differs among clones. This pattern is also seen in natural populations of *D. magna* (Ebert, unpublished data).

Variation in juvenile growth

Juvenile increment is the other major factor that determines the maturation phenotype in *Daphnia*. But how much variation is there for juvenile growth within and between environments? Figure 5, which was produced by vary-

Table 1. Reported body length ranges of newborn from the genus *Daphnia*

Species	Sample-size	Body-length min/max	Ratio** max/min	Reference
D. ambigua	1225	0.26–0.43	1.66	Lei and Armitage (1980)
D. hyalina	84*	0.52–0.84	1.61	Hrbáčková (1971)
D. pulex	155	0.56–0.80	1.43	Hrbáčková-Esslová (1962)
D. pulicaria	217	0.52–0.96	1.85	Hrbáčková-Esslová (1962)
D. curvirostris	109	0.50–0.61	1.22	Hrbáčková-Esslová (1962)
D. obtusa	93	0.48–0.62	1.29	Hrbáčková-Esslová (1962)
D. magna	996	0.61–1.01	1.66	Green (1954)
D. magna	> 2000	0.74–1.36	1.84	Ebert (1991) (15°C)
D. magna	162	0.70–1.18	1.68	Ebert (1994) (20°C)

*   values obtained from graphic presentation.
** ratio: maximum length/minimum length.

ing food quantity over four orders of magnitude, shows a range for juvenile increment from about 1.15 to nearly 1.35. Comparison with other studies, particular those which list individual instar lengths (Banta, 1939; Anderson and Jenkins, 1942) shows that the range might be as large as 1.05 to 1.40 in extreme cases.

For most laboratory experiments it is desirable to reduce environmental variation as much as possible, which generally leads to a strong reduction in variation of growth. To estimate how much variation in growth is still present when environmental conditions were controlled, I calculated juvenile growth measured as mean juvenile increment for each of the 273 females from the 88-clone experiment. Variation in clonal means of juvenile increment (calculated until the adolescent instar) was low: mean = 1.296, SD = 0.011, CV = 0.68%. The broad-sense heritability of the mean increment until the adolescent instar was 24% ($p < 0.05$). There was virtually no covariance between length at birth and juvenile growth. The genetic correlation between length at birth and the juvenile increment was -0.01 ($p > 0.5$) (total correlation $r = -0.05$, non-genetic correlation $r = -0.10$).

In summary, juvenile increment can vary strongly across different environmental conditions, but varies little under controlled (laboratory!) conditions. Thus, the main source of variation of the maturation phenotype within environments is caused by variability in size at birth, while across environmental conditions variation is caused by both juvenile increment and size at birth. Variation in size at birth and juvenile increment seem to be independent from each other.

The distribution transformer

Combining what has been said about the origin of variation and the mechanisms of maturation in *Daphnia*, it is possible to make predictions of variances and covariances of life-history traits and draw conclusions about their evolution. To illustrate this, I return again to the data set of the 88 clones, since these data were collected prior to the development of the maturation-threshold model. However, the results of this experiment provoked my consideration of a mechanistic model of maturation in *Daphnia*.

The 88 clones of this experiment were a random sample of clones hatched from ephippia in the laboratory. In 21 clones, some females matured in five and others in six instars, indicating non-genetic variance for instar number at maturity. Within each instar group, there was a positive correlation of age and length at maturity with the length at birth (Ebert, 1991). A comparison of the 21 clones which had females in both instar groups revealed that in only one clone was a female with six instars at maturity larger at birth then a female with five instars at maturity. But the difference was only 0.02 mm, which is about the limit of measurement precision. Thus, large sizes at birth were correlated with less instars at maturity. I conclude that there is little evidence that the threshold mechanism is not rigid, but that clones differed in either growth or the threshold size.

The data structure of the 88-clone experiment facilitates analysis of total, genetic, and non-genetic correlations for various subsamples. In Table 2, correlations are given for the total data set, for females born in the first or the second clutch of their mother (second-clutch females are larger at birth!), and for females with five or six instars at maturity. Within instar

Table 2. Total, genetic and non-genetic correlations between length at birth, age at maturity and length at maturity

Correlation/ prediction	All females	Clutch 1 females	Clutch 2 females	Instar-group	
				5	6
birth length with length at maturity					
total	0.055	− 0.485 ***	0.328 ***	0.651 ***	0.549 **
genetic	0.079	− 0.511 **	0.318 **	0.643 ***	0.546 **
non-genetic	0.032	− 00.371	0.344 ***	0.663 ***	0.603 *
birth length with age at maturity					
total	− 0.191 **	− 0.466 ***	0.032	0.394 ***	0.732 ***
genetic	− 0.129	− 0.463 *	0.082	0.499 ***	0.797 ***
non-genetic	− 0.247 ***	− 0.481 *	− 0.028	0.345 ***	− 0.109
length at maturity with age at maturity					
total	0.618 ***	0.850 ***	0.525 ***	0.254 ***	0.448 **
genetic	0.589 ***	0.834 ***	0.517 ***	0.136	0.468 *
non-genetic	0.651 ***	0.929 ***	0.541 ***	0.376 ***	0.032
sample size					
total	271	50	223	240	33
genetic	88	27	88	87	22
non-genetic	186	24	136	154	12

groups, most correlations are significantly positive. The correlation coefficients obtained from the analysis of the total data set and those obtained from the analysis of the clutch 1 and the clutch 2 females, however, are positive, negative or close to zero. This lack of a consistency of the sign of covariance is because in these three data sets females from different instar groups (five and six instars at maturity) occur in different proportions.

The positive correlations of age and length at maturity with body length at birth suggest positive correlations between age and length at maturity. Since age and length at maturity are affected in a similar way by variation in the instar number at maturity, this holds across instar groups (Table 2) (Note that this correlation is positive within environments, but negative across environments. Negative norm of reaction!).

The origin of the differences in the correlations of clutch 1 and clutch 2 females is illustrated in Figure 8. The graphs show the distribution of the length at birth and of length and age at maturity. Black bars indicate those females which matured in six instars. They were smaller at birth. The distributions of adult phenotypes of first-clutch females clearly show a bimodal distribution, although, the length-at-birth distribution was nearly normal (actually even leptokurtic). For females from second clutches, length at

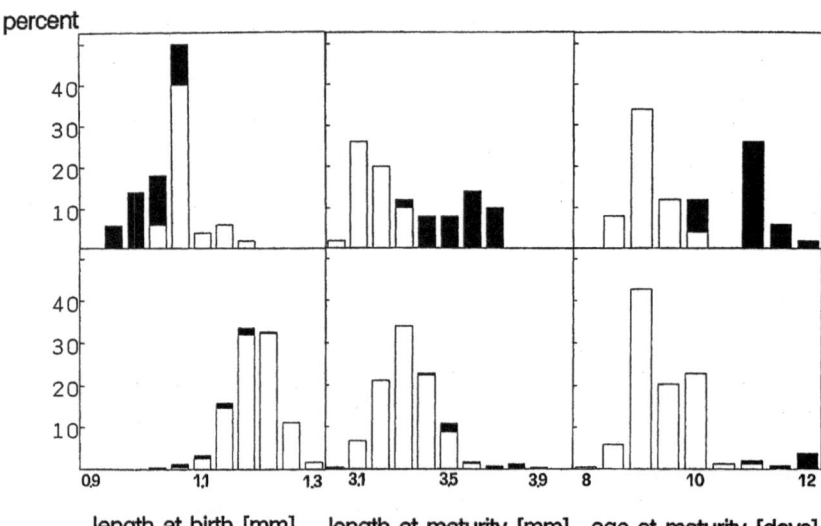

Figure 8. Frequency distributions for body-length at birth, body-length at maturity and for age at maturity for females derived from first clutches of their mother (upper graphs) and females derived from second-clutches (lower graphs). Note that the second clutch females are larger at birth, but not larger at maturity. Black bars: instar number at maturity was 6 ($n = 50$); white bars: instar number at maturity was 5 ($n = 223$). Data from the carp pond population from Ebert (1991).

birth was slightly skewed to the left, while age and length at maturity were skewed to the right. For both first- and second-clutch females, the smaller females at birth end up as the larger and older at maturity. This reversal shifts the correlation coefficients toward negative values. The change in the distribution of lengths at birth into lengths at maturity has been called a "distribution transformer" by Jones et al. (1991).

Also visible in Figure 8 is the canalizing effect of the threshold. Although females from the first and second clutch of their mothers differed in their length at birth, their lengths at maturity were approximately the same (Fig. 8).

Discussion

Maturation marks a crucial step in the life-history of any organism. Life history theory has provided ample examples for the evolution of optimal maturation phenotypes, while experiments have offered convincing support for the predictions made. Surprisingly, we know very little about the mechanisms which mediate the shift of resources from growth to reproduction. The maturation threshold model proposed for *Daphnia* is a small step in this direction, since it considers aspects of development, physiology, and ecology, and thus might contribute to a more integrated concept for *Daphnia* life-history evolution.

At this point I want to mention that what I labeled a "size threshold" might in reality be a different type of threshold. However, this other type of threshold is clearly correlated with body size. Body size could therefore be a "morphological marker" for the real threshold.

Support for the threshold model

The discussed experiments qualitatively support the proposed threshold model. The main points are: Larger size at birth and higher juvenile increments leads to maturation in fewer instars. Within instar groups, both age and length at maturity are positively correlated with size at birth. Females with nearly equals lengths at birth but different instar numbers at maturity differ in their length and age at maturity. Length at maturity is buffered against large variation in length at birth.

Many of these observations described here for *D. magna* are also known for other species of *Daphnia*. For example, variation in the instar number at maturity in *Daphnia* is a phenomenon commonly observed (Anderson, 1932; Anderson and Jenkins, 1942; Green, 1954, 1956; Porter et al., 1983; Urabe, 1988; Ketola and Vuorinen, 1989; see Bottrell, 1975 for a review). Porter et al. (1983) and Urabe (1988) have shown that variation in instar number at maturity can be found under various food conditions, and that the average number of instars at maturity increases when food decreases. Green (1956) demonstrated for seven (!) *Daphnia* species that females that were small at birth need more instars to reach maturity than females born larger. Lei and Armitage (1980) observed that *D. ambigua* with nearly the same length at birth but different numbers of juvenile instars can differ markedly in their length at maturity.

No correlations between size at birth and adult traits for *Daphnia* have been published. The here reported positive correlation between age and size at maturity within environments was also found by Lynch (1984). Note that across food levels, length and age at maturity are negatively correlated, i.e., under good feeding conditions females are large and young at maturity, while they are older and smaller at poor feeding conditions (Weglenska, 1971; Tillmann and Lampert, 1984; Taylor, 1985; Lynch, 1989; Enserink et al., 1995). This negatively-sloped reaction norm (sensu Stearns and Koella, 1986) for age and size at maturity is predicted by the threshold model as well.

The distribution transformer

Large variation in body-length at birth can generate surprising relations between length at birth and age or length at maturity (Fig. 8). A given distribution of length at birth can be transformed into different distributions at maturity. The dependence of the covariance between length at birth and the adult phenotype on the distribution of lengths at birth can have interesting effects during size-selective predation.

During recent years it has been argued that the genetic covariance matrix of life-history traits is of fundamental importance in shaping the evolutionary potential of populations (Lande, 1982; Maynard Smith et al., 1985; Spitze et al., 1991). The distribution transformer has implications for the correlated response to size-dependent selection on birth or maturation length. Depending on the sign of the covariance between length at birth and at maturation, selection against small newborn (e.g., due to starvation (Gliwicz, 1990) or invertebrate predation (Elser et al., 1987)) could indirectly select for larger or smaller lengths at maturity. For example, consider the case shown in Figure 8. Selection against the smallest newborn would alter the distribution of lengths at maturity and decrease the mean length at maturity. High mortality in newborn less than 1 mm long would reduce the frequency of adults which take one instar more to mature and are therefore large at maturity (the black bars in Fig. 8). Clearly this type of selection on length at maturity is limited because length at maturity is somewhat buffered against large variation in length at birth. Whatever the range and variance of length at birth, the body-length at maturity will always stay within a defined range (Fig. 8), because the fixed threshold size regulates the initiation of the maturation process (Ebert, 1992).

The microevolutionary consequences of the distribution transformer on the age and size structure of a *Daphnia* population are difficult to understand, given that genetic, environmental, and individual variation in length at birth, growth, age and threshold length must be considered. The factors that influence genetic covariances between length at birth and age or length at maturity are manifold, e.g., the growth conditions or the clutch number females are born in. The distribution of length at birth depends strongly on the age distribution of mothers, and thus depends on the mortality regime of adult females. Given this complexity, it is difficult to formulate general predictions regarding the sign and strength of correlation coefficients between birth length and maturation length in natural environments.

The model (Figs 1, 2) predicts that the lower the threshold size, the lower the minimum and maximum length at maturity. Thus, given that genetic variance for a threshold size exists, continuous selection pressure favoring a smaller length at maturity would select clones with a lower threshold size and vice versa. The distribution transformer slows down this process because the actual size of a daphnid is the target of selection, not the threshold size. The correlation of length at maturity and the threshold value is weak, since the realized length at maturity falls somewhere within a defined range of lengths.

It is not clear what role the distribution transformer plays in a population, in which the threshold size is adapted to local conditions. A possible adaptive explanation could be that the transformer buffers the effects of rapidly fluctuating size-dependent mortality regimes (e.g., Lynch, 1983; Elser et al., 1987). The production of a range of phenotypes per genotype, and the instability of the correlation matrix between length at birth and adult characters could be advantageous in temporally unstable selection regimes (Gillespie, 1973; Seger and Brockmann, 1987; Philippi and Seger, 1989). For a daphnid, non-genetic variability in body-length of offspring might increase the chances that at least some proportion of the offspring will survive under unpredictable conditions. Such polyphenism is known as "adaptive-coin-flipping" (Cooper and Kaplan, 1982; Kaplan and Cooper, 1984) or "stochastic polyphenism" (Walker, 1986). Each single female daphnid produces a wide range of offspring lengths and given the distribution transformer, an unpredictable correlation with the adult phenotype.

While this discussion deals mainly with the influence of size-dependent selection on *Daphnia* life-history, one might emphasize age-dependent selection rather than size-dependent selection as the most important factor in *Daphnia* life-history evolution. However, the distribution transformer does not only change the size distribution, but in a similar way the distribution of age at maturity (Fig. 8). Therefore, the effects of indirect selection seem to be of similar nature for age- and size-dependent selection. Combining both types of selection increases the complexity of the analysis beyond the scope of this paper.

Is the size threshold the result of selection for canalization?

Canalization, here defined as "an abstract term that describes unknown developmental mechanisms that reduce phenotypic variation" (Stearns and Kawecki, 1994) is assumed to be the result of stabilizing selection. Canalization would be advantageous if it constrains a trait closer to its optimum (Rendel, 1967). In other words, canalization is a process which reduces the fitness loss resulting from deviations from the optimum. Stearns and Kawecki (1994) used this concept to predict that traits more important to fitness should be more strongly canalized than traits with less impact on fitness. They also presented evidence that differential canalization of life-history traits exists in *Drosophila melanogaster*, and that the degree of canalization is correlated with the sensitivity of the trait to small deviations from its mean.

With respect to the maturation-size-threshold of *Daphnia*, three types of maturation thresholds have been considered: maturation occurs at a given age (Paloheimo et al., 1982), in a given instar (Banta, 1939) or at a particular size (McCauley et al., 1990; Ebert, 1991). As body size can be considered to have the highest impact on fitness in *Daphnia*, canalization would be expected to reduce variation in body size.

Although each trait should be canalized to deviate as little as possible from its optimum, this might not be possible because traits are not independent from each other. Therefore, canalization of one trait might prevent canalization of another trait. Evolution should canalize those traits in which variance is costly at the expense of increased variance in traits in which variance is less costly.

A convenient measure to compare variation of different traits is the coefficient of variation, (cv). McKee (1995) and McKee and Ebert (1996a) tested the hypothesis that length at maturity is canalized in *Daphnia*, as opposed to age or instar number at maturity. It was predicted that the cv of length at maturity should, in any environment, be smaller or equal to the cv's of age and instar number at maturity. This prediction was tested and confirmed for three temperature conditions (McKee and Ebert, 1996a). It was concluded that the evolution of a size threshold, as opposed to an age or instar threshold, is adaptive, since it presumably reduces deviations from the optimal fitness to the least extent (McKee and Ebert, 1996a). Further, canalization of size at maturity in *Daphnia* increases the variance for age and instar number a maturity.

Canalization in other arthropods

One might argue that the size threshold is the result of the arthropod-specific stepwise growth pattern. However, a comparison of different arthropod groups shows that the threshold criterion varies strongly across groups. Variation in the number of juvenile instars is found in many groups, including spiders (Deevey, 1949), decapods (Hartnoll, 1985), lepidopterans (Clare and Singh, 1991) and orthopterans (Uvarov, 1966). By contrast, the number of pre-adult instars is fixed in other groups, e.g., dipterans, hymenopterans, and coleopterans. A fixed (canalized) number of pre-adult instars would result in a smaller size at maturity if growth is poor, as is typical for fruit flies and bumble bees (Plowright and Jay, 1968; Gebhardt and Stearns, 1988). The most prominent example for a canalized age at maturity are the periodic cicadas, in which some populations of each of three known species are canalized to mature in 13 years, while other populations are canalized to mature in 17 years (Yoshimura, 1996). In other cases, combinations of traits might be important and two traits might be canalized at the expense of a third. For example, the tobacco horn worm, *Manduca sexta*, always pupates in its fifth instar. However, if growth is poor, maturation is delayed considerably until the fifth instar larva weights about five grams (Nijhout and Williams, 1974). In this case, it seems as if instar number and weight at maturity are canalized, while costs are paid in terms of high variability in time to maturity. In summary, canalization of life-history traits among arthropods is highly variable. It is therefore reasonable to conclude that canalization in arthropods reflects an adaptation rather than a constraint. Understanding the evolutionary pathways that determine which traits will be canalized is one of the most exciting aspects in current life-history research.

How important is the maturation size threshold for the experimentalist?

The results reviewed above are all based on laboratory experiments with *D. magna*. Among the better-studied *Daphnia* species, *D. magna* is the most phenotypically plastic and therefore is very suitable for an experimental approach to study the mechanics of this plasticity. For example, maturation in *D. magna* can occur as early as the fourth instar or as late as the ninth instar. Such large variation must have consequences for other life-history traits. Smaller *Daphnia* species mature usually within five or six instars (for a review see Bottrell, 1975). In some studies, deviations from five instars at maturity were excluded from the data set because they were considered to be odd (e.g., Banta, 1939).

Researchers working with smaller *Daphnia* species, might not always be aware of the additional variance caused by variable numbers of instars at maturity. I believe that for many questions this does not influence the results of experiments, but knowledge of the causes of variances and covariances of life-history traits would certainly help to understand unexplained variance.

The problem of large unexplained variation in life-history traits (not only in *D. magna*) has long been recognized and a variety of approaches have been used to deal with this problem. For example, it has been common practice to minimize noise in the data by choosing offspring of certain clutches only (usually not the very variable first clutch) such that the variance of length at birth was minimized (e.g., Bell, 1983; Lynch and Ennis, 1983; Goulden et al., 1987; Glazier, 1992). Others presented their data separately for each instar group (Anderson and Jenkins, 1942; Ketola and Vuorinen, 1989). I have mentioned already that *Daphnia* with deviant numbers of instars have been excluded from some data sets to manage the unexplained variance (Banta, 1939). W. Lampert (personal communication) refers to unusually small and early-maturing female daphnids (those with one instar less at maturity) as "Lolita's", supposedly a reference to Vladimir Nabokov's master novel. Lolita's are characterized as being statistical outliers that cause problems when parametric statistics are used (W. Lampert, personal communications). The problem that needs to be addressed is whether the experimental reduction of variance, e.g., using only newborn from the same clutch, changes the generality of our results, or whether it allows us to gain a deeper insight into *Daphnia* life-history mechanics without seriously compromising our conclusions.

Conclusions

The study of canalization of phenotypic traits might prove to be a fruitful approach in the study of cladoceran evolution. Every trait studied so far is embedded in a covariance matrix with other traits. This matrix is not only important for life-history traits (Spitze et al., 1991), but also for behavioral traits (De Meester, 1993), predator defense (Tollrian, 1995) and hybridization (Spaak, 1995). Understanding the trade-offs that favor canalization of one trait over other traits would be a major step in the study of cladoceran evolution.

   "It is an old maxim of mine," said Sherlock Holmes, "that when you have excluded the impossible, whatever remains, however improbable, must be the truth." "Perhaps", said Watson, skeptically, "you may have convinced me as to the motive, but you are yet to explain how it is done."

Acknowledgments
I thank M. Boersma, T. Little, C. Lively, Th. Städler and J. Wearing-Wilde for reading and improving the manuscript. Any mistakes left in the paper are certainly my own fault. The work was supported by Swiss National Fond Grants Nr. 3100-43093.95 and Nr. 3.643.0.87.

References

Anderson, B.G. (1932) The number of pre-adult instars, growth, relative growth, and variation in *Daphnia magna. Biol. Bull.* 63:81–98.
Anderson, B.G. and Jenkins, J.C. (1942) A time study of events in the life span of *Daphnia magna. Biol. Bull.* 83:260–272.

Banta, A.M. (1939) Studies on the physiology, genetics, and evolution of some Cladocera. *Carnegie Institution of Washington* Paper No. 39.

Bell, G. (1983) Measuring the cost of reproduction III. The correlation structure of the early life history of *Daphnia pulex*. *Oecologia* 60:378–383.

Bergmann, C. (1847) Über die Verhältnisse der Wärmeökonomie der Thiere zu ihrer Grösse. *Göttinger Studien* 1:595–708.

Boersma, M. (1995) The allocation of resources to reproduction in *Daphnia galeata:* Against the odds? *Ecology* 76:1251–1261.

Botsford, L.W. (1985) Models of growth. *In:* A.M. Wenner (ed.): *Factors in Adult Growth*. Balkema, Rotterdam, pp 101–128.

Bottrell, H.H. (1975) Generation time, length of life, instar duration and frequency of moulting, and their relationship to temperature in eight species of Cladocera from the river Thames, Reading. *Oecologia* 19:129–140.

Brooks, J.L. and Dodson, S.I. (1965) Predation, body size, and composition of plankton. *Science* 150:28–35.

Clare, G.K. and Sing, P. (1991) Variation in the number of larval instars of the brownheaded leafroller, *Ctenopseustis obliquana* (Lepidoptera: Tortricidae) at constant laboratory temperatures. *N.Z.J. Zool.* 17:141–146.

Confer, J.L. and Blades, P.I. (1975) Omnivorus zooplankton and planktivorus fish. *Limnol. Oceanogr.* 20:571–579.

Cooper, W.S. and Kaplan, R.H. (1982) Adaptive 'coin-flipping': A decision-theoretic examination of natural selection for random individual variation. *J. Theor. Biol.* 94:135–151.

Darwin, C. (1859) *The Origin of Species by Means of Natural Selection*. John Murray, London.

De Bernardi, R. and Giusani, G. (1975) Population dynamics of three cladocerans of Lago Maggiore related to predation pressure by a planktophagous fish. *Verh. Internat. Verein. Limnol.* 19:2906–2912.

De Meester, L. (1993) Genotype, fish-mediated chemicals, and photoactic behavior in *Daphnia magna. Ecology* 74:1467–1474.

Deevey, G.B. (1949) The developmental history of *Latrodectus mactans* (Fabr.) at different rates of feeding. *Amer. Midland Naturalist* 42:189–219.

Ebert, D. (1991) The effect of size at birth, maturation threshold and genetic differences on the life history of *Daphnia magna. Oecologia* 86:243–250.

Ebert, D. (1992) A food independent maturation threshold and size at maturity in *Daphnia magna. Limnol. Oceanogr.* 37:878–881.

Ebert, D. (1993) The trade-off between offspring size and number in *Daphnia magna:* The influence of genetic, environmental and maternal effects. *Arch. Hydrobiol.* Suppl. 90:453–473.

Ebert, D. (1994) A maturation size threshold and phenotypic plasticity of age and size at maturity in *Daphnia magna. Oikos* 69:309–317.

Ebert, D., Yampolsky, L.Y. and Stearns, S.C. (1993a) Genetics of life-history traits in *Daphnia magna:* 1. Heritabilities in two food levels. *Heredity* 70:335–343.

Ebert, D., Yampolsky, L.Y. and van Noordwijk, A.J. (1993b) Genetics of life-history traits in *Daphnia magna:* 2. Phenotypic plasticity. *Heredity* 70:344–352.

Elser, M.M., von Ende, C.N., Sorrano, P. and Carpenter, S.R. (1987) *Chaoborus* populations: Response to food web manipulation and potential effects on zooplankton communities. *Can. J. Zool.* 65:2846–2852.

Enserink, E.L., Kerkhofs, M.J.J., Baltus, C.A.M. and Koeman, J.H. (1995) Influence of food quantity and lead exposure on maturation in *Daphnia magna*: evidence for a trade-off mechanism. *Func. Ecol.* 9:175–185.

Gebhardt, M.D. and Stearns, S.C. (1988) Reaction norms for developmental time and weight at eclosion in *Drosophila mercatorum. J. evol. Biol.* 1:335–354.

Giles, B.E. (1990) The effects of variation in seed size on growth and reproduction in the wild barley *Hordeum vulgare* ssp. *Heredity* 64:239–250.

Gillespie, J.H. (1973) Polymorphism in random environments. *Theoret. Pop. Biol.* 4:193–195.

Glazier, D.S. (1992) Effects of food, genotype, and maternal size and age on offspring investment in *Daphnia magna. Ecology* 73:910–926.

Gliwicz, Z.M. (1990) Food thresholds and body size in Cladocerans. *Nature* 343:638–640.

Gliwicz, Z.M. and Guisande, C. (1992) Family planning in *Daphnia*: Resistance to starvation in offspring born to mothers grown at different food levels. *Oecologia* 91:463–468.

Goulden, C.E., Henry, L. and Berrigan, D. (1987) Egg size, postembryonic yolk, and survival ability. *Oecologia* 72:28–31.

Green, J. (1954) Size and reproduction in *Daphnia magna* (Crustacea: Cladocera) *Proc. zool. Soc. Lond.* 124:535–545.

Green, J. (1956) Growth, size and reproduction in *Daphnia* (Crustacea: Cladocera). *Proc. zool. Soc. Lond.* 126:173–204.

Guisande, C. (1993) Reproductive strategy as population density varies in *Daphnia magna* (Cladocera). *Freshwater Biol.* 29:463–467.

Hall, D.J., Threlkeld, S.T., Burns, C.W. and Crowley, P.H. (1976) The size-efficiency hypothesis and the size structure of zooplankton communities. *Ann. Rev. Ecol. Syst.* 7:177–208.

Hartnoll, R.G. (1982) Growth. *In:* L.G. Abele (ed.): *The Biology of Crustacea, Vol. 2.* Academic Press, New York, pp 111–196.

Hartnoll, R.G. (1985) Growth, sexual maturity and reproductive output. *In:* A.M. Wenner (ed.): *SR Crustacean Issues 3, Factors in adult growth.* Balkema, Rotterdam, pp 101–128.

Hrbácek, J. and Hrbácková-Esslova, M. (1960) Fish stock as a protective agent in the occurence of slow developing dwarf species and strains of the genus *Daphnia. Int. Revue ges. Hydrobiol.* 45:355–358.

Hrbácková, M. (1971) The size and distribution of neonates and growth of *Daphnia hyalina* Leydig (Crustacea, Cladocera) from Lake Maggiore under laboratory conditions. *Mem. Ist. Ital. Idrobiol.* 27:357–367.

Hrbácková-Esslova, M. (1962) Postembryonic development of cladocerans I. *Daphnia pulex* group. *Vest. Cest. Spol. Zool.* 26:212–233.

Hutchinson, G.E. and MacArthur, R. (1959) A theoretical ecological model of size distributions among species of animals. *Amer. Nat.* 93:117–126.

Jones, J.S., Ebert, D. and Stearns, S.C. (1991) Life-history constraints and the mechanics of reproduction in genes, cells and *Daphnia. In:* R.J. Berry, T.J. Crawford and G.M. Hewitt (eds): *Genes in Ecology,* Blackwell Scientific Publications, Oxford, pp 393–404.

Kaplan, R.H. and Cooper, W.S. (1984) The evolution of developmental plasticity in reproductive characteristics: An application on the 'adaptive coin-flipping' principle. *Amer. Nat.* 123:393–410.

Kerfoot, W.C. and Sih, A. (1987) *Predation.* University Press of New England, Hanover.

Ketola, M. and Vuorinen, I. (1989) Modification of life-history parameters of *Daphnia pulex* Leydig and *D. magna* by the presence of *Chaoborus* sp. *Hydrobiologia* 179:149–155.

Lack, D. (1947) *Darwin's Finches.* Cambridge University Press, Cambridge.

Lande, R. (1982) A quantitative genetic theory of life-history evolution. *Ecology* 63:607–615.

Lei, C. and Armitage, K.B. (1980) Growth, development and body size of field and laboratory populations of *Daphnia ambigua. Oikos* 35:31–48.

Lynch, M. (1977) Fitness and optimal body size in zooplankton populations. *Ecology* 58:763–774.

Lynch, M. (1980) Predation, enrichment, and the evolution of cladoceran life histories: A theoretical approach. *In:* W.C. Kerfoot (ed.): *Evolution and Ecology of Zooplankton Communities.* University Press of New England, Hanover, NH, pp 367–376.

Lynch, M. (1983) Estimation of size-specific mortality rates in zooplankton populations by periodic sampling. *Limnol. Oceanogr.* 28:533–545.

Lynch, M. (1984) The limits to life history evolution in *Daphnia. Evolution* 38:465–482.

Lynch, M. (1989) The life history consequences of resource depression in *Daphnia pulex. Ecology* 70:246–256.

Lynch, M. and Ennis, R. (1983) Resource availability, maternal effects, and longevity. *Exp. Gerontol.* 18:147–165.

Makrushin, A.V. (1981) Ovary cycles of *Daphnia pulex* and *Moina macrocope. Hydrobiological Journal* 17:66–70 (in Russian).

Maynard Smith, J., Burian, R., Kauffman, S., Alberch, P., Campell, J., Goodwin, B., Lande, R., Raup, D. and Wolpert, L. (1985) Developmental constraints and evolution. *Quart. Rev. Biol.* 60:265–287.

McCauley, E., Murdoch, W.W., Nisbet, R.M. and Gurney, W.S.C. (1990) The physiological ecology of *Daphnia* I. The importance of stage-specific patterns of growth. *Ecology* 71:703–715.

McKee, D. (1995) *The Influence of Temperature on Some of the Life History, Behaviour and Population Characteristics of Daphnia magna.* PhD thesis, Imperial College, London.

McKee, D. and Ebert, D. (1996a) The effect of temperature on maturation threshold length in *Daphnia magna. Oecologia* 108:627–630.

McKee, D. and Ebert, D. (1996b) The interactive effects of temperature, food level and maternal phenotype on offspring size in *Daphnia magna* and a test of the growth rate-offspring size hypothesis. *Oecologia* 107:189–196.

Murtaugh, P.A. (1981) Size-selective predation on *Daphnia* by *Neomysis mercedis. Ecology* 62:894–900.

Neill, W.E. (1975) Experimental studies of microcrustacean competition, community composition and efficiency of resource utilization. *Ecology* 56:809–826.

Nijhout, H.F. and Williams, C.M. (1974) Control of moulting and metamorphosis in the tobacco hornworm, *Manduca sexta* (L.): Growth of the last-instar larva and the decision to pupate. *J. Exp. Biol.* 61:481–491.

Pace, M.L., Porter, K. and Feig, Y.S. (1984) Life history variation within a parthenogenetical population of *Daphnia parvula* (Crustacea: Cladocera). *Oecologia* 63:43–51.

Paloheimo, J.E., Crabtree, S.J. and Taylor, W.D. (1982) Growth model of *Daphnia. Can. Fish. Aquat. Sci.* 39:598–606.

Philippi, T. and Seger, J. (1989) Hedging one's evolutionary bets, revisited. *Trends Ecol. Evol.* 4:41–44.

Plowright, R.C. and Jay, S.C. (1968) Caste differentiation in bumblebees (*Bombus* Latr.: Hym.) I. The determination of female size. *Insectes Soc.* 15:171–192.

Porter, K.G., Orcutt, J.D. and Gerritsen, J. (1983) Functional response and fitness in a generalist filter feeder, *Daphnia magna* (Cladocera: Crustacea). *Ecology* 64:735–742.

Rendel, J.M. (1967) *Canalization and Gene Control.* Logos Press, London.

Roff, D.A. (1992) *The Evolution of Life Histories.* Chapman and Hall, New York.

Scott, M.A. and Murdoch, W.W. (1983) Selective predation by the backswimmer *Notonecta. Limnol. Oceanogr.* 28:352–366.

Seger, J. and Brockmann, H.J. (1987) What is bethedging? *Oxford Surveys Evol. Biol.* 4: 182–211.

Spaak, P. (1995) Sexual reproduction in *Daphnia*: interspecific differences in a hybrid species complex. *Oecologia* 104:501–507.

Spitze, K., Burnson, J. and Lynch, M. (1991) The covariance structure of life-history characters in *Daphnia pulex. Evolution* 45:1081–1090.

Stearns, S.C. (1976) Life-history tactics: A review of the ideas. *Quart. Revi. Biol.* 51:3–47.

Stearns, S.C. (1992) *The Evolution of Life Histories.* Oxford University Press, Oxford.

Stearns, S.C. and Kawecki, T.J. (1994) Fitness sensitivity and the canalization of life-history traits. *Evolution* 48:1438–1450.

Stearns, S.C. and Koella, J.C. (1986) The evolution of phenotypic plasticity in life-history traits: Predictions of reaction norms for age and size at maturity. *Evolution* 40:893–913.

Stirnadel, H.A. and Ebert, D. (1997) Prevalence, host specificity and impact on host fecundity of microparasites and epibionts in three sympatric *Daphnia* species. *J. Anim. Ecol.* 66: 212–222.

Taylor, B.E. (1985) Effects of food limitation on growth and reproduction of *Daphnia. Arch. Hydrobiol. Beih. Ergeb. Limnol.* 21:285–296.

Tessier, A.J. and Consolatti, N.L. (1989) Variation in offspring size in *Daphnia* and consequences for individual fitness. *Oikos* 56:269–276.

Tessier, A.J., Henry, L.L., Goulden, C.E. and Durand, M.W. (1983) Starvation in *Daphnia*: energy reserves and reproductive allocation. *Limnol. Oceanogr.* 28:667–676.

Threlkeld, S.T. (1976) Starvation and the size structure of zooplankton communities. *Freshwater Biol.* 6:489–496.

Tillmann, U. and Lampert, W. (1984) Competitive ability of differently sized *Daphnia* species: An experimental test. *J. Freshwater Ecol.* 2:311–323.

Tollrian, R. (1995) Predator-induced morphological defenses: Costs, life history shifts, and maternal effects in *Daphnia pulex. Ecology* 76:1691–1705.

Urabe, J. (1988) Effect of food conditions on the net production of *Daphnia galeata*: Separate assessment of growth and reproduction. *Bull. Plankt. Soci. Japan* 35:159–174.

Uvarov, B. (1966) *Grasshoppers and Locusts.* Cambridge University Press, Cambridge.

Vidtmann, S. (1993) The peculiarities of prevalence of microsporidium *Larssonia daphniae* in natural *Daphnia pulex* population. *Ekologija* 1:61–69 (in Russian).

Walker, T.J. (1986) Stochastic polyphenism: Coping with uncertainty. *Florida Entomol.* 69:46–62.

Weglenska, T. (1971) The influence of various concentrations of natural food on the development, fecundity and production of planktonic crustacean filtrators. *Ekologia Polska* 19:427–472.

Yoshimura, J. (1997) The evolutionary origins of periodical cicadas during ice ages. *Amer. Nat.* 149:112–124.

Zaffagnini, F. (1987) Reproduction in *Daphnia. Mem. Ist. Ital. Idrobiol.* 45:245–284.

Evolutionary Ecology of Freshwater Animals
ed. by B. Streit, T. Städler and C. M. Lively
© 1997 Birkhäuser Verlag Basel/Switzerland

Optimal energy allocation tactics and indeterminate growth: Life-history evolution of long-lived bivalves

J. Jokela

ETH-Zürich, Experimental Ecology, ETH-Zentrum NW, CH-8092 Zürich, Switzerland

Summary. The theory of optimal energy allocation among maintenance, somatic growth and reproduction forms the backbone of life-history theory. Allocation to each of these functions need not be equally important with respect to fitness. For example, for young individuals the allocation to maintenance (survival) may be more important than allocation to somatic growth or reproduction. The priority rank of allocation targets may depend on a multitude of factors, both external (e.g., seasonal environment) and internal (e.g. age) to the organism. Individuals that can identify the changes in the expected priority rank of allocation targets and respond accordingly will have a selective advantage. In other words, selection favors individuals whose allocation pattern is closer to the theoretical optimum than the mean allocation pattern of the population. A match of the observed and expected allocation patterns may be investigated by manipulating the availability of resources and measuring the response of the organism. I review the literature of energy allocation in long-lived, iteroparous, indeterminately growing bivalves. I discuss the relevant time scales of allocation decisions (among-season vs. within-season), the role of storage allocation in seasonal environments, and review the theory of optimal resource allocation for this type of life histories. Empirical and experimental studies suggest that priority ranks among allocation targets are an important part of life-history strategy in these long-lived bivalves.

Introduction

Allocation of energy among maintenance, growth and reproduction is a central part of life-history strategy (Cody, 1966; Levins, 1968; Sibly and Calow, 1986; Williams, 1966a, b). Growth at the juvenile stage, the onset of reproduction, survival, and reproductive output may all be described as functions of energy allocation. It is not an overstatement to suggest that the theory of optimal energy allocation forms the backbone of life-history theory (see Stearns, 1992). Optimal energy allocation may be defined as the pattern of energy allocation that maximizes the fitness of an organism for its expected life span. Energy allocation, of course, is a dynamic process depending on a multitude of ecological and physiological factors (Sibly and Calow, 1986; Stearns, 1992).

In this paper, I will focus on the pattern of energy allocation in long-lived, iteroparous species that grow indeterminately. The "allocation pattern" may be envisioned as a set of coadapted, covarying physiological traits that determine which of the "competing" functions of the organism are supported at which time (Bradley et al., 1991; Glazier and Calow, 1992;

Gurney et al., 1990; Jokela and Mutikainen, 1995; McCauley et al., 1990; Perrin, 1992). Iteroparity and indeterminate growth are life-history terms, referring to species which reproduce several times, and grow all their life (Stearns, 1992). Such species are numerous, including, for example, most fish.

In the first section of this paper, I will briefly review the evolution of iteroparity and indeterminate growth in the light of life-history theory. Then, I will present a conceptual framework for problems of energy allocation that iteroparity and indeterminate growth bring about, and how these problems are approached with optimality models. In the latter part of the paper I will review the results of some case studies of energy allocation in marine and freshwater bivalves. I chose to limit my view to life histories represented by large bivalves, therefore, I do not attempt to cover but a thin slice of the vast literature on optimal energy allocation. Large bivalves are suitable for my purpose, because they are long-lived, iteroparous, and have indeterminate growth. In these organisms assessment of growth (using the annual growth rings) and reproductive allocation (production of larvae) is relatively easy, allowing studies of energy allocation in natural populations.

Evolution of iteroparity and indeterminate growth

In a seminal paper, Cole (1954) suggested that for a semelparous species, the gain in the rate of increase from switching to iteroparity is equal to increasing the average semelparous clutch size by one offspring. This result seemed paradoxical ("Cole's paradox"), since so many organisms are iteroparous (Roff, 1992; Stearns, 1992), and lead to serious attempts to understand the evolution of iteroparity.

Cole's paradox was later solved by Charnov and Schaffer (1973) who found that Cole's result was true only when juvenile survival was equal to adult survival. More recently, Charlesworth (1980) showed that in a constant environment, iteroparity may be favored if adult survival is high, and the intrinsic growth rate, r, is low. This result may be understood through energy allocation. When adult survival is high, a semelparous strategist has to allocate a very high amount of energy to reproduction to match the average lifetime reproductive allocation of an iteroparous strategist (Roff, 1992). However, when the rate of increase is high, early allocation has a higher pay-off, and it is easier for a semelparous strategists to match the allocation of an iteroparous strategist. Hence, life-history theory predicts that, in a constant environment, iteroparous strategies are associated with more or less stable populations and high adult survival (Charlesworth, 1980).

Another avenue for the evolution of iteroparity is stochastic variation in offspring survival, i.e., temporal unpredictability of the environment (Murphy, 1968). In an unpredictable environment a single catastrophic year may wipe out all semelparous strategists from the population, therefore, spreading the offspring production among several years may be a more favorable strategy (bet-hedging; Stearns, 1976). In other words, in temporally stochastic environments, where variation in offspring survival is high, an iteroparous strategy may yield higher geometric mean fitness than a semelparous strategy.

The evolution of indeterminate growth has not been studied as extensively as the evolution of iteroparity. However, indeterminate growth has been associated with two life-history characteristics: (1) increasing fecundity with size, and (2) iteroparous lifestyle (Roff, 1992). Obviously, allocation to growth after the onset of reproduction is simply wasted if it does not lead to a corresponding increase in fecundity (or corresponding decrease in adult mortality). It is noteworthy that organisms following an indeterminate growth pattern also have an iteroparous life style. Hence, the evolution of indeterminate growth is often discussed together with the evolution of risk spreading tactics, and iteroparity.

In their thorough review of optimal energy allocation models, Perrin and Sibly (1993) concluded that indeterminate growth may result from two alternative allocation patterns: (1) either energy is allocated simultaneously to growth and reproduction, or (2) the life-history includes multiple switches from allocation to growth to allocation to reproduction. Of these, the latter is a plausible model for iteroparous organisms. Furthermore, Perrin and Sibly (1993) concluded that linear allocation trade-offs between growth and fecundity are not likely to produce indeterminate growth in a constant environment. However, annual environments, seasonality, and even unpredictable stochasticity in the environment may select for multiple switches between growth and reproduction.

Therefore, the conclusion is that the outlines of the evolution of iteroparity and indeterminate growth are well covered in current life-history theory (Engen and Saether, 1994; Perrin and Sibly, 1993; Roff, 1992; Stearns, 1992). What remains unclear is how the energy allocation pattern is shaped by natural selection to support this kind of life history.

The role of storage allocation in a seasonal environment

In iteroparous organisms seasonality plays an important role in the allocation of energy. Here, seasonality may be defined as any regular periodicity in the fluctuation of resources, not solely as annual sequences of favorable periods. Seasonality imposes temporal constraints on the reproductive allocation; reproduction is favorable, or possible, only within a specific time frame.

Seasonality has important consequences for energy allocation. Seasonality not only constrains the reproductive allocation, and allocation to growth, but also the risks of mortality may be higher during the less favor-

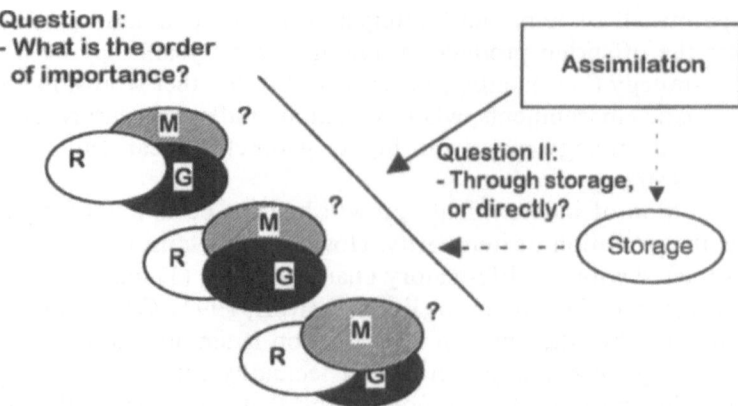

Figure 1. Schematic presentation of two different pathways the assimilated energy may be directed to maintenance (M), reproduction (R) and growth (G). Often storage allocation is associated with maintenance, but stored energy may also be used for other functions. The role of storage is to buy time; often the availability and need of energy do not match.

able season. One "solution" for the periods of negative energy balance is the evolution of traits allowing storage of energy. Storage may be envisioned as "the fourth allocation target"; stored energy may later be used for maintenance, reproduction or growth (Fig. 1). However, independent of the final target, storage allocation always serves the same function: it moves energy from times of plenty to times of scarcity, and provides flexibility to energy allocation (Ankney and MacInnes, 1978; Perrin et al., 1990; Perrin and Sibly, 1993; Reznick and Braun, 1987).

It is important to pay attention to two different time scales when "optimality" of allocation is considered, namely, the patterns of energy allocation observed among- and within seasons. The life history of an organism may be more or less completely outlined by describing the among-season allocation decisions. For example, age at maturity, number of reproductive bouts, and longevity are all outcomes of changes in the among-season allocation pattern.

However, it is the within-season allocation decisions that define the reproductive output, growth and survival for each particular season. These within-season decisions are connected to the among-season pattern. In fact, the among-season pattern may be best seen as a compilation of the within-season decisions (Kozlowski and Wiegert, 1987). Lifetime scale, of course, is the one where evolutionary competence is measured (Stearns, 1982), but to understand what leads to the observed life history, it is essential to look at the dynamics of the within-season energy allocation. This is emphasized when the constraints that define the evolution of among-season allocation patterns are considered (Stearns, 1989). However, questions of optimal energy allocation may be, and have been, studied successfully at both these time scales.

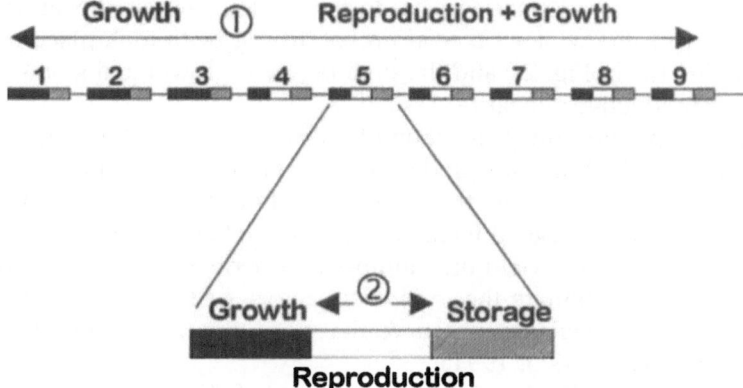

Figure 2. A scheme of an iteroparous life history in a seasonal environment. Numbered rectangles represent consecutive seasons. The first level (top) represents changes in the allocation pattern observed at the among-season time scale. An example of change in the allocation pattern at this scale is the switch ①, determining the onset of reproduction. The location of this switch has important fitness consequences, and may be under strong selection. The second level of energy allocation is the within-season pattern (bottom). Decisions at this level determine the reproductive output, growth and survival of the organism within each season. An example of this is the sequence ②, determining the proportion of reproductive allocation. Note that optimal allocation among the functions within the season may depend on age. In this scheme, storage allocation is thought to be used for maintenance during winter (winter is represented by thin lines between boxes).

Figure 2 is an attempt to describe a typical life-history of a long-lived iteroparous organism that has the potential for indeterminate growth. The purpose of the figure is to illustrate the two time scales discussed above. In Figure 2, the onset of reproductive allocation at age 4 represents an obvious among-season change in the allocation pattern. However, after reproduction has been commenced, the reproductive allocation has to be settled within each season with respect to growth and storage to ensure survival until the next favorable season. Here it is assumed that fecundity increases with size. The solution to this within-season "optimization problem" depends on age and survival probabilities to the next season. Hence, fitness of an organism can be viewed as a composite outcome of both among and within season decisions of energy allocation.

"Kozlowski-model" for optimal energy allocation

Kozlowski (1991) constructed an optimal energy allocation model for the type of life history depicted in Figure 2. His goal was to find the optimal allocation pattern within each season for different ages, and to simulate the resulting growth trajectory. The model included two switches of allocation within each season: First from growth to reproduction, and second from reproduction to storage (see Fig. 2). In the model, storage allocation was associated with the winter survival probability; growth took place before reproduction (as in Fig. 2), and the time between growth and storage was used for reproductive allocation.

The model predicts the proportion of allocation to each function by age (Kozlowski, 1991). When the prediction was compared to the best available field data on the relative reproductive allocation by age (Vahl's data on reproductive effort by age in Iceland scallops, see below and Vahl, 1981a), the model matched the field data almost in a perfect manner. The model was also used to simulate the growth trajectory of the scallops, and the result was a sigmoidal growth curve that very closely matched the true growth curve (Kozlowski, 1991).

The predictions of Kozlowski's model were twofold: Allocation to growth decreases with age, and the proportion devoted to storage increases with the increasing survival probability to the next season (Kozlowski, 1991). Furthermore, if survival probability was high, the model predicted that optimal age at maturity should be delayed. Kozlowski, with a very good reason, suggested that these types of allocation models are very powerful tools for life-history studies (Kozlowski, 1991).

In the discussion of the predictions of the model, Kozlowski (1991) devoted some thoughts to how organisms could possibly follow the "optimal" schedule of switching the allocation from growth to reproduction to storage within a season. He pointed out that when solving the optimization problem with an analytical model, one has to proceed backwards, from the last year to the earlier ones. Of course, this is not the way organisms face allocation decisions. Therefore, he proposed that either organisms "… 'decide' about energy allocation in a forward manner or *use some rigid decision rules how to proceed under given conditions*. Although such rules cannot lead exactly to the optimal solution, they can help to behave almost optimally" (italics mine; Kozlowski, 1991). Next I will make an attempt to identify these "rigid decision rules", and show how they may be studied. I propose that "priority rank of allocation targets" may be a sufficient guide for optimal energy allocation. Below I focus on how the priority rank of allocation targets may change, and how organisms best respond to these changes.

Priority rank of allocation targets (PRAT) in a seasonal environment

As the name implies, the "priority rank" indicates the marching order of allocation targets at a specific point in time. Expected priority rank of allocation targets (EPRAT) may be defined as the relative importance of allocation targets with respect to fitness (Fig. 3). For example, if resources for 1 year are very poor, and fecundity increases with size, it may be more beneficial for a young individual to cut the allocation to reproduction for

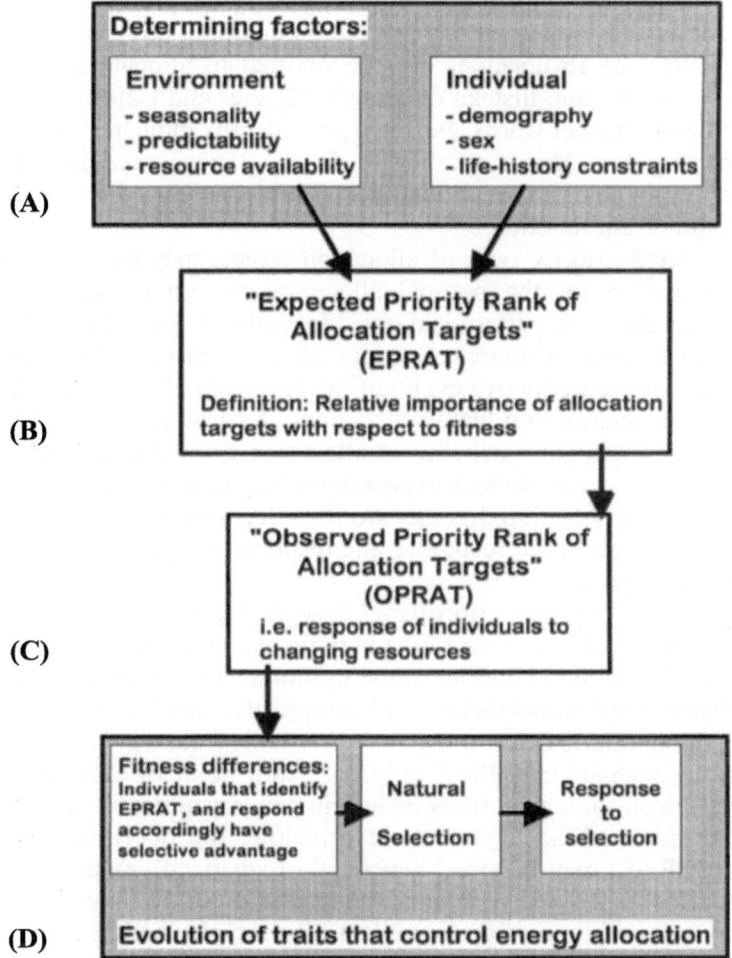

Figure 3. Summary of factors that are important for the evolution of traits that control energy allocation. The graph is divided into four logically connected sections indicated with capital letters (A, B, C, and D). Determining factors (A) are components that determine the relative importance of allocation targets at a specific point in time. These determining factors include the type of environmental variation and resource availability, as well as individual factors that are associated with allocation targets. These include demographic factors like survival probability, age, reproductive schedule, and general life-history constraints that limit the strategy set of an individual. Using this information, hypothesis on the expected priority rank of allocation targets (EPRAT) can be formulated (B). The relative importance of allocation targets is measured in the currency of fitness. (C) The observed priority rank of allocation targets (OPRAT) may be measured, for example, by manipulating the resource so that the individuals are forced to respond. Resource manipulation simulates naturally occurring changes in the environment. Evolution of the traits that control energy allocation (D) takes place if individuals differ in their ability to "recognize" and respond to EPRAT. This leads to selection on the traits involved and, may lead to evolutionary change in the allocation pattern. Note that allocation traits may be genetically canalized, developmentally converged, or phenotypically modulated (see text).

that year, and instead allocate all available energy to somatic growth. Of course, for an old individual of the same population, the priority rank of allocation may be reversed: it may be optimal to allocate all available energy to reproduction instead of growth (Glazier and Calow, 1992; Williams, 1966b). In other words, expected priority rank indicates which of the allocation targets is least important to fitness. If resources decrease, the "correct" response of an individual would be to decrease the allocation to the least important function.

The expected priority rank of allocation targets may be envisioned to form a guideline for the optimal allocation pattern of the individual (Fig. 3). This expected pattern is derived from theory (see above). EPRAT is based on external (resources, season, environment) and internal properties (age, size, reproductive cycle) of the organism (determining factors, Fig. 3). These determining factors also guide the changes in the allocation pattern of the organism (switches in allocation from one function to the other). Needless to say, we may expect the ability to respond "correctly" to evolve by natural selection. In other words, selection may favour individuals that have the ability to recognize and respond to expected priority rank of allocation targets.

In Figure 3 the expected priority rank is separated from observed priority rank (OPRAT). A close match of the individual's response with the expectation takes the individual close to the optimum. The traits that are under selection are called "traits that control energy allocation" in Figure 3. Evolution of these traits is determined by the heritable fitness differences in the response to changes in EPRAT (Fig. 3). From point of view of natural selection, the physiological traits underlying energy allocation may be like any other traits of the organism. In principle, the observed allocation pattern (OPRAT) may be fixed genetically (canalized), fixed during the early development (developmental conversion), or it may be flexible throughout the lifetime of the organism (phenotypic modulation) (Jokela and Mutikainen, 1995).

How can the priority rank of allocation targets be assessed? Both EPRAT and OPRAT are concepts that have to be approached both through information on the environment and information on the individual. In Figure 3 these are called determining factors (A). These factors determine the expected optimal allocation pattern for an individual at any point in time. Kozlowski's model (see above, and Kozlowski, 1991) is a prime example how this kind of information may be processed and how the expectations may be derived (see also Kozlowski, 1996).

In seasonal environments the expected priority rank of allocation targets may change within the season (Fig. 2). Therefore, assessing the allocation priorities early in the season may give quite different result than the same survey late in the season. In other words, the expected fitness returns of allocation to a specific function may depend on the timing of allocation within the season. Late in the season the allocation to growth or reproduction may not give as large returns in fitness as does allocation to maintenance (e.g., through storage).

The relative allocation to reproduction may also change with age (e.g., as predicted by the Kozlowski model). Classical demographic life-history theory predicts that reproductive effort increases with age, i.e., reproductive allocation should become relatively more important than allocation to growth or maintenance with increasing age (Fisher, 1930; Williams, 1966b). This is because the reproductive value of an organism decreases with age. In other words, higher "risks" may be taken in reproduction at older age, because less is at stake. This prediction, however, has to be considered carefully with respect to density dependence and to the relationship between fecundity and age, as pointed out by Charlesworth (1980).

Below I review studies in which the observed priority rank of allocation targets has been assessed in natural populations of large bivalves.

Empirical studies of energy allocation in natural populations of marine and freshwater bivalves

The natural history of large bivalves gives some hints of the "allocation problems" these organisms face during their life. Marine bivalves primarily have external fertilization, pelagic larvae, and a specific period within the favourable season when spawning takes place. Often larvae contain lipids for buoyancy and fuel. Large freshwater bivalves, on the other hand, have parasitic larvae that are brooded in the gill blades of female clams, and released at a species-specific period within the favorable season.

The timing of reproduction is an important determinant of fitness for large bivalves. The timing of reproduction as such is a complex life history phenomenon. Depending on the type of larva (pelagic or parasitic) and the dynamics of the environment, the favorable period for release may vary from anywhere between spring and winter. Of course, the favorable period for release of larvae does not always coincide with the favorable period of resources. For example, in freshwater clams of the genus *Anodonta*, the release of parasitic larvae takes place in very early spring (Jokela et al., 1991; Negus, 1966; Ökland, 1963). This marks the end of several months of scarce resources (winter). However, at this time the host availability is very high due to spawning schools of host fish in the habitats occupied by the clams (Jokela and Palokangas, 1993).

Another example of the mismatch between the time of favorable resources and the best timing of reproduction may be when the larvae need a good start, i.e., larvae have to be released just before the growth season so that they can take the full benefit of it. Indeed, external determinants of juvenile survival may lead to a considerable variation in the within-season energy allocation patterns among species. This emphasizes the importance to work out the natural history before asking questions about the optimal allocation pattern, or the priority rank of allocation targets.

It is important to separate "reproduction" from "reproductive alloca-tion", because these can take place at different times. For example, larvae may be stored for long periods, or the energy for the production of larvae may be stored earlier in the season. Temporal mismatch between favorable periods of resources and favorable periods for reproduction may be solved with allocation to storage; therefore it is not surprising that both marine and freshwater clams store energy as fat and/or glycogen.

Vahl studied the reproductive energy allocation of Iceland scallop (*Chlamys islandica*) in a series of detailed studies (Vahl, 1978; Vahl, 1981 a, b, c). This indeterminately growing species becomes reproductively mature at 3 to 6 years of age, and is very long-lived (maximum longevity more than 20 years). These scallops divide the favorable season so that growth takes place from early March to June, while spawning takes place in July. Vahl set out to study the reproductive allocation with respect to age. His studies were motivated by the prediction of higher reproductive alloca-tion with decreasing residual reproductive value with age (Fisher, 1930; Williams, 1966b). However, he found no correlation between reproductive effort and age, until at very advanced age (> 13 years) (Vahl, 1981 a, b, c).

But he found something else that was very interesting. He had estimated the metabolic rates of scallops of different ages. When compiling the avail-ability of resources and energy-budgets for different age classes, Vahl (1981 a, b, c) came to the conclusion that the old age classes actually have a relatively shorter favorable season than the younger age classes. He concluded that because of this, the older age classes were more constrained by resources, and that the priority rank of allocation targets at these older age classes was: maintenance > reproduction > growth. Because of the exponential increase in fecundity with size and high survival, even the older age classes had relatively high residual reproductive values; in fact, reproductive values started to decrease at the age of 13 years. Therefore, Vahl suggested that for most adult ages somatic and reproductive alloca-tion should be adjusted to a level that leads to a low risk of mortality (Vahl, 1981 a, b, c).

These conclusions are supported by a series of studies of energy al-location of the giant scallop, *Placopecten magellanicus* (MacDonald and Bayne, 1993; MacDonald and Thompson, 1985 a, b; MacDonald and Thompson, 1986; MacDonald et al., 1987; Thompson and MacDonald, 1990). These studies used a decrease in resource availability by depth as a "natural experiment". In the deep habitat scallops grew slower and allocat-ed less energy to reproduction (MacDonald et al., 1987). No obvious "con-flict" between growth, maintenance and reproduction were observed until at old age, corroborating the observation of Vahl (see above). However, variation in reproductive allocation was higher than in growth suggesting that for most ages reproduction was compromised in favor of maintenance and somatic growth (maintenance = growth > reproduction). This may be adaptive if losses in reproduction during one season are less than fecundity loss in the future due to smaller size (MacDonald and Thompson, 1985b). At the older ages the conclusion was similar to Vahl's, growth was traded to reproduction at old age (MacDonald et al., 1987).

Moreover, the within-season decisions of energy allocation have been studied in natural populations of bivalves. When studying the growth of freshwater zebra mussels, *Dreissena polymorpha,* Smit et al. (1992) found evidence of within-season differences in the allocation priorities. They found the peak growth to occur in late May, although peak productivity at the study habitats took place later. They suggested that slow growth during early and mid-summer was due to allocation to reproduction at this period. This was supported by the observation that the growth of the zebra mussels increased later in the season when the reproductive period was passed. Furthermore, Jokela et al. (1993) found that the growth rate of female freshwater clams, *Anodonta piscinalis,* was faster in the early season, before fertilization and development of larvae, than in the late season, when larvae were developed and brooded in the gill blades.

Together these studies of energy allocation patterns in natural bivalve populations indicate that both within- and among-season variation in allocation patterns may be found (Tab. 1). This variation may covary with the timing of the reproductive schedule, suggesting that the observed priority rank of allocation targets may indeed respond to changes in the environment.

Priority rank of allocation targets in bivalves: Experimental evidence

Two aspects of energy allocation are difficult to study without experimental manipulation in a natural setting: the role of flexible phenotypic responses in adjusting the energy allocation to changing environments (phenotypic plasticity), and the existence of priority rank among allocation targets. Furthermore, the appropriate time scale should be taken into account when designing such experiments. Unfortunately, not many experimental manipulations exist. However, Peterson and Fegley (1986), Jokela and Mutikainen (1995) and Jokela (1996) have conducted field experiments to study phenotypic plasticity and priority rank of allocation targets in large bivalves.

For a 1-year reciprocal transplant experiment with the freshwater clam *Anodonta piscinalis,* Jokela and Mutikainen (1995) chose six study sites that differed with respect to resource availability. They started the experiment (transplanted groups of clams reciprocally) in autumn, right after the development of larvae was completed. At this time larvae were still brooded in the gill blades of females. The purpose of the experiment was to stock clams at sites in a reciprocal design at higher than normal density, and

Table 1. Summary of studies of energy allocation in natural populations of bivalves. M = maintenance allocation, R = reproductive allocation, G = growth

Species	Measured characteristics	Observed priority rank of allocation targets		Ref.
Chlamys islandica	Growth, metabolic rate, energy budget, and reproductive allocation of different age classes	Old age classes: M > R > G		1
Placopecten magellanicus	Growth, reproductive and maintenance allocation of different age classes	Old age classes: M > R > G	other age classes: M = G > R	2
Dreissena polymorpha	Within-season variation in growth	early season: R>G	late season: G > R	3
Anodonta piscinalis	Within-season variation in growth	early season: G > R	late season: R > G	4

References:

1. Vahl, 1978; Vahl, 1981a, 1981b; 1981c.
2. MacDonald and Bayne, 1993; 1985a; MacDonald and Thompson, 1985b; MacDonald and Thompson, 1986; MacDonald et al., 1987; Thompson and MacDonald, 1990.
3. Smit et al., 1992.
4. Jokela et al., 1993.

monitor the effect of the treatment on survival, growth and reproduction. The design created a wide array of past-present environments, useful in revealing the phenotypic responses of clams to changing resources.

Jokela and Mutikainen (1995) drew two main conclusions based on the results of the experiment. First, the allocation pattern of the clams was very flexible. Shell growth, somatic mass, percent body fat, proportion of females reproducing and reproductive output were all affected more by transplant site than by origin, indicating that clams phenotypically adjusted their energy allocation to the prevailing conditions. However, origin had a statistically significant effect on reproductive investment (reproductive output in relation to body mass), somatic mass and shell growth, indicating that also genetic differences in the relative allocation to different functions exist among populations.

Their second conclusion was that maintenance allocation had the top priority in this 1-year experiment (Jokela and Mutikainen, 1995). If resources allowed, reproduction was the next target. Clearly, shell growth was given up first if resources were poor. That the allocation to shell growth was subordinate to reproductive allocation was evident from a comparison of undisturbed control clams to experimental clams kept at high density. The experiment clearly affected the growth of the clams, but not the reproductive output. Jokela and Mutikainen (1995) also found a negative correlation with the proportion of reproductive females and mortality. This suggests that at those treatment combinations that suffered the most physiological stress (as indicated by high mortality), a higher proportion of females that survived had shut off the reproductive allocation altogether.

In the second experiment, Jokela (1996) used reciprocal transplants to study the flexibility of energy allocation of *Anodonta* clams within seasons. In these clams, fertilization takes place in mid-summer, after which the larvae develop and are brooded in the gill blades of females (Jokela et al., 1991). The most intensive period of larval development is during July. To reveal the within-season differences in the priority rank of allocation targets, Jokela (1996) conducted reciprocal transplants for 2 months in different parts of the season.

The results of the experiment indicated that early-season changes (May – June, before fertilization) in the environment had no detectable effect on the reproductive output of clams, but for some origins a response in the late season (November) fat content and body mass was observed. However, transplants in the late season had the predicted effects; clams transplanted to productive sites increased their reproductive output, while at non-productive sites the reproductive output decreased. Late transplants also clearly affected the somatic condition of the clams. Together these results suggest that clams were able to reproductively compensate for early-season changes in the environment, but were not able to compensate for changes that occurred later in the season (Jokela, 1996).

Interestingly, the late-season transplants had no effect on percent body fat. This led to the suggestion that in the late season clams allocated energy to storage in preparation for winter. The general conclusion of the experiment was that the priority rank of allocation targets varied within the season. More specifically, in these clams allocation to maintenance (storage) seems to override the allocation to reproduction in the late season, but in the early season reproductive allocation seems to be the top priority. Overall, the experiment paints a picture of restrained reproductive tactic (Jokela, 1996; Sibly and Calow, 1984), and suggests that within-season switches of allocation exist, as assumed by the Kozlowski model (Kozlowski, 1991).

The existence of within-season switches in allocation is also supported by the results of a field experiment conducted by Peterson and Fegley (1986). They enclosed juvenile and adult hard clams (*Mercenaria mercenaria*) in field enclosures and observed seasonal differences in the growth rates of the adult clams when compared to the juveniles. Clams were individually marked, and care was taken to be able to correct for seasonal variation in the resources and reproductive cycle. They observed that in the late autumn the allometric growth rate of the juveniles was much faster than the growth rate of the adults. These differences were most likely due to increased allocation to storage by the adults, when they prepared for the coming reproductive period. Therefore, the study showed that adult *Mercenaria* clams may change their energy allocation among growth and reproduction with respect to seasonal demands, and that juvenile and adult allocation patterns may differ (Peterson and Fegley, 1986).

These three field experiments suggest that priorities among allocation targets exist in large bivalves (Jokela and Mutikainen, 1995), and that both demographic differences and seasonal changes have to be taken into account when a complete picture of energy allocation patterns is needed (Jokela, 1996; Peterson and Fegley, 1986). The results of these studies are summarized in Table 2.

Concluding remarks

The models of optimal allocation to reproduction and growth for iteroparous organisms of different ages suggest a very good match with empirical data (Kozlowski, 1991; 1992; 1996). The importance of these models is that finally there is a theory that explains convincingly how indeterminate growth pattern evolves. The simple explanation is that increase in the relative allocation to reproduction at older age, at the expense of growth, is the optimal pattern of energy allocation. This result advocates that the expected priority rank of allocation targets changes during the lifetime of the organism. Organisms that respond accordingly by adjusting their allocation pattern will have a selective advantage.

Table 2. Summary of experimental studies of energy allocation in bivales. M = maintenance allocation, R = reproductive allocation, G = growth. PRAT = Priority rank of allocation targets

Species	Type of experiment	Measurements	Observed rank of allocation targets	Ref.
Anodonta piscinalis	One year reciprocal transplant experiment among sites of different quality	Survival, growth reproduction and fat content	Allocation pattern responded strongly to environmental variation. M>R>G	1
Anodonta piscinalis	Within-season reciprocal transplant experiments	Growth, reproductive allocation and fat content	Clams were able to compensate for early season resource shortage later in the season (reproductive allocation not affected). Storage allocation had high priority late in the season. PRAT changed within-season	2
Mercenaria mercenaria	Enclosures of juvenile and adult clams	Within-season variation in somatic growth, shell growth and gonad mass	Within-season differences in PRAT among demographic groups, old clams allocated more to storage late in the season	3

References:

1. Jokela and Mutikainen, 1995.
2. Jokela, 1996.
3. Peterson and Fegley, 1986.

The empirical evidence of energy allocation patterns in long-lived bivalves seems sufficient for some general conclusions to be drawn. The expected difference in priority rank of allocation targets with age has been found in two species, *Placopecten magellanicus* (MacDonald et al., 1987) and *Mercenaria mercenaria* (Peterson and Fegley, 1986). Furthermore, it is obvious that storage allocation plays an important role in long-lived bivalves. It may be quite common that the most favorable period for reproduction does not coincide with the most productive period in terms of resources. Storage allocation allows temporal transport of energy, but it also may include costs (transformation efficiency may be low). Clearly, more studies on the changes in the priority rank of allocation targets with age and on the role of storage allocation are needed.

Another conclusion from studies of natural populations is that the season may be divided into periods that favor allocation to different functions. Growth may precede reproduction, which may precede storage allocation. However, care must be taken in generalizing the result. The timing of reproduction is a very complex life-history decision that may covary with unexpected biotic or abiotic factors of the environment. The bottom line is that the priority rank of allocation targets may vary within seasons.

The priority rank of allocation targets is best revealed by manipulating the environment, and measuring the corresponding response of the organisms (Jokela, 1996; Jokela and Mutikainen, 1995; Peterson and Fegley, 1986). The three experimental studies reviewed here may be justly criticized for their rather indirect approach to energy allocation; no attempt was made to actually measure the energy balance of clams in the different treatments. Direct measurement of assimilation and allocation would have allowed a much more detailed assessment of the allocation pattern. However, practical problems in conducting such experiments in natural populations are overwhelming. What these large-scale field experiments show is that indirect information may be successfully used to explore energy allocation patterns. The results of the experiments emphasize that when conclusions regarding the adaptivity of energy allocation patterns are made, it is important to take note of the time scale addressed with the experiments.

In conclusion, large long-lived iteroparous bivalves seem to be prime organisms to study the evolution of indeterminate growth and priority rank of allocation targets. At the moment there are good grounds to suggest that allocation patterns may change as a response to age and seasonal fluctuation of resources, and that these changes may "help" in adjusting the energy allocation to an optimal direction.

Acknowledgments
I thank J. Kozlowski, P. Mutikainen, T. Städler and J. Tuomi for valuable comments on the manuscript. I was funded by Academy of Finland while writing this review.

References

Ankney, C.D. and MacInnes, C.D. (1978) Nutrient reserves and performance of female lesser snow geese. *Auk* 95:459–471.

Bradley, M.C., Perrin, N. and Calow, P. (1991) Energy allocation in the cladoceran *Daphnia magna* Straus, under starvation and refeeding. *Oecologia* 86:414–418.

Charlesworth, B. (1980) *Evolution in Age-Structured Populations*. Cambridge University Press, Cambridge.

Charnov, E.L. and Schaffer, W.M. (1973) Life history consequences of natural selection: Cole's result revisited. *Amer. Nat.* 107:791–793.

Cody, M.L. (1966) A general theory of clutch size. *Evolution* 20:174–184.

Cole, L.C. (1954) The population consequences of life history phenomena. *Quart. Rev. Biol.* 29: 103–137.

Engen, S. and Saether, B.E. (1994) Optimal allocation of resources to growth and reproduction. *Theor. Popul. Biol.* 46:232–248.

Fisher, R.A. (1930) *The Genetical Theory of Natural Selection*. Clarendon Press, Oxford.

Glazier, D.S. and Calow, P. (1992) Energy allocation rules in *Daphnia magna* – Clonal and age differences in the effects of food limitation. *Oecologia* 90:540–549.

Gurney, W.S.C., McCauley, E., Nisbet, R.M. and Murdoch, W.W. (1990) The physiological ecology of *Daphnia:* A dynamic model of growth and reproduction. *Ecology* 71:716–732.

Jokela, J. (1996) Within-season reproductive and somatic energy allocation in a freshwater clam, *Anodonta piscinalis*. *Oecologia* 105:167–174.

Jokela, J. and Mutikainen, P. (1995) Phenotypic plasticity and priority rules for energy allocation in a freshwater clam: a field experiment. *Oecologia* 104:122–132.

Jokela, J. and Palokangas, P. (1993) Reproductive tactics in *Anodonta* clams – parental host recognition. *Anim. Behav.* 46:618–620.

Jokela, J., Valtonen, E.T. and Lappalainen, M. (1991) Development of glochidia of *Anodonta piscinalis* and their infection of fish in a small lake in northern Finland. *Arch. Hydrobiol.* 120: 345–355.

Jokela, J., Uotila, L. and Taskinen, J. (1993) Effect of castrating trematode parasite *Rhipidocotyle fennica* on energy allocation of freshwater clam *Anodonta piscinalis*. *Funct. Ecol.* 7: 332–338.

Kozlowski, J. (1991) Optimal energy allocation models – an alternative to the concepts of reproductive effort and cost of reproduction. *Acta Oecol.* 12:11–33.

Kozlowski, J. (1992) Optimal allocation of resources to growth and reproduction: Implications for age and size at maturity. *Trends Ecol. Evol.* 7:15–19.

Kozlowski, J. (1996) Optimal allocation of resources explains interspecific life-history patterns in animals with indeterminate growth. *Proc. R. Soc. Lond.* 263:559–566.

Kozlowski, J. and Wiegert, R.G. (1987) Optimal age and size at maturity in annuals and perennials with determinate growth. *Evol. Ecol.* 1:231–244.

Levins, R. (1968) *Evolution in Changing Environments*. Princeton University Press, Princeton, New Jersey.

MacDonald, B.A. and Bayne, B.L. (1993) Food availability and resource allocation in senescent *Placopecten magellanicus:* Evidence from field populations. *Funct. Ecol.* 7:40–46.

MacDonald, B.A. and Thompson, R.J. (1985a) Influence of temperature and food availability on the ecological energetics of the giant scallop *Placopecten magellanicus* I. Growth rates of shell and somatic tissue. *Mar. Ecol.-Progr. Ser.* 25:279–294.

MacDonald, B.A. and Thompson, R.J. (1985b) Influence of temperature and food availability on the ecological energetics of the giant scallop *Placopecten magellanicus* II. Reproductive output and total production. *Mar. Ecol. Progr. Ser.* 25:295–304.

MacDonald, B.A. and Thompson, R.J. (1986) Production, dynamics and energy partioning in two populations of the giant scallop *Placopecten magellanicus*. *J. Exp. Mar. Biol. Ecol.* 101: 285–299.

MacDonald, B.A., Thompson, R.J. and Bayne, B.L. (1987) Influence of temperature and food availability on the ecological energetics of the giant scallop *Placopecten magellanicus*. IV. Reproductive effort, value and cost. *Oecologia* 72:550–556.

McCauley, E., Murdoch, W.W., Nisbet, R.M. and Gurney, W.S.C. (1990) The physiological ecology of *Daphnia:* Development of a model of growth and reproduction. *Ecology* 71: 703–715.

Murphy, G.I. (1968) Pattern in life history and the environment. *Amer. Nat.* 102:391–403.

Negus, C.L. (1966) A quantitative study of growth and production of unionid mussels in the river Thames at Reading. *J. Anim. Ecol.* 35:513–532.

Ökland, J. (1963) Notes on population density, age distribution, growth, and habitat of *Anodonta piscinalis* Nilss. (Moll., Lamellibr.) in a eutrophic Norwegian lake. *Nytt. Mak. Zool.* 11:19–43.

Perrin, N. (1992) Optimal resource allocation and the marginal value of organs. *Amer. Nat.* 139:1344–1369.

Perrin, N. and Sibly, R.M. (1993) Dynamic models of energy allocation and investment. *Annu. Rev. Ecol. Syst.* 24:379–410.

Perrin, N., Bradley, C. and Calow, P. (1990) Plasticity of storage allocation in *Daphnia magna*. *Oikos* 59:70–74.

Peterson, C.H. and Fegley, S.R. (1986) Seasonal allocation of resources to growth of shell, soma, and gonads in *Mercenaria mercenaria*. *Biol. Bull.* 171:597–610.

Reznick, D.N. and Braun, B. (1987) Fat cycling in the mosquitofish (*Gambusia affinis*): Fat storage as a reproductive adaptation. *Oecologia* 73:401–413.

Roff, D.A. (1992) *The Evolution of Life Histories; Theory and Analysis.* Chapman & Hall, New York.

Sibly, R. and Calow, P. (1984) Classification of habitats by selection pressures: A synthesis of life-cycle and r/k theory. *In:* R.M. Sibly and R.H. Smith (eds): *Behavioural Ecology. Ecological Consequencies of adaptive behaviour. 25th Symp. British Ecol. Soc.* Blackwell, Oxford, pp 75–97.

Sibly, R.M. and Calow, P. (1986) *Physiological Ecology of Animals.* Blackwell, Oxford.

Smit, H., Devaate, A.B. and Fioole, A. (1992) Shell growth of the zebra mussel (*Dreissena polymorpha* (Pallas)) in relation to selected physicochemical parameters in the lower Rhine and some associated lakes. *Arch. Hydrobiol.* 124:257–280.

Stearns, S.C. (1976) Life-history tactics: A review of the ideas. *Quart. Rev. Biol.* 51:3–47.

Stearns, S.C. (1982) On fitness. *In:* D. Mossakowski and G. Roth (eds): *Environmental Adaptation and Evolution.* Gustav Fischer; Stuttgart, New York, pp 3–17.

Stearns, S.C. (1989) Trade-offs in life-history evolution. *Funct. Ecol.* 3:259–268.

Stearns, S.C. (1992) *The Evolution of Life Histories.* Oxford University Press, Oxford.

Thompson, R.J. and MacDonald, B.A. (1990) The role of environmental conditions in the seasonal synthesis and utilization of biochemical energy reserves in the giant scallop, *Placopecten magellanicus*. *Can. J. Zool.* 68:750–756.

Vahl, O. (1978) Seasonal changes in oxygen consumption of the Iceland scallop *Chlamys islandica* (O.F. Müller) from 70°N. *Ophelia* 17:143–154.

Vahl, O. (1981a) Age-specific residual reproductive value and reproductive effort in the Iceland scallop, *Chlamys islandica* (O.F. Müller). *Oecologia* 51:53–56.

Vahl, O. (1981b) Energy transformations by the Iceland scallop *Chlamys islandica* (O.F. Müller) from 70°N. I. The age-specific energy budget and net growth efficiency. *J. Exp. Mar. Biol. Ecol.* 53:281–296.

Vahl, O. (1981c) Energy transformations by the Iceland scallop *Chlamys islandica* (O.F. Müller) from 70°N. II. The population energy budget. *J. Exp. Mar. Biol. Ecol.* 53:297–303.

Williams, G.C. (1966a) *Adaptation and Natural Selection.* Princeton University Press, Princeton, New Jersey.

Williams, G.C. (1966b) Natural selection, the costs of reproduction, and a refinement of Lack's principle. *Amer. Nat.* 100:687–690.

Population biology and reproductive modes

Evolutionary Ecology of Freshwater Animals
ed. by B. Streit, T. Städler and C.M. Lively
© 1997 Birkhäuser Verlag Basel/Switzerland

Ecology and genetics of interspecific hybridization in *Daphnia*

K. Schwenk[1] and P. Spaak[2]

[1] *Netherlands Institute of Ecology, Centre for Limnology, Rijksstraatweg 6,*
NL-3631 AC Nieuwersluis, The Netherlands
[2] *Max-Planck-Institut für Limnologie, Abteilung Ökophysiologie, Postfach 165,*
D-24302 Plön, Germany
Present address: EAWAG/ETH, Department of Limnology, Überlandstrasse 133,
CH-8600 Dübendorf, Switzerland

Summary. A central question in evolutionary biology concerns the ecological and genetic processes by which species hybridize. We analyze the evolutionary consequences of hybridization, biogeographic patterns and fitness comparisons in the crustacean *Daphnia* (Branchiopoda) within the conceptual framework of theories on interspecific hybridization. In contrast to most cases of hybridization among animals, *Daphnia* species and hybrids do not form interpopulational transition zones (hybrid zones), but rather patchy distributions of hybrids and parentals have been found. In addition, due to ameiotic parthenogenetic reproduction, hybrid breakdown can be avoided and hybrids can reach higher abundances than parental species. Hybrids within the *D. galeata* complex occur across broad geographic ranges, and lakes vary significantly in species and hybrid composition over time. Species differ in traits related to predator avoidance, such as size at maturity, intrinsic rate of increase (r), behaviour, and induction of helmets and spines. Hybrids tend to exhibit a combination of characters which enables them to persist in ecological niches which arise seasonally due to changes in predation and food regimes. Hybrids seem to be of recent and multiple origin and seem to form backcrosses occasionally, but no mitochondrial and only low levels of nuclear introgression have been detected. Interspecific matings are non-random, which leads to directionality of hybridization and introgression. Environmental settings seem to facilitate hybridization (*temporal hybrid superiority model*), whereas evolutionary consequences may arise from repeated backcrossing, which in some cases results in patterns of reticulate evolution.

The shape of the land forms a number of places
   which allow the survival of different races.
When enclaves advance with the ice in retreat
   some form hybrid zones where the two ranges meet.
Such regions are common and not very wide
   so the mixing of genes affects neither side.
They divide up the range in a patchwork of pieces
   which echoes and glimpses on the nature of species.
A brief rendez-vous and the ice comes again.   Godfrey M. Hewitt, 1993

Introduction

A fundamental aim of evolutionary biology is to provide proximate and ultimate explanations for the variation and diversity of organisms in space and time. Because of the complex interactions of different processes like genetic drift, mutation and selection, it is difficult to estimate the significance of one specific evolutionary factor. Compared with these evolutionary processes, interspecific hybridization provides a prominent window to the genetic changes in populations (Hewitt, 1988; Harrison, 1990). Hybrids possess a combination of genes from two different parental gene pools. In contrast to the relatively slow rate of change in gene frequencies due to selection or genetic drift, a hybridization event can be a *dramatic* process. F_1 hybrids differ markedly from their parental species, since, in a single generation, two originally separated gene pools may create significantly more combinations of genes than recombination within a given species should provide. Comparing parental and hybrid taxa offers the possibility to test evolutionary theories of selection, adaptation and gene flow (Harrison, 1990, 1993).

The establishment of hybrids among sympatric species depends on both the possibilities for interspecific cross-fertilization (e.g., spatial/temporal synchronization, field frequencies, mate choice) and its success in a specific habitat (i.e., relative fitness of hybrids). The outcome of hybridization has frequently been considered as a race between fusion and speciation, depending on the fitness of hybrid and the initial level of positive assortative mating (Harrison, 1990, 1993). Studies of natural interspecific hybridization have addressed mainly five issues: 1) Origin of hybridization (sympatric or secondary contact of formerly allopatric or parapatric taxa), 2) hybrid zone dynamics (maintenance of hybridization, stability of hybrid zones), 3) genetic and evolutionary consequences of hybridization (introgression, reticulate evolution, phylogeny), 4) speciation and genetic/reproductive isolation (species concepts, assortative mating, mating barriers) and 5) fitness comparisons (habitat associations, ecological isolation).

In the following, we first provide a framework of theories on interspecific hybridization and then call attention to recent work on the *Daphnia galeata* multi-species complex. The genus *Daphnia* consists of two main groups, the *D. pulex* and the *D. longispina* group; the latter can be further divided into the *D. galeata mendotae* (North America) and *D. galeata* (Europe) multi-species complexes. The *D. galeata* complex comprises five commonly studied species (*D. galeata, D. cucullata, D. hyalina, D. rosea* and *D. longispina*) and five interspecific hybrids. Three of these species frequently co-occur with their interspecific hybrids in large permanent lakes (Fig. 1). Using genetic and ecological data, we address the following issues in the *D. galeata* species complex: 1) the population structure of species complexes, 2) biogeographic patterns, 3) fitness comparisons, and 4) evolutionary consequences of interspecific hybridization. In general, we aim to compare and contrast interspecific hybridization among *Daphnia* species with other species complexes among animals, and discuss these results in the context of models on hybrid maintenance.

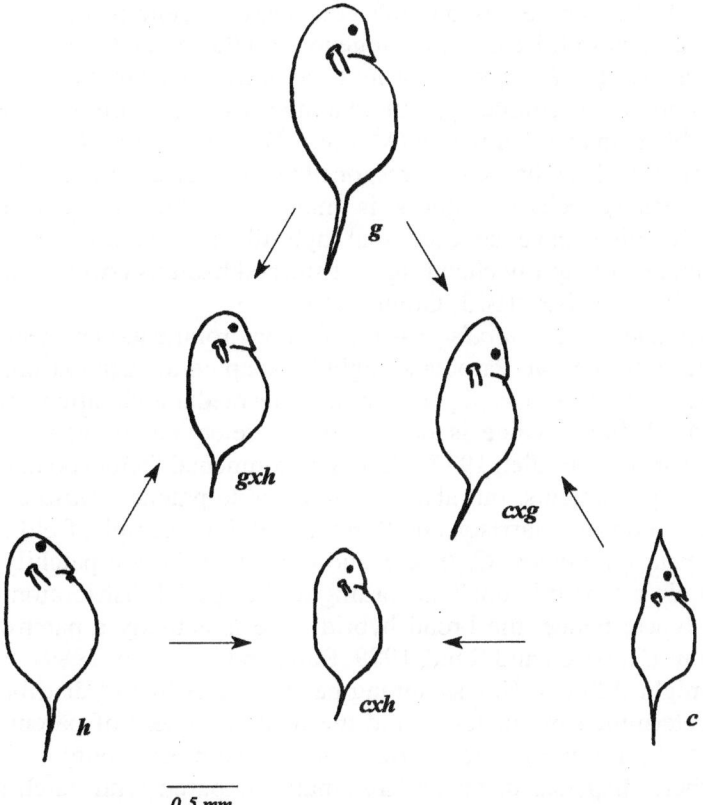

0.5 mm

Figure 1. Typical morphs of the European *D. galeata* species complex, redrawn from Flößner and Kraus (1986) and Flößner (1993). Species are: **g**: *D. galeata*, **c**: *D. cucullata*, **h**: *D. hyalina*, and three interspecific hybrids (**gxh**: *D. galeata x hyalina*, **cxh**: *D. cucullata x hyalina*, **cxg**: *D. cucullata x galeata*).

Conceptual framework of theories on interspecific hybridization

The majority of theoretical work on hybrid zones has concentrated on hybrid zone origin (secondary contact vs. primary intergradation), and on the maintenance of hybridization and speciation (Moore, 1977; Barton and Hewitt, 1985, 1989; Bullini, 1985; Hewitt, 1988; Harrison, 1990, 1993; Abbott, 1992; Arnold, 1992). According to their basic assumptions, models of hybrid zone maintenance can be grouped into two classes. One group of models is based on the assumption that the relative fitness of genotypes is largely determined by ecological factors, hence selection regimes vary along spatial gradients (e.g., *bounded hybrid superiority model, mosaic models, gradient and ecotone models*; Endler, 1977; Moore, 1977). Alternatively, *tension zone models* assume no ecological components responsible for selection processes, but only consider endogenous factors, such as genetic incompatibilities of parental genomes (Barton and Hewitt, 1985).

The *bounded hybrid superiority model* assumes that parentals are adapted to different environments, but that hybrids are more successful in certain habitats than either parent (Moore, 1977; Moore and Koenig, 1986). The location of these habitats is responsible for species and hybrid distributions, because relative fitness is mainly determined by ecological factors. Botanists have stressed that hybridization is often associated with unstable and rapidly changing or disturbed habitats (Anderson, 1948; Stebbins, 1950; Heiser, 1973; Grant, 1981).

Gradient and *ecotone models* assume that environmental gradients cause differences in selection. Parentals might be adapted to each extreme habitat, whereas the hybrid is superior in an intermediate situation (habitat). Thus, hybrid maintenance is dependent on selection gradients and the extent of dispersal (Endler, 1977). Since environmental factors do not always change along gradients, but also among discrete patches, *mosaic models* have been proposed (Harrison and Rand, 1989). For example, field crickets (*Gryllus pennsylvanicus, G. firmus* and their hybrids) are patchily distributed within a "hybrid zone", according to the spatial distribution of certain soil types; hence, the broad hybrid zone is actually a patchwork of populations (Harrison and Rand, 1989; Rand and Harrison, 1989). Parental species might differ in fitness among patches, thus hybrid distribution is largely determined by dispersal and the relative fitness of parentals and hybrids. The structure of the hybrid zone will be mosaic only in environments where dispersal distances are small compared with patch size. If dispersal distances are large compared with patch size, then a gradient pattern (not a mosaic) would be expected (Harrison and Rand, 1989). Several cases of hybrid taxa that are ecologically differentiated have been found, e.g., species distributions according to abiotic factors such as soil type, altitude or moisture (Bert and Harrison, 1988; Rand and Harrison, 1989) and biotic factors, such as host plant differences (Sperling, 1987).

In contrast to the models mentioned above, *tension zone models* assume that environmental factors are not involved in the maintenance of hybrid zones (Barton and Hewitt, 1985). Theoretical and empirical studies have shown that a dynamic equilibrium, resulting from a balance between reduced hybrid fitness and dispersal, can account for hybrid zone maintenance (Barton and Hewitt, 1985). Selection operates through reduced fitness of hybrids, e.g., due to (partial) genetic incompatibility of parental genomes. Several examples have been reported where hybrids are sterile (Hewitt et al., 1987), have reduced viability (Barton, 1980), or have relatively higher embryonic mortality (Szymura and Barton, 1986). One of the strongest arguments supporting tension zone models is the observation of concordant clines among independent characters (such as morphology, nuclear and mitochondrial DNA markers). Such patterns are difficult to explain solely by ecology-mediated selection, since this would imply that all characters exhibit an identical response to environmental factors.

In addition to numerous examples of narrow hybrid or tension zones, some examples are known where parentals and hybrids occur in sympatry across a broad geographic scale (Hovanitz, 1943; Heiser, 1947). Although hybridization can occur on a continental scale in these cases, parental taxa are able to maintain their distinctness, probably through ecological differentiation. Whether hybrid zone maintenance and geographical patterning are adequately explained by ecological factors or by an equilibrium between selection against hybrids and dispersal is a matter of ongoing research and debate (e.g., Bert and Arnold, 1995). Since tension zones are not linked to ecological factors, they presumably are able to move (Kraus and Miyamoto, 1990). Hewitt and Barton (1980) pointed out that the movement of tension zones and their tendency to become "frozen" along natural barriers complicates the discrimination between the alternative hypotheses, because the association of ecotones with species and hybrid distributions could be coincidental. Although it is not our intention to evaluate different theories of hybrid zone maintenance, we would like to emphasize their significance in studying the relationship between ecological parameters and selection in hybrid zones, e.g., through fitness comparisons.

Besides studies on the maintenance of hybridization, several, mainly empirical, inquiries have focused on evolutionary and genetic consequences of interspecific hybridization, such as the directionality of hybridization, phylogenetic relationships and introgression. In line with Rieseberg and Wendel (1993), we define introgression as gene exchange between species, subspecies, races, or any other set of differentiated population systems. In general, the process of introgression comprises cytoplasmic introgression (mitochondrial or chloroplast DNA) and nuclear DNA introgression. These processes are presumed to be important evolutionary factors, since they might lead to an increase in genetic diversity, or to the origin or transfer of adaptations (for review see Rieseberg and Wendel, 1993).

Furthermore, introgression has substantial consequences for phylogenetic relationships (Arnold, 1992), for relative fitness values (Lewontin and Birch, 1966; Levin and Bulinska-Radomska, 1988; Arnold and Hodges, 1995) and for adaptive properties of introgressants, in particular in disturbed or changing environments where new ecological niches become available (Stebbins and Daly, 1961). Introgression has been studied extensively in plant species (for reviews see Heiser, 1973; Rieseberg and Wendel, 1993) and during the last decade also has gained prominence in animal studies (e.g., Ferris et al., 1983; Avise et al., 1984; Szymura and Barton, 1986; Harrison et al., 1987; Aubert and Solignac, 1990; Forbes and Allendorf, 1991; Sperling and Spencer, 1991; Scribner, 1993). If one disregards the mechanism of introgression as such, and focuses on the evolutionary consequences, in particular reticulate patterning of phylogenies, then several additional questions can be raised (Arnold, 1992). For example, do ecological, genetic or historical factors determine the spatial distribution of introgression? Is introgression asymmetric or bi-directional? How long has introgression proceeded: a few or many generations?

Interspecific hybridization in *Daphnia*

Despite extensive empirical as well as theoretical studies of hybrid zones among animals, in particular in terrestrial habitats (e.g., Barton and Hewitt, 1985; Hewitt, 1989; Harrison, 1990, 1993; Grant and Grant, 1992), very little information has been derived from freshwater habitats, especially lakes. Hybridization among aquatic organisms has only recently been studied (Avise et al., 1984; Bert and Harrison, 1988; Forbes and Allendorf, 1991; DeMarais et al., 1992; Konkle and Philipp, 1992; Dowling and DeMarais, 1993; Scribner, 1993; Scribner and Avise, 1993). Terrestrial habitats frequently show a gradient in ecological parameters like temperature, humidity and vegetation structure. Hybrids tend to occur in intermediate environments and can exist either in sympatry or parapatry with the parental species. In contrast, freshwater habitats, particularly lakes and ponds, are characterized by a patchy to linear spatial distribution, by homogeneity in abiotic and biotic factors, and by distinct "island-like" isolation from each other.

Among aquatic organisms, members of the microcrustaceans (Branchiopoda: Anomopoda, Ctenopoda, Haplopoda and Onychopoda) exhibit several unique features related to the study of interspecific hybridization. Species in this group reproduce both sexually and parthenogenetically (via obligate or cyclic ameiotic parthenogenesis), therefore F_1 hybrids can potentially circumvent possible deleterious effects of reduced sexual fertility by propagating parthenogenetically. Zooplankton species are confronted with low dispersal rates but relatively stable environments. Species which produce resting eggs can be dispersed via other organisms or wind. However, they face two basic problems: first, dispersal is passive and thus dependent on the availability of potential vectors, and second, the likelihood of reaching a suitable new habitat is low. Parthenogenetic reproduction and relatively short generation times (e.g., the genus *Daphnia*) facilitate the detailed study of life-history variation and quantitative genetics (e.g., Ebert et al., 1993; Spitze, 1993). These characteristics allow fitness comparisons and the possibility of testing assumptions of hybrid maintenance with regard to selection and ecological differentiation (Weider, 1993; Boersma, 1994; Spaak and Hoekstra, 1995). Since ecological and genetic aspects of interspecific hybridization among cladocerans have mainly been studied in the genus *Daphnia*, we focus in the following on species complexes in the genus *Daphnia* and particular on the *D. galeata* complex (Tab. 1).

The *D. carinata* complex in Australia, the *D. obtusa, D. pulex* and *D. galeata mendotae* complexes in North America, and the *D. galeata*

Table 1. Publications concerning interspecific hybridization among *Daphnia* species complexes

Multi-species complex	Systematics and taxonomy	Population genetics	Ecology
D. carinata (Australia)	Hebert and Wilson (1994)	Hebert (1985) Hebert and Wilson (1994)	
D. obtusa (North America)	Hebert (1995) Hebert and Finston (1996) Colbourne and Hebert (1996)		
D. pulex (North America)	Hebert (1995) Colbourne and Hebert (1996)	Hebert et al. (1989a) Hebert et al. (1993) Van Raay and Crease (1995)	
D. galeata mendotae (North America)	Taylor and Hebert (1994) Hebert (1995) Taylor et al. (1996)	Taylor and Hebert (1992, 1993 a–c)	
D. galeata (Europe)	Flößner and Kraus (1986) Flößner (1993) Schwenk (1993) Taylor et al. (1996)	Wolf and Mort (1986) Gießler (1987) Wolf (1987) Hebert et al. (1989b) Weider and Stich (1992) Müller (1993) Schwenk (1993) Spaak and Hoekstra (1993) Müller and Seitz (1995) Spaak (1996) Ender et al. (1996)	Mort (1990) Weider and Wolf (1991) Wolf and Weider (1991) Weider (1993) Müller and Seitz (1993, 1994) Spaak (1994) Boersma (1994, 1995) Boersma and Vijverberg (1994a, b) Spaak (1995a, b) Spaak and Hoekstra (1995, 1997)

complex in Europe together comprise more than 20 hybridizing species within the genus *Daphnia* (see Tab. 1 for references). Several hypotheses have been proposed to explain why some *Daphnia* species complexes have such high levels of hybridization (with regard to frequency of hybridization and numbers of species), whereas other daphnids and other zooplankton species in general do not hybridize, or only at very low frequen-

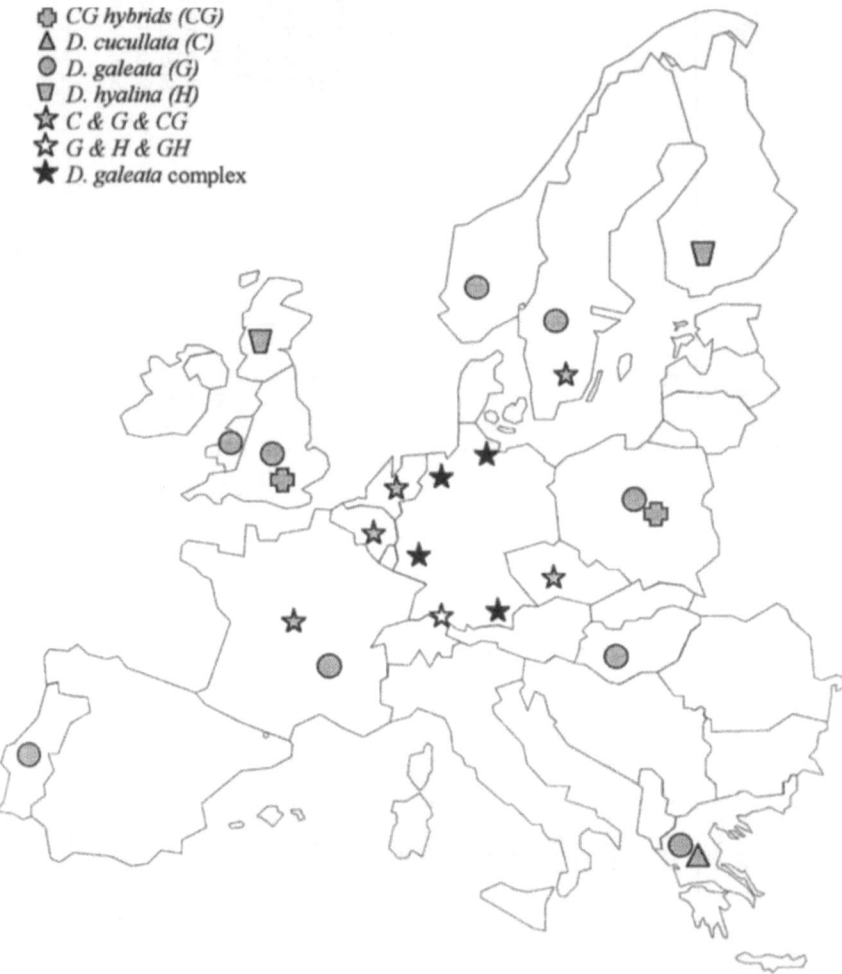

Figure 2. The distribution of three *Daphnia* species (*D. galeata, D. hyalina, D. cucullata*) and their interspecific hybrids (*D. galeata x hyalina, D. cucullata x galeata* and *D. cucullata x hyalina*) in Europe. Stars represent locations where either all three species and three hybrids co-occur (black stars), or two species with their interspecific hybrids (gray and white stars) are found. Other symbols represent single species or hybrids found within a lake district. Species and hybrid determination are either based on population genetic studies (e.g., Wolf and Mort, 1986; Hebert et al., 1989b; Gießler, 1987; Müller and Seitz, 1995) or nuclear and mitochondrial DNA markers (K. Schwenk, unpublished data).

cies. For example, littoral zone daphnids show limited hybridization compared with pelagic species. Hann (1987) assumed that short-term survival of hybrids and behavioural specialization (e.g., different habitat preferences) is responsible for this pattern. If hybrids reproduce only parthenogenetically and have to cope with rapid changes in environmental conditions (including droughts), which require the production of resting eggs, then the success of clonal hybrid lineages will be minimized. This is of course only true for species which require sexual reproduction to produce resting eggs. In large temperate lakes, zooplankton species do not face extreme environmental perturbations, such as droughts and strong temperature fluctuations. Thus hybrids, once produced, may persist as clonal lineages via parthenogenetic reproduction. Although it seems that interspecific hybridization among large-lake *Daphnia* species is more common (frequency of hybridization) than among temperate pond species, information about interspecific hybridization is still insufficient to draw general conclusions. However, environmental conditions might influence the likelihood of hybridization or the persistence of hybrids, but hybridization *per se* is facilitated by genetic and historic factors (level of reproductive isolation, mode of speciation).

Hebert (1985) suggested that the incidence of interspecific hybridization and temporarily high abundances of hybrids among *Daphnia* and rotifers might be due to their ability to reproduce parthenogenetically. Parthenogenetic reproduction certainly facilitates the occasionally higher abundances of hybrids, compared with parental species. But the occurrence of interspecific hybridization among *Daphnia* and rotifer species is most likely based on semi-permeable reproductive barriers which allow interspecific crosses (Schwenk, 1993; Taylor and Hebert, 1993a). However, the phenomenon of interspecific hybridization among cyclic parthenogenetic animals has to be examined first for the occurrence of hybridization *per se*, and second, for the ecological success and persistence of hybrid clones (e.g., Arnold and Hodges, 1995). Hence, in addition to the "classical" evolutionary questions with regard to interspecific hybridization mentioned above, other factors such as the consequences of parthenogenetic reproduction and the "evolutionary age" of hybrid clones have to be considered as well.

The *Daphnia galeata* species complex

Biogeographic patterns

Daphnia hybrids (*D. galeata x hyalina*, *D. cucullata x galeata* and *D. cucullata x hyalina*) have been found in several geographic regions across Europe (Fig. 2). Interspecific hybridization has been discovered using species-specific morphological characters and allozyme markers (fixed

Table 2. Diagnostic allozyme loci used to discriminate between *Daphnia* species and hybrids in the European *D. galeata* complex

Taxon	Loci and alleles						
	sAAT	AO	PGI	PGM	MDH	PEP	GPDH
D. galeata	f	f	m, f	s, m, f, f$^+$	s, f	f, m	f
D. hyalina	s	s	m	m, f, f$^+$, f^{++}	f, f$^+$	s	s
D. cucullata	s$^-$	f	s$^-$, s, m, f	s, m	s, f, f$^+$	−	−

Allele designations reflect the relative anodal migration distances as follows: f^{++}: super fast, f$^+$: very fast, f: fast, m: medium, s: slow, s$^-$: very slow, −: not determined (Wolf and Mort, 1986; Weider, 1993). Loci are either fixed or differ significantly in allele frequencies between species, sAAT: aspartate aminotransferase, EC 2.6.1.1, AO: aldehyde oxidase, EC 1.2.3.1., PGI: phosphoglucose isomerase, EC 5.3.1.9, PGM: phosphoglucomutase, EC 2.7.5.1, MDH: malate dehydrogenase, EC 1.1.1.40, PEP: peptidase, EC 3.4.11, GPDH: glycerol-3-phosphate dehydrogenase, EC 1.1.1.8

alleles). F$_1$ hybrids are characterized through intermediate morphology (compared with parentals) and heterozygous genotypes at marker loci (Tab. 2). In the following, the term "hybrid" is used in the sense of mixed ancestry, therefore the group of hybrids may comprise F$_1$ hybrids, F$_2$ hybrids and backcross genotypes. *Daphnia* hybrids have been detected in lake areas in southern (Bavaria), northern (Plön), and western Germany (Eifel), as well as The Netherlands, Sweden, France and Poland (Wolf and Mort, 1986; Gießler, 1987; Müller and Seitz, 1994; Spaak and Hoekstra, 1995; K. Schwenk, unpublished data). In addition, *D. cucullata x galeata* and *D. galeata x longispina* hybrids are present in several lakes in Bohemia, Czech Republic (Hebert et al., 1989b). Since only a few population genetic studies have been performed, and sampling locations appear to be associated with limnological institutes rather than covering representative areas, any generalization of biogeographic patterns seems premature. In addition samples were taken by different methods and during different times of the year, thus the true diversity was most likely underestimated. However, across central Europe within the *D. galeata* complex, the biogeographic pattern indicates that species and hybrids occur in a rather patchy distribution with large differences in taxa composition among and within lake districts (Figs 2 and 3). Thus, either *Daphnia* species form broad hybrid zones, or species and hybrid distributions primarily reflect the patchy structure of the environment. In North America, a similar pattern was found for hybrids of *D. galeata mendotae* and *D. rosea*. Both species also occur across a relatively large geographic area, ranging from lakes in northern Indiana, USA, to the province of Ontario, Canada and into New England (Taylor and Hebert, 1992, 1993b). The geographically widespread hybridization among these species complexes is a rare phenomenon when compared with other animal species, which typically form rather narrow hybrid zones (Barton and Hewitt, 1985).

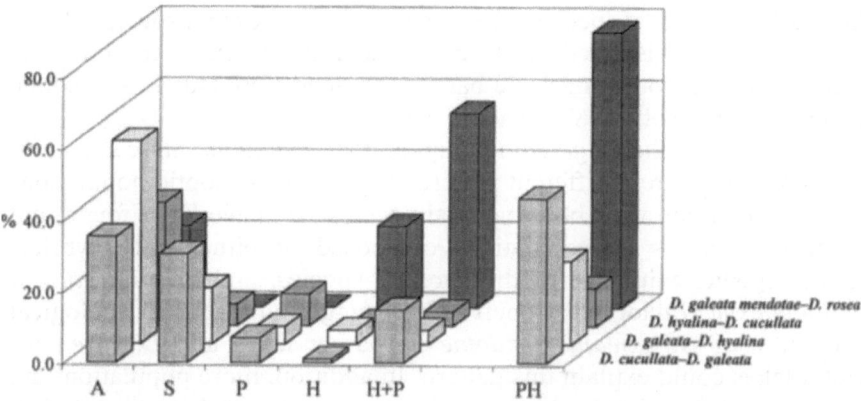

Figure 3. Species and hybrid compositions of 124 populations of the *D. galeata* and the *D. galeata mendotae* species complexes divided into pairs of hybridizing species and five classes of populations. A: only one parental species per lake, S: hybrids occur together with both parental species (syntopy), P: both parental species without hybrids, H: only hybrids, H+P: hybrids together with only one of the parental species. Bars on the far right show the percentage of populations in which hybrids were detected (PH).

Population structure

To examine the composition of populations we divided the *D. galeata* complex into its three species pairs plus their hybrids (associations) and classified each lake into one of five categories: *A*) only one of the parental species present, *S*) all three taxa occur together (syntopy), *P*) both parental species without hybrids, *H*) only hybrids, or *H+P*) hybrids together with only one of the parental species (Fig. 3). This analysis is based on data from 102 lakes across Europe and 22 lakes in North America (Wolf and Mort, 1986; Gießler, 1987; Hebert et al., 1989b; Taylor and Hebert, 1992; Müller, 1993; Schwenk, unpublished data).

Around 30% of the *D. cucullata–D. galeata* associations were composed of both parental species and hybrids (synoptic), whereas the *D. galeata–D. hyalina* and the *D. hyalina–D. cucullata* associations exhibited lower values. Since not all lakes were sampled more than once per year, which may cause a bias in the estimation of the number of taxa present, we expect the proportion of synoptic populations to be even higher. In all three associations however, the majority of populations is composed of only one parental species, *D. galeata* within the *D. cucullata–D. galeata* and *D. galeata–D. hyalina* associations, and *D. cucullata* within the *D. hyalina* –*D. cucullata* association. Whether these patterns are caused by ecological or historical factors remains an open question. However, evidence for size-selective predation and the influence of food quality and quantity on species composition has been adduced (Weider, 1993; Spaak and Hoekstra, 1995). Interestingly, no hybrids have been detected in some populations, although both parental species are present. This implies that species/clones are reproductively isolated either by asynchronous sexual reproduction of parentals or by other intrinsic barriers to gene flow (such as genomic incompatibility; Wolf, 1987; Spaak, 1996).

Compared with the *D. galeata* complex, the *D. galeata mendotae* complex shows an entirely different pattern (Fig. 3). No synoptic populations (*S*), or populations with both parental species (*P*) have been found, but similar proportions of populations composed of either only hybrids, hybrids together with one of the parental species, and only *D. galeata mendotae* exist. Taylor and Hebert (1992) suggested that clear ecological preferences of *D. galeata mendotae* for larger lakes and *D. rosea* for smaller lakes could explain this pattern. In addition, more populations are composed solely of hybrids than of only one parental species. Taken together, all these patterns suggest that hybrids cope with both parental "habitat types", and in addition successfully exploit an intermediate ecological niche. Although hardly any synoptic populations of both parentals and hybrids occur, the proportion of hybrid populations (> 70%, Fig. 3) and the local abundance of hybrids (> 40%) is very high. These proportions are smaller for populations of the *D. galeata* complex (10–45% and 12–26%, respectively), although synoptic populations are frequently found (30%). This pattern might be explained by assuming that species of the *D. galeata mendotae* complex are spatially separated through their ecological differentiation, whereas species of the *D. galeata* complex are separated primarily by various levels of reproductive isolation, e.g., through seasonally asynchronous sexual reproduction or intrinsic barriers to gene flow.

Life-history evolution and fitness comparisons

One of the most effective ways to reveal the potential evolutionary significance of hybridization and introgression is to measure the relative fitness of different hybrid and parental genotypes (Arnold, 1992; Arnold and Hodges, 1995). This might be especially illuminating for cyclic parthenogens like *Daphnia*, which can circumvent the disadvantage of reduced sexual fertility. In principle, hybrid genotypes that bear combinations of advantageous traits can be maintained indefinitely in a population through parthenogenetic reproduction.

Compared with genetic studies, fitness differences of *Daphnia* hybrids and their parental species have been less investigated. The available studies are restricted to the *D. galeata* complex in which life-history characteristics of hybrids and parental species were compared with respect to food level and temperature (e.g., Weider, 1993; Spaak and Hoekstra, 1995). The comparison of intrinsic rates of increase (*r*) of parental and hybrid genotypes can be used as a measurement of fitness. Although *r* represents only one component of fitness and a definition of fitness among cyclic parthenogens is not straightforward, the estimation of *r* together with other life-history characteristics reveals the capability of clones to respond to environmental fluctuations (ecological success, environmental sensitivity). In addition, size-related traits must be taken into account, since both invertebrate and vertebrate predation on *Daphnia* is size-selective (Zaret, 1975).

Hybrids and parentals were found to be significantly different in most life-history characters. The intrinsic rate of increase (*r*) of *D. galeata x hyalina* was significantly higher compared with the parental species at low temperature (14°C), while at high temperature (20°C) the parental species performed better (Wolf and Weider, 1991). In their extended study, Weider and Wolf (1991) showed that *D. cucullata x galeata* hybrids have an *r* that differs significantly from *D. cucullata* but does not differ significantly from *D. galeata*. The size of *Daphnia* hybrids at birth and maturity is in most cases intermediate to the parental species, however, only one or two clones per taxon were used.

In a study on the niche breadth of *D. galeata x hyalina* and its parental species, using five clones of each taxon raised at four food levels ranging from 0.2 to 2 mg C mL^{-1}, Weider (1993) found significant taxon *x* food level interactions for *r* as well as for size-related traits. At extreme food levels (0.2 and 2 mg C mL^{-1}), the *r* of hybrids did not differ significantly from *D. galeata* but both had a significantly higher *r* when compared with *D. hyalina*. Differences in niche breadth could not be shown between the hybrids and parentals; only *D. galeata* and *D. hyalina* differed significantly in this respect. A similar pattern has been discovered for *D. cucullata, D. galeata* and their hybrids; both species compete with the hybrids (food), but not among each other, which suggests limited niche overlap of species (Boersma, 1995).

It might be that fluctuating environmental conditions, e.g., for temperature, food levels and predation regimes, create conditions under which hybrids temporarily have a higher fitness than the parental species. This hypothesis is supported by the temporal distribution of syntopically occurring hybrids and parentals; in certain periods of the year hybrids numerically dominate the species complexes (Wolf, 1987; Spaak and Hoekstra, 1993; Spaak 1996).

The hypothesis that *Daphnia* hybrids are maintained through temporarily higher fitness (*temporal hybrid superiority model*) was tested by Spaak and Hoekstra (1995) with *D. galeata, D. cucullata* and hybrid clones from Lake Tjeukemeer, The Netherlands. In the laboratory, mean *r* of the hybrids did not differ significantly from *D. galeata*, but was significantly higher than in *D. cucullata* (Fig. 4(A)). Furthermore, life-history comparisons of hybrids and parentals demonstrated that *r* values for hybrids are higher at high food conditions than those of parentals (Boersma and Vijverberg, 1994a). Because the hybrids are significantly smaller (offspring size and size at maturity) than the larger parental *D. galeata* (Fig. 4(A)) and predation by fish is positively size-selective, they will have an advantage in the field. Fish predation on *Daphnia* is very high in this lake (Lammens et al., 1985;

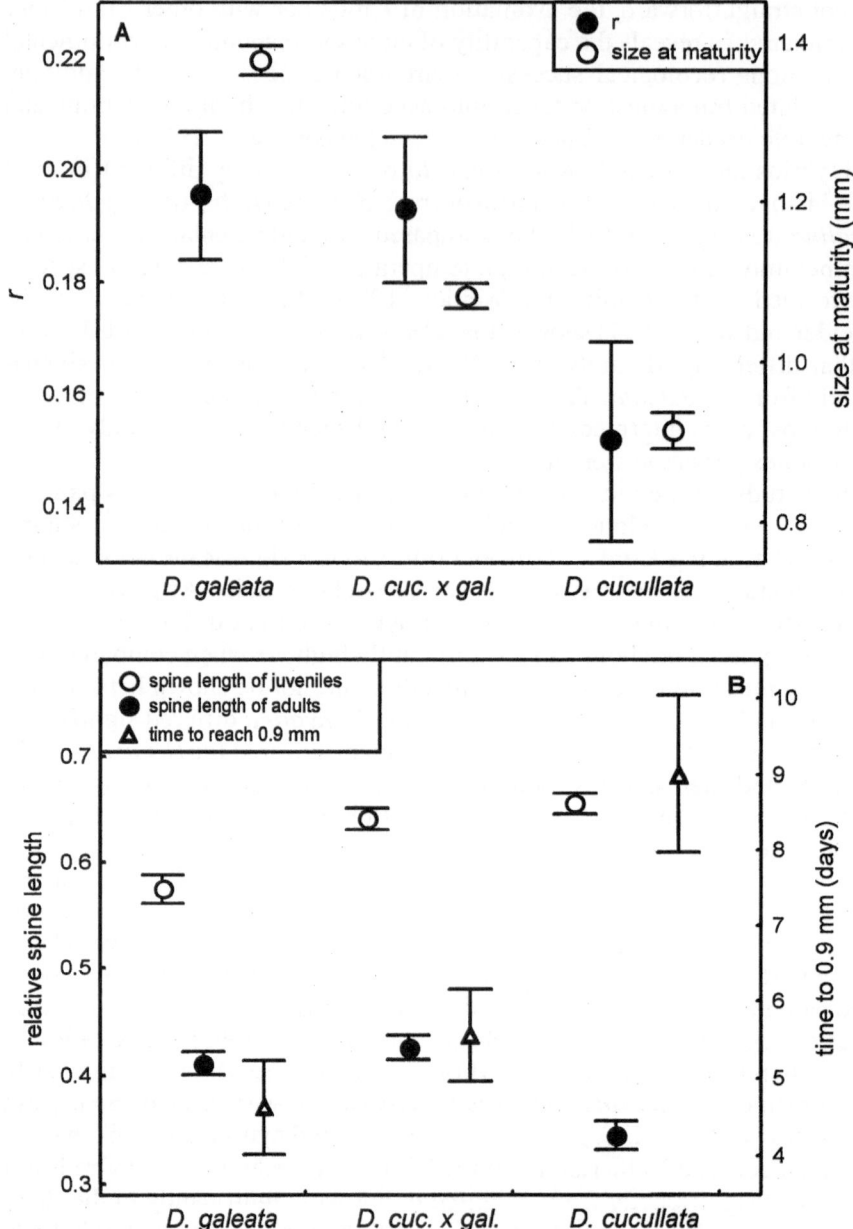

Figure 4. Interaction plots of means and 95% confidence intervals for four life-history characters of *D. galeata, D. cucullata* and *D. cucullata x galeata*. (A) The intrinsic rate of increase (*r*) and size at maturity are shown for the three taxa (Spaak and Hoekstra, 1995). (B) The means and confidence intervals of the relative tail spine lengths are plotted for the first juvenile instars (white circle) and the fourth adult instars (black circle). Triangles represent the average number of days required for each taxon to reach 0.9 mm (Spaak, 1995a).

Vijverberg et al., 1990). Boersma et al. (1991) showed that smelt is a size-selective predator on *Daphnia*: in June, smelt selected for smaller *Daphnia*, and in the rest of the year for larger daphnids, especially individuals with a body length >1 mm. In addition, during certain periods of the year size-dependent death rates reach 10% d^{-1} for daphnids around 1 mm body length to >60% d^{-1} for daphnids around 1.6 mm body length (Vijverberg and Richter, 1982). These data show that the smaller size at maturity for *D. cucullata* and the hybrids can be selectively advantageous.

The observed differences in size and fecundity between the taxa may explain why *D. galeata* and the hybrids dominate the *Daphnia* assemblage in Lake Tjeukemeer early in the season (higher *r*, low predation risk, high food concentrations), and *D. cucullata* in the summer with high fish preda-tion (lower vulnerability to fish predation, lower food concentrations). For example, at the end of May 1989, both *D. galeata* and the hybrids reached peak densities (65 ind./L), high abundances lasted for about 1 week for *D. galeata*, but the hybrid population did not collapse until August (Spaak and Hoekstra, 1997). Thereafter, dominance of *D. cucullata* can thus be explained by size-related differences in vulnerability to visual predators (fishes). In a 3-year study of the population dynamics of the *D. cucullata*– *D. galeata* association in Lake Tjeukemeer, Spaak and Hoekstra (1997) provided evidence in support of the *temporal hybrid superiority* model. Deviations from the mean instantaneous rate of increase on a certain date (r_r) showed a positive relationship with fish predation for *D. cucullata*, a negative one for *D. galeata*, and no relationship for the hybrids. This sug-gests a higher influence of fish predation on the *r* of the larger *D. galeata* compared with the smaller hybrids and *D. cucullata*.

Niche segregation and maintenance of hybrid lineages

A large number of factors may determine whether a trait can be advan-tageous for hybrid taxa. In lakes with strong positive size-selective fish predation, small daphnids have an advantage compared to larger ones (Boersma et al., 1991, 1996). But the same trait can be a disadvantage under resource (food) limitation, since larger daphnids have been shown to withstand food shortage better compared with smaller ones (Gliwicz, 1990). Therefore, studies on the fitness consequences of hybridization of *Daphnia* can only indicate trends for the hybrids as a "taxon." It depends on local circumstances if a certain combination of traits can be considered as advantageous or disadvantageous for *Daphnia* hybrids. In addition to laboratory studies on fitness, field data related to condition, mortality, predation and competition have significantly improved our understand-ing of niche segregation and hybrid maintenance (e.g., Boersma and Vijverberg, 1994b; Boersma, 1995; Boersma et al., 1996; Spaak and Hoekstra, 1997).

Table 3. Ecological characteristics of *D. galeata*, *D. cucullata* and *D. cucullata x galeata* .

Taxon	Predation risk fish	Predation risk invertebrate	r	Interspecific competition
D. galeata	high	low	high	low
D. cucullata	low	high	low	low
D. cucullata x galeata	intermediate	intermediate	high	high

r represents the intrinsic rate of increase.

With respect to invertebrate predation, Spaak (1995a) found that the *D. cucullata x galeata* hybrids have features which are similar to those of *D. galeata*, resulting in a reduction of predation risk by invertebrates. Invertebrate predators like *Leptodora kindtii* prefer daphnids smaller then 0.9 mm (Herzig and Auer, 1990). Thus, for a daphnid it appears to be important to reach the size of 0.9 mm as soon as possible. Although *D. cucullata x galeata* hybrids have a smaller size at maturity compared with *D. galeata*, they do not significantly differ from *D. galeata* in the time they need to attain 0.9 mm (Fig. 4(B)). A second important trait related to invertebrate predation are structures which increase the handling time by predators or enhance escape and thus reduce the predation risk (Pijanowska, 1990). Hybrids do not differ significantly from the favorable traits of *D. cucullata* during the first instars in relative spine length, when the animals are still small and experience high predation risk from invertebrates (Fig. 4(B)). Similar to the example of vertebrate predation, *Daphnia* hybrids seem to represent a blend of advantageous features, which most likely originate from selection on predation-related traits (Tab. 3; Fig. 4(A), (B)).

Studies on the temporal variation in hybrid and parental species occurrence have shown that the relative abundances of these taxa differ significantly over time (Wolf, 1987; Weider and Stich, 1992; Spaak, 1996). In Lake Constance, *D. galeata* was more abundant from May to September, compared with *D. hyalina*, while the proportion of hybrids varied between 20 and 40% during the whole year (Weider and Stich, 1992). A 3-year study in Lake Tjeukemeer showed that large among-year differences in species composition can occur (Spaak, 1996). For example, in the summer of 1990 the daphnid population was composed of 30% *D. cucullata x galeata* hybrids, but in 1991 they decreased to less than 10%. These results, together with data from experimental studies, suggest that the relative abundance of taxa is determined by the interaction of life-history features and environmental factors, e.g., food availability, temperature and predation.

Besides variation in life histories, differences in the vertical distribution of *Daphnia* hybrids and parental species can lead to niche segregation. Weider and Stich (1992) found that *D. galeata* was the dominant taxon in the upper 20 m of Lake Constance during summer and early autumn, and *D. hyalina* and the hybrids were most abundant below 30 m. A comparable pattern was found for *D. galeata* and the hybrid *D. cucullata x galeata* in

Lake Neuhofener Altrhein (Germany) by Müller and Seitz (1993). Day depth of *D. galeata* was confined to the border between the meta- and hypolimnion, whereas the smaller hybrids showed a more heterogeneous distribution. This result might reflect the lower predation risk for the smaller hybrids compared with the larger *D. galeata*.

Life-history characteristics of hybrids, such as *r*, interspecific competition and vulnerability to predation (fish and invertebrates) show significant differences from the parental species (Tab. 3). This pattern suggests niche segregation, and might explain the temporarily high abundance (ecological success) of hybrid clones. If hybrids reproduce preferentially via parthenogenesis, hybrid clones can be maintained because of their ecological differentiation from parental species. The mode of reproduction in combination with the variation of environmental factors seems to be the essential presupposition for the ecological success of *Daphnia* hybrids; *Daphnia* hybrids might be characterized by high clonal reproductive success but low sexual fertility.

Genetic and evolutionary consequence of interspecific hybridization

Consequences of natural hybridization may comprise: 1) merging of populations, 2) reinforcement of reproductive barriers, 3) introgression (Anderson and Hubricht, 1938), or 4) speciation (Bullini, 1985). The origin and dynamics of the hybridization process is dependent on level of reproductive isolation, ecological differentiation between hybridizing populations and factors such as genetic drift, migration, assortative mating and selection on hybrids (Arnold, 1992). Some of these issues have been investigated using the *D. galeata* and the *D. galeata mendotae* complexes (Gießler, 1987; Taylor and Hebert, 1992, 1993a, b; Schwenk, 1993; Spaak and Hoekstra, 1995). In the following, we discuss four aspects: 1) the origin of hybrid lineages, 2) mitochondrial and nuclear DNA introgression, 3) the directionality of hybridization and 4) phylogenetic relationships.

The origin of hybrid lineages
Considering cyclically parthenogenetic reproduction and hybridization, two important questions arise: 1) How did hybrids originate, and 2) Are they maintained through ameiotic parthenogenesis as "evolutionary old" asexual lineages? Population genetic investigations of hybrid populations of the *D. galeata* complex have revealed the highest level of clonal diversity (negative logarithm of Simpson's index of concentration) for *D. cucullata*, intermediate values for the hybrids and lowest values for *D. galeata* (Spaak, 1996). Müller and Seitz (1995) found a similar pattern, however, some hybrid populations had even higher values of genetic diversity then parentals. In contrast, hybrid populations of the *D. galeata mendotae* complex are characterized by relatively low values of genetic diversity compared

with parentals (Taylor and Hebert, 1992); this pattern might be explained by the distribution of parental species. As shown in Figure 3, species of the *D. galeata* complex frequently occur syntopically, whereas species of the *D. galeata mendotae* complex coexist infrequently. However, the possibility remains that reproductive isolation, strong selection (such as size-selective predation), or different levels of sexual reproduction among species may be responsible for the numbers of multilocus genotypes (= clones) detected.

Since *Daphnia* species are able to reproduce parthenogenetically, hybrids of the *D. galeata* complex could be maintained as clonal lineages. One approach to reveal whether hybrids constitute distinct clonal lineages is to compare mitochondrial DNA sequences of hybrid clones and clones of their maternal species. Within the *D. galeata* complex, nucleotide divergence of hybrids and maternal species is virtually identical to divergence among clones within species (Schwenk, 1993). In addition, frequencies of hybrid genotypes not only vary over seasons, but also between years (Spaak, 1996). On average, hybrid populations exhibit equal or higher levels of genetic variation than parentals and hybridization is found in several locations (see references in Tab. 1). Taken together, these findings suggest that hybrids are of multiple origin, i.e., that they do not form old independent parthenogenetic lineages, but are rather produced continuously. If hybridization occurs frequently and sexual reproduction between parental species and hybrids occurs occasionally (Hebert et al., 1989b; Taylor and Hebert, 1992), a further question is raised: does genetic material cross species boundaries via repeated backcrossing of recombinants, i.e., does introgression occur?

Mitochondrial and nuclear DNA introgression
Among *Daphnia* species the occurrence of introgression has been investigated using morphological characters (Lieder, 1983, 1987). Since this approach is based on morphological traits whose genetic bases are unknown, additional markers such as allozymes, nuclear and mitochondrial DNA have recently been applied (Ender 1993; Schwenk, 1993; Taylor and Hebert, 1993a; Spaak, 1996). *Daphnia* backcrosses have been detected occasionally at high frequencies (Gießler, 1987; Hebert et al., 1989b; Taylor and Hebert, 1992; Müller, 1993), indicating the possibility of gene flow between species. For example, the frequency of backcrosses per lake ranges form <9% in northern Germany (Wolf and Mort, 1986) to <11% in southern Germany (S. Gießler, personal communication) to <5–85% in Bohemia (Hebert et al., 1989b). Recent crossing experiments between *D. galeata* and *D. cucullata*, and between *D. cucullata x galeata* hybrids and *D. cucullata* revealed unambiguously the sexual reproductive capacity of hybrids (K. Schwenk and M. Bijlsma, unpublished data).

In neighbouring sympatric populations of *D. galeata mendotae* and *D. rosea*, alleles have been detected which are absent or are in allopatric reference populations. This pattern has been interpreted as introgression of nuclear alleles, mainly from *D. rosea* to *D. galeata mendotae*. A second example of introgression revealed gene flow from *D. cucullata* to *D. galeata* (Spaak, 1996). Some alleles of high frequency in *D. cucullata* were not present in allopatric populations of *D. galeata*, but present in low frequency in synoptic populations of *D. galeata*. In addition, an UPGMA analysis of more than 5000 individuals collected over a period of 3 years showed that *D. cucullata x galeata* hybrids are more similar to *D. galeata* (difference < 0.04; Nei's genetic distances), than to *D. cucullata* (difference 0.30; Spaak, 1996). A similar investigation which applied numerous nuclear DNA markers (*random amplified polymorphic DNA*; RAPD analysis) has revealed a similar pattern (Ender, 1993). Since it is difficult to discriminate between patterns of introgression and maintained polymorphisms from a common ancestor, or mutational events, comparisons of independent markers such as morphological characters, nuclear and mitochondrial DNA are needed to falsify the hypothesis of introgression (Harrison, 1990; Arnold, 1992).

Mitochondrial DNA introgression has been investigated using species-specific mitochondrial DNA markers within the *D. galeata* and the *D. galeata mendotae* complexes. In neither case could mitochondrial introgression be detected whereas nuclear introgression has been discovered (Schwenk, 1993; Taylor and Hebert, 1993a). Since it has been found that mitochondrial DNA crosses species boundaries more easily than nuclear DNA among animals (e.g., Ferris et al., 1983; Aubert and Solignac, 1990), it is possible that the sample size was too low to detect small amounts of mitochondrial gene flow.

Species within the *D. galeata* complex exhibit relatively high value of nucleotide divergence in the mitochondrial cyt *b* gene, although species are considered to be closely related, and interspecific hybridization and back-crossing occurs (Schwenk, 1993). This pattern might be a result of the differential contributions of hybrid and backcross females and males to back-crossing. For example, if F_1 hybrid males cross repeatedly with *D. galeata* females, we expect nuclear, but no mitochondrial introgression to occur.

The directionality of hybridization
Another aspect to interspecific crossing concerns the directionality of hybridization. Are hybrids always formed by females of one parental species (unidirectional hybridization), or are reciprocal crosses possible (bidirectional hybridization)? Several studies have demonstrated that directionality can be influenced by 1) abundances of parental species (Avise and Sanderson, 1984; Dowling et al., 1989), 2) asymmetric interspecific mate choice or reproductive isolation (Lamb and Avise, 1986; Harrison et al., 1987; Arntzen and Wallis, 1991; Bradley et al., 1991; Dowling and Hoeh, 1991; Patton and Smith, 1993), or 3) differential hybrid viability (Herke et al., 1990; McGowan and Davidson, 1992).

D. galeata mendotae x rosea hybrids were found to exhibit mitochondrial DNA from both parental species (Taylor and Hebert, 1993a). This pattern was found to be frequency dependent; in most cases only the dominant parental species passed on its mitochondrial DNA, which could be attributed to environmental induction of sexual reproduction in *Daphnia*. Under this hypothesis, the production of males and sexual females differs slightly in quality and quantity of necessary stimuli. Sexual females can be induced through changes in photoperiod, low food concentration and crowding, whereas males are induced by changes in photoperiod or a chemical cue which is produced during crowding of daphnids (Hobæk and Larsson, 1990; Larsson, 1991). Furthermore, this chemical cue has been sued successfully to induce males in different species, thus it does not seem to be species-specific (Hobæk and Larsson, 1990). Given these conditions, Taylor and Hebert (1993a) proposed the following scenario: Sexual *D. galeata mendotae* females are produced because of the high density of *D. galeata mendotae* individuals, whereas due to low density of *D. rosea* individuals only *D. rosea* males are formed (because of chemical induction by *D. galeata mendotae*), or *vice versa*. Thus, the dominant species in such a system would determine the directionality of hybridization.

A different pattern of directionality of hybridization has been reveled among species of the *D. galeata* complex. Based on limited mtDNA data, it appears that unidirectional hybridization is prevalent (Schwenk, 1993), although in some lakes bidirectional hybridization has been detected (K. Schwenk, unpublished data). Since in Europe species and hybrids of the *D. galeata* complex occur in syntopy, and also in similar proportions, the null hypotheses would be that hybrids exhibit mitochondrial DNA of both parentals in similar frequencies. But the data generated so far seem to reject this hypothesis because significantly more *D. cucullata x galeata* hybrids exhibit mitochondrial DNA of *D. cucullata*; similar results hold for *D. galeata x hyalina* hybrids (K. Schwenk, unpublished data). One possible explanation might be that sexual females and males of both parental species rarely co-occur at the same time during the year (Wolf, 1987; Spaak, 1995b). These observations suggest that either temporal differentiation of sexual reproduction or asymmetrical reproductive isolation contribute to nonrandom mating among parental species.

Phylogenetic relationships. Another consequence of interspecific hybridization and introgression is the reticulation of phylogenetic relationships, rather than hierarchical patterns or lineage diversification. Occasional horizontal gene flow between lineages might cause estimated phylogenies based on morphological, nuclear DNA, and cytoplasmic DNA data to be discordant (e.g., Solignac and Monnerot, 1986; Dowling and Brown, 1989; Kraus and Miyamoto, 1990; Lehman et al., 1991; Arnold 1992; McDade, 1992; Moore, 1995).

Due to phenotypic plasticity, cyclomorphosis and interspecific hybridization of species in the *D. longispina* group, morphological investigations used to reconstruct phylogenetic relationships have been highly controversial (Frey, 1982; Benzie, 1986; Flößner and Kraus, 1986; Hrbáček, 1987; Lieder, 1987; Flößner, 1993). Phenetic and cladistic analysis using 42 morphological traits among 43 *Daphnia* species even failed to establish the existence of a *D. longispina* group (Benzie, 1986). Alternatively, Hrbáček (1987) grouped *D. galeata, D. hyalina* and *D. cucullata* was well as *Daphnia longispina* and *Daphnia rosea* together as one clade. In addition, Wolf and Mort (1986) and Hebert et al. (1989b) suggested that the occurrence of interspecific hybridization as well as electrophoretic data indicate that these species are closely related. However, more genetic information is needed to evaluate species affinities, since morphological criteria alone are not sufficient to resolve the phylogenetic relationships among closely-related *Daphnia* species. Recent reappraisals of the phenotypic variation in *D. galeata mendotae, D. galeata galeata, D. hyalina* and *D. cucullata*, using additional genetic markers (allozyme and mtDNA sequence analysis), have revealed cryptic species complexes and frequent interspecific hybridization (Wolf and Mort, 1986; Taylor and Hebert, 1992, 1993b; Schwenk, 1993; Taylor et al., 1996). Thus, taxonomic uncertainties in the past were most likely caused by the production of intermediate forms and several "recombinations" of species-specific traits. However, within the *D. galeata* complex coherent phylogenetic relationships have been found using morphological traits, allozyme and nuclear DNA markers (Ender, 1993; Müller, 1993; S. Gießler, personal communication). *D. galeata* and *D. hyalina* seem to be more closely related to each other than either of them is to *D. cucullata*. However, an alternative branching order was suggested based on mitochondrial DNA information (Schwenk, 1993; Schwenk et al., 1995; Taylor et al., 1996). *D. galeata* and *D. cucullata* appear to be more closely related than either of them is to *D. hyalina*. If the tendency of discordance between nuclear and mitochondrial phylogenesis is supported by further studies (and lineage sorting can be excluded), we may infer incidences of reticulate evolution in this group.

Synopsis

The most distinct characteristics that separate cladoceran, particularly *Daphnia* species complexes, from most other animal species complexes investigated so far are: 1) the island-like nature of lake habitats, 2) the occurrence of hybrids and parentals in sympatry and even syntopy across large geographic areas, and 3) cyclical parthenogenesis which provides the possibility for hybrids to circumvent any deleterious effect of reduced sexual fertility. The first characteristic facilitates local hybridization; hybrids from different lakes vary in genotypic characteristics: they are probably not dispersed, but once they are produced they are able to persist locally via parthenogenetic reproduction (Taylor and Hebert, 1993b). In the *D. galeata mendotae* complex, only a few hybrid genotypes are found, which reflects either strong selection against certain genotypes or rare hybridization events. Since the parental species are rarely found to coexist syntopically, the latter explanation may hold true. Lakes seem to be markedly different in their ecological settings, and species have clear preferences for certain habitats, thus the island-like character of lakes and low dispersal rates are responsible for a small number of interspecific crosses. However, if intermediate habitats are available ("hybridization of the habitat"; Anderson, 1948), then opportunities for hybridization increase. In the *D. galeata* species complex the proportions of habitats in which species co-occur is significantly higher, and more hybrid genotypes are found. In addition, variation in genotype frequencies between years (Spaak, 1996), and variation of species and hybrid compositions across lake regions (Weider, 1993) suggest that the maintenance of hybrid genotypes largely depends on suitable ecological parameters of lakes, such as predation and food regimes. Among species of the *D. galeata* complex, the genotype composition of hybrid populations are highly diverse and variable between years, but some lakes are known to be dominated by only one hybrid genotype for longer periods (Spaak and Hoekstra, 1993). These observations raise the question of how hybrid lineages are maintained. Hybrid fitness can be differentiated into two aspects: first, the comparison of fitness values of hybrids and parentals *per se* (parthenogenetic phase of life cycle; e.g., predation risk, r), and second the sexual contribution of hybrids to the next generation (sexual phase of life cycle; lifetime reproductive success). With regard to the first issue, empirical data support the view that hybrids could be, at least temporarily, superior to parental species. Once produced, hybrids might occupy a distinct ecological niche, either through differential diel vertical migration strategies (Weider and Stich, 1992; Müller and Seitz, 1993), or through a combination of traits which reduce predation risks (Spaak and Hoekstra, 1995). However, if hybrids reproduce only parthenogenetically, then their potential for genetic adaptation depends solely on mutation. As a consequence, hybrids could be locally and within an "ecological time scale" very important, but of minor or no importance on an "evolutionary time scale". In principle, hybrids are able to follow several evolutionary pathways (Fig. 5). They can either persist as a parthenogenetically reproducing clonal lineage (a), become extinct (b), or reproduce sexually with parentals or other hybrid clones (c). Information on life-history characteristics helps to elucidate the short-term ecological success and the niche breadth of hybrid clones, whereas the evolutionary significance of interspecific hybridization depends on (sexual) reproductive success of hybrids (F_2 or backcross genotypes).

One way of maintaining advantageous traits that have become established through successful hybridization is to switch from cyclical to obligate parthenogenesis. Certain clones of the *D. pulex* group are able to

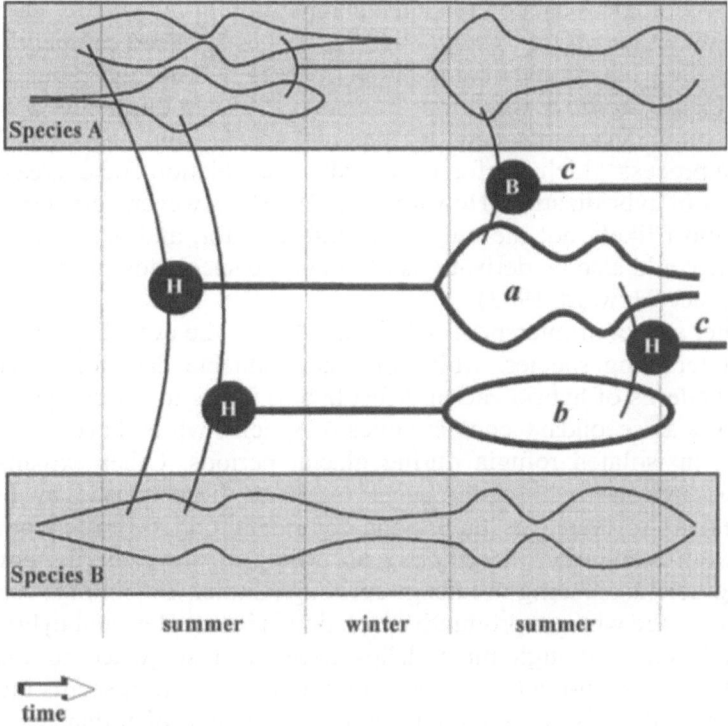

Figure 5. Model of *Daphnia* interspecific hybridization with regard to reproductive modes and clonal abundances. Envelopes (width = abundance) symbolize parthenogenetic reproduction, lines sexual reproduction and subsequent resting egg stages, and black circles (H) hybridization and backcross events (B). If hybrids are produced they might persist for one season (or even longer) as resting eggs before they hatch and reproduce via ameiotic parthenogenesis (*a*), or they become extinct after a certain time of ameiotical reproduction (*b*). If hybrids are able to reproduce sexually, they might form backcrosses or F₂ hybrids (*c*).

produce resting eggs parthenogenetically, hence they evolve independently from their sexual ancestors (Hebert and Crease, 1980; Crease et al., 1989). In contrast, no obligatory parthenogenetic clones have been detected so far in the *D. galeata* complex, thus hybrid or backcross clones can only maintain their advantageous traits by subsequent parthenogenetic reproduction. However, to what extent hybrids are limited in their capability to reproduce sexually because of deleterious effects during meiosis (i.e., hybrid sterility) remains an open question. Evidence of nuclear introgression (Taylor and Hebert, 1993a; Spaak, 1996) together with the directionality of hybridization and backcrossing (Schwenk, 1993; Taylor and Hebert, 1993a) suggest that probably hybrid males are involved in sexual reproduction. However, data from crossing experiments are needed to compare the level of sexual reproduction and fitness of hybrids with parentals.

Recent ecological data on resource competition in *D. galeata, D. cucullata* and the *D. cucullata x galeata* hybrids suggest reduced competition between species, but stronger competition between hybrids and each parental species (*hemispecific competition;* Boersma, 1995). In this context, hybridization seems disadvantageous for parental species and one could hypothesize a process which reinforces reproductive isolation and decreases the frequency of hybridization (Howard et al., 1993). However, competition for food is most likely not the only competitive factor, and barriers to gene exchange could also be derivations of allopatric speciation prior to secondary contact (Howard, 1993).

Another aspect of interspecific hybridization is the evolutionary history of the interacting species, which provides valuable data for explaining current patterns of hybridization. Most hybrid zones in Europe have been interpreted as secondary contact zones of species which have previously diverged in isolated refugia during glacial periods. Other explanations consider environmental disturbances (such as those caused by human activities) as relevant for facilitating hybridization. A third hypothesis assumes differentiation in sympatry according to, for example, environmental gradients, leading to lineage splitting (Endler, 1977). This scenario includes a stage where reproductive isolation is incomplete and hybrids are still produced. Although many lakes have been subjected to massive changes due to man-made eutrophication and pollution, it seems unlikely that hybridization is caused solely by environmental disturbances. Since interspecific hybridization comprises several *Daphnia* species and occurs on different continents (North America, Europe and Australia), other than ecological factors seem to be involved as well. For example, interspecific hybridization among various *Daphnia* species occurs despite significant genetic divergence of hybridizing taxa, indicating reduced or absent reproductive isolation among members of this genus (Schwenk 1993; Taylor et al., 1996: Colbourne and Hebert, 1996).

Future perspectives

Although initial results on *Daphnia* species complexes have shown ecological and genetic patterns which describe and explain the evolutionary process of interspecific hybridization, some central questions remain. Is interspecific hybridization among *Daphnia* species an important evolutionary factor, or nothing more than inconsequential evolutionary noise?

Both premating isolation (e.g., ecological and temporal) and postmating isolation (genetic incompatibility) require further investigation. If one assumes that hybrids rarely reproduce sexually and are incapable of resting egg production via apomixis (which might be true for most of the hybrids within the *D. galeata* complex), we expect that hybrids are able to persist until extinction in their lake of origin, since no dispersal of resting eggs is possible. Consequently, a comparison of hybrid and parental species genotypes of various isolated sites should reveal higher levels of genetic divergence between hybrid populations than between species populations. The other extreme would be that hybrids reproduce sexually, which could lead to speciation via hybridization or, more likely, to fusion of parental taxa (Bullini and Nascetti, 1990). However, data on hybrid classes (such as F_1, F_2 and backcrosses) and induction of sexual females and males of parental species could provide information about the level of reproductive isolation. Recently, Spaak (1995b, 1996) reported that three *Daphnia* species and their interspecific hybrids differed in their reaction to sexual stimuli (production of males and sexual females). This variation might constitute a necessary condition for mechanisms of reproductive isolation.

The study of interspecific hybridization among zooplankton species (cyclic parthenogens) provides two important additional features compared with "classical" examples among animals. First, it is possible to separate long-term evolutionary and short-term ecological consequences of interspecific hybridization. Although hybrids might exhibit very low sexual fertility, they can nevertheless dominate zooplankton communities due to their capacity for parthenogenetic reproduction. Hybrids within the *D. galeata* complex form genetically diverse groups, thus various combinations of parental genes will be subjected to selection. Comparisons of life-history characteristics and predation regimes are bound to reveal which parental characters are under hard selection (e.g., predation and resource competition). Second, zooplankton species are relatively easy to cultivate in laboratories and have short generation times, which facilitates experiments on life-history variation, selection and predation.

Some extremely interesting but difficult questions that need to be addressed are related to the phenomenon of resting egg formation. What is the ecological and evolutionary significance of resting egg pools? And how frequent are dispersal events of *Daphnia* resting eggs between habitats? Since it is known that hybrids can hatch successfully from ephippia (Carvalho and Wolf, 1989), it would be interesting to know the relative frequencies, hatching rates and the maximum duration of diapause. Production of resting eggs can serve as a predator-avoidance mechanism, or as a process which maintains high genetic diversity (e.g., Hairston and Dillon, 1990). Since several lakes have been found in which only hybrids are present, the question arises whether these populations were founded by dispersed ephippia or whether the parental species were displaced by hybrids and went extinct. Genetic analysis of resting eggs from lake sediments could provide a unique record of the population structure in the past, and could facilitate the estimation of dispersal rates of species and hybrids.

Acknowledgments
Generous and detailed advice from Lawrence Weider and Thomas Städler significantly improved the presentation of these ideas. We thank Jos van Damme, Ramesh Gulati, Riks Laanbroek, Wolf Mooij, Jacobus Vijverberg and Bill DeMott for comments on an earlier version of this manuscript. The comments of Derek Taylor, Richard Harrison and two anonymous reviewers also substantially improved the manuscript. We appreciated many stimulating discussions and support by Maarten Boersma, Michaela Brehm and Eva Mader. We are grateful to Sabine Gießler, Rita deMelo, Paul Hebert, John Colbourne and Derek Taylor for supplying unpublished data. This is publication 2122 from the Centre for Limnology (CL), Netherlands Institute for Ecology (NIE), Nieuwersluis, The Netherlands.

References

Abbott, R.J. (1992) Plant invasions, interspecific hybridization and the evolution of new plant taxa. *Trends Ecol. Evol.* 7:401–405.

Anderson, E. (1948) Hybridization of the habitat. *Evolution* 2:1–9.

Anderson, E. and Hubricht, L. (1938) Hybridization in *Tradescantia*. III. The evidence for introgressive hybridization. *Am. J. Bot.* 25:396–402.

Arnold, M.L. (1992) Natural hybridization as an evolutionary process. *Annu. Rev. Ecol. Syst.* 23:237–261.

Arnold, M.L. and Hodges, S.A. (1995) Are natural hybrids fit or unfit relative to their parents? *Trends Ecol. Evol.* 10:67–71.

Arntzen, J.W. and Wallis, G.P. (1991) Restricted gene flow in a moving hybrid zone of the newts *Triturus cristatus* and *T. marmoratus* in western France. *Evolution* 45:805–826.

Aubert, J. and Solignac, M. (1990) Experimental evidence of mitochondrial DNA introgression between *Drosphila* species. *Evolution* 44:1272–1282.

Avise, J.C. and Saunders, N.C. (1984) Hybridization and introgression among species of sunfish (*Lepomis*): Analysis by mitochondrial DNA and allozyme markers. *Genetics* 108:237–255.

Avise, J.C., Bermingham, E., Kessler, L.G. and Saunders, N.C. (1984) Characterization of mitochondrial DNA variability in a hybrid swarm between subspecies of bluegill sunfish (*Lepomis macrochirus*). *Evolution* 38:931–941.

Barton, N.H. (1980) The fitness of hybrids between two chromosomal races of the grasshopper *Podisma pedestris*. *Heredity* 45:47–59.

Barton, N.H. and Hewitt, G.M. (1985) Analysis of hybrid zones. *Annu. Rev. Ecol. Syst.* 16:113–148.

Barton, N.H. and Hewitt, G.M. (1989) Adaptation, speciation and hybrid zones. *Nature* 341:497–503.

Benzie, J.A.H. (1986) Phenetic and cladistic analyses of the phylogenetic relationships within the genus *Daphnia* worldwide. *Hydrobiologia* 140:105–124.

Bert, T.M. and Arnold, W.S. (1995) An empirical test of predictions of two competing models for maintenance and fate of hybrid zones. Both models are supported in a hard-clam hybrid zone. *Evolution* 49:276–289.

Bert, T.M. and Harrison, R.G. (1988) Hybridization in western Atlantic stone crabs (genus *Menippe*): Evolutionary history and ecological context influence species interactions. *Evolution* 42:528–544.

Boersma, M. (1994) *On the Seasonal Dynamics of Daphnia Species in a Shallow Eutrophic Lake*. Ph.D. thesis, University of Amsterdam, The Netherlands.

Boersma, M. (1995) Competition in natural populations of *Daphnia*. *Oecologia* 103:309–318.

Boersma, M. and Vijverberg, J. (1994a) Resource depression in *Daphnia galeata, Daphnia cucullata* and their interspecific hybrid: Life history consequences. *J. Plankton Res.* 16:1741–1758.

Boersma, M. and Vijverberg, J. (1994b) Seasonal variations in the condition of two *Daphnia* species and their hybrid in a eutrophic lake: Evidence for food limitation. *J. Plankton Res.* 16:1793–1809.

Boersma, M., van Densen, W.L.T. and Vijverberg, J. (1991) The effect of predation by smelt (*Osmerus eperlanus*) on *Daphnia hyalina* in a shallow eutrophic lake. *Verh. internat. Verein. Limnol.* 24:2438–2442.

Boersma, M., van Tongeren, O.F.R. and Mooij, W.M. (1996) Seasonal pattern in the mortality of *Daphnia* species in a shallow lake. *Can. J. Fish. Aquat. Sci.* 53:1–11.

Bradley, R.D., Baker, R.J. and Davis, S.K. (1991) Genetic control of premating-isolating behavior: Kaneshiro's hypothesis and asymmetrical sexual selection in pocket gophers. *J. Hered.* 82:192–196.

Bullini, L. (1985) Speciation by hybridization in animals. *Boll. Zool.* 52:121–137.

Bullini, L. and Nascetti, G. (1990) Speciation by hybridization in phasmids and other insects. *Can. J. Zool.* 68:1747–1760.

Carvalho, G.R. and Wolf, H.G. (1989) Resting eggs of lake-*Daphnia*. I. Distribution, abundance and hatching of eggs collected from various depths in lake sediments. *Freshw. Biol.* 22: 459–470.

Colbourne, J.K. and Hebert, P.D.N. (1996) The sytematics of North American *Daphnia* (Crustacea: Anomopoda): A molecular phylogenetic approach. *Phil. Trans. Roy. Soc. Lond.* B 351:349–360.

Crease, T.J., Stanton, D.J. and Hebert, P.D.N. (1989) Polyphyletic origins of asexuality in *Daphnia pulex*. II. Mitochondrial-DNA variation. *Evolution* 43:1016–1026.

DeMarais, B.D., Dowling, T.E., Douglas, M.E., Minckley, W.L. and Marsh, P.C. (1992) Origin of *Gila seminuda* (Teleostei, Cyprinidae) through introgressive hybridization: Implications for evolution and conservation. *Proc. Natl. Acad. Sci. USA* 89:2747–2751.

Dowling, T.W. and Brown, W.M. (1989) Allozymes, mitochondrial DNA, and level of phylogenetic resolution among four minnow species (*Notropis*: Cyprinidae). *Syst. Zool.* 38: 126–143.

Dowling, T.E. and DeMarais, B.D. (1993) Evolutionary significance of introgressive hybridization in cyprinid fishes. *Nature* 362:444–446.

Dowling, T.E. and Hoeh, W.R. (1991) The extent of introgression outside the contact zone between *Notropis cornutus* and *Notropis chrysocephalus* (Teleostei: Cyprinidae). *Evolution* 45:944–956.

Dowling, T.E., Smith, G.R. and Brown, W.M. (1989) Reproductive isolation and introgression between *Notropis cornutus* and *Notropis chrysocephalus* (family Cyprinidae): Comparison of morphology, allozymes, and mitochondrial DNA. *Evolution* 43:620–634.

Ebert, D., Yampolsky, L. and Stearns, S.C. (1993) Genetics of life history in *Daphnia magna*. 1. Heritabilities at two food levels. *Heredity* 70:335–343.

Ender, A. (1993) *Identifizierung von Nuklearen DNA-Markern in Daphnia-Hybrid-Komplexen: RAPD-Analyse*. Diploma thesis, J.W. Goethe-Universität, Frankfurt/Main, Germany.

Ender, A., Schwenk, K., Städler, T., Streit, B. and Schierwater, B. (1996) RAPD identification of microsatellites in *Daphnia*. *Mol. Ecol.* 5:437–441.

Endler, J.A. (1977) *Geographic Variation, Speciation, and Clines*. Princeton University Press, Princeton.

Ferris, S.D., Sage, R.D., Huang, C.H., Nielsen, J.T., Ritte, U. and Wilson, A.C. (1983) Flow of mitochondrial DNA across a species boundary. *Proc. Natl. Acad. Sci. USA* 80:2290–2294.

Flößner, D. (1972) *Kiemen- und Blattfüsser, Branchiopoda, Fischläuse, Branchiura*. Die Tierwelt Deutschlands 60:1–501, VEB Gustav Fischer Verlag, Jena.

Flößner, D. (1993) Zur Kenntnis einiger *Daphnia*-Hybriden. *Limnologica* 23:71–79.

Flößner, D. and Kraus, K. (1986) On the taxonomy of the *Daphnia hyalina-galeata* complex (Crustacea: Cladocera). *Hydrobiologia* 137:97–115.

Forbes, S.H. and Allendorf, F.W. (1991) Associations between mitochondrial and nuclear genotypes in cutthroat trout hybrid swarms. *Evolution* 45:1332–1349.

Frey, D.G. (1982) G.O. Sars and the Norwegian Cladocera: A continuing frustration. *Hydrobiologia* 96:267–293.

Gießler, S. (1987) *Mikroevolution und Populationsgenetik im Daphnia galeata/hyalina/cucullata-Komplex*. Ph.D. thesis, Ludwig Maximilians-Universität München, Germany.

Gliwicz, Z.M. (1990) Food thresholds and body size in cladocerans. *Nature* 341:638–640.

Grant, P.R. and Grant, B.R. (1992) Hybridization of bird species. *Science* 256:193–197.

Grant, V. (1981) *Plant Speciation*, Second Edition. Columbia University Press, New York.

Hairston, N.G., Jr. and Dillon, T.A. (1990) Fluctuating selection and response in a population of freshwater copepods. *Evolution* 44:1796–1805.

Hann, B.J. (1987) Naturally occurring interspecific hybridization in *Simocephalus* (Cladocera, Daphniidae): Its potential significance. *Hydrobiologia* 145:219–224.

Harrison, R.G. (1990) Hybrid zones, windows on evolutionary process. *Oxford Surv. Evol. Biol.* 7:69–128.

Harrison, R.G. (ed.) (1993) *Hybrid Zones and the Evolutionary Process.* Oxford University Press, New York.

Harrison, R.G. and Rand, D.M. (1989) Mosaic hybrid zones and the nature of species boundaries. *In*: D. Otte and J.A. Endler (eds): *Speciation and its Consequences.* Sinauer Associates, Sunderland, Massachusetts, pp 111–133.

Harrison, R.G., Rand, D.M. and Wheeler, W.C. (1987) Mitochondrial DNA variation in field crickets across a narrow hybrid zone. *Mol. Biol. Evol.* 4:144–158.

Hebert, P.D.N. (1985) Interspecific hybridization between cyclic parthenogens. *Evolution* 39:216–220.

Hebert, P.D.N. (1995) *The Daphnia of Northern America: An Illustrated Fauna.* CD-ROM, distributed by the author. Department of Zoology, University of Guelph, Guelph, Ontario.

Hebert, P.D.N. and Crease, T.J. (1980) Clonal coexistence in *Daphnia pulex* (Leydig): Another planktonic paradox. *Science* 207:1363–1365.

Hebert, P.D.N. and Finston, T.L. (1996) A taxonomic reevaluation of North American *Daphnia* (Crustacea: Cladocera). II. New species in the *D. pulex* group from the South-Central United States and Mexico. *Can. J. Zool* 74:632–653.

Hebert, P.D.N. and Wilson, C.C. (1994) Provincialism in plankton: Endemism and allopatric speciation in Australian *Daphnia. Evolution* 48:1333–1349.

Hebert, P.D.N., Beaton, M.J., Schwartz, S.S. and Stanton, D.J. (1989a) Polyphyletic origins of asexuality in *Daphnia pulex.* 1. Breeding-system variation and levels of clonal diversity. *Evolution* 43:1004–1015.

Hebert, P.D.N., Schwartz, S.S. and Hrbáček, J. (1989b) Patterns of genotypic diversity in Czechoslovakian *Daphnia. Heredity* 62:207–216.

Hebert, P.D.N., Schwartz, S.S., Ward, R.D. and Finston, T.L. (1993) Macrogeographic patterns of breeding system diversity in the *Daphnia pulex* group. 1. Breeding systems of Canadian populations. *Heredity* 70:148–161.

Heiser, C.B., Jr. (1947) Hybridization between the sunflower species *Helianthus annuus* and *H. petioloaris. Evolution* 1:249–262.

Heiser, C.B., Jr. (1973) Introgression re-examined. *Bot. Rev.* 39:347–366.

Herke, S.W., Kornfield, I., Morgan, P. and Moring, J.R. (1990) Molecular confirmation of hybridization between northern pike (*Esox lucius*) and chain pickerel (*E. niger*). *Copeia* 1990:846–850.

Herzig, A. and Auer, B. (1990) The feeding behaviour of *Leptodora kindtii* and its impact on the zooplankton community of Neusiedler See (Austria). *Hydrobiologia* 198:107–117.

Hewitt, G.M. (1988) Hybrid zones – natural laboratories of evolutionary studies. *Trends Ecol. Evol.* 3:158–167.

Hewitt, G.M. (1989) The subdivision of species by hybrid zones. *In:* D. Otte and J.A. Endler (eds): *Speciation and Its Consequences.* Sinauer Associates, Sunderland, Massachusetts, pp 85–110.

Hewitt, G.M. (1993) Postglacial distribution and species substructure: Lessons from pollen, insects and hybrid zones. *In:* D.R. Lees and D. Edwards (eds): *Evolutionary Pattern and Processes.* Linnean Society Symposium Series 14, Academic Press, London, pp 97–123.

Hewitt, G.M. and Barton, N.H. (1980) The structure and maintenance of hybrid zones as exemplified by *Podisma pedestris. In:* R.L. Blackman, G.M. Hewitt and M. Ashburner (eds): *Insect Cytogenetics.* Royal Entomological Society of London Symposia 10, Blackwell, Oxford, pp 159–169.

Hewitt, G.M., Butlin, R.K. and East, T.M. (1987) Testicular dysfunction in hybrids between parapatric species of the grasshopper *Chorthippus parallelus. Biol. J. Linn. Soc.* 31:25–34.

Hobæk, A. and Larsson, P. (1990) Sex determination in *Daphnia magna. Ecology* 71:2255–2268.

Hovanitz, W. (1943) Hybridization and seasonal segregation in two races of a butterfly occurring together in two localities. *Biol. Bull.* 85:44–51.

Howard, D.J. (1993) Reinforcement: Origin, dynamics, and fate of an evolutionary hypothesis. *In:* R.G. Harrison (ed.): *Hybrid Zones and the Evolutionary Process.* Oxford University Press, New York, pp 46–69.

Howard, D.J., Waring, G.L., Tibbets, C.A. and Gregory, P.G. (1993) Survival of hybrids in a mosaic hybrid zone. *Evolution* 47:789–800.

Hrbáček, J. (1987) Systematic and biogeography of *Daphnia* species in the northern temperate regions. *In:* R.H. Peters and R. de Bernardi (eds): *Daphnia*. Memorie dell'Instituto Italiano di Idrobiologia, Vol. 45, Verbania Pallanza, Pallanza, pp 37–76.

Konkle, B.R. and Philipp, D.P. (1992) Asymmetric hybridization between two species of sunfishes (*Lepomis*: Centrarchidae). *Mol. Ecol.* 1:215–222.

Kraus, F. and Miyamoto, M.M. (1990) Mitochondrial genotype of a unisexual salamander of hybrid origin is unrelated to either of its nuclear haplotypes. *Proc. Natl. Acad. Sci. USA* 87:2235–2238.

Lamb, T. and Avise, J.C. (1986) Directional introgression of mitochondrial DNA in a hybrid population of tree frogs: The influence of mating behavior. *Pro. Natl. Acad. Sci. USA* 83: 2526–2530.

Lammens, E.H.R.R., de Nie, J.W., Vijverberg, J. and van Densen, W.L.T. (1985) Resource partitioning and niche shifts of bream (*Abramis brama*) and eel (*Anguilla anguilla*) mediated by predation of smelt (*Osmerus esperlanus*) on *Daphnia hyalina*. *Can. J. Fish. Aquat. Sci.* 42:1342–1351.

Larsson, P. (1991) Intraspecific variability in response to stimuli for male and ephippia formation in *Daphnia pulex*. *Hydrobiologia* 225:281–290.

Lehman, N., Eisenhawer, A., Hansen, K., Mech, L.D., Peterson, R.O., Gogan, P.J.P. and Wayne, R.K. (1991) Introgression of coyote mitochondrial DNA into sympatric North American gray wolf populations. *Evolution* 45:104–119.

Levin, D.A. and Bulinska-Radomska, Z. (1988) Effects of hybridization and inbreeding on fitness in *Phlox*. *Am. J. Bot.* 75:1632–1639.

Lewontin, R.C. and Birch, L.C. (1966) Hybridization as a source of variation for adaptation to new environments. *Evolution* 20:315–336.

Lieder, U. (1983) Introgression as a factor in the evolution of polytypical plankton Cladocera. *Int. Revue ges. Hydrobiol.* 68:269–284.

Lieder, U. (1987) The possible origin of *Daphnia cucullata procurva* POPPE 1887 in the lakes of the Pomeranian Lakeland by hybridization in the past. *Hydrobiologia* 149:201–211.

McDade, L.A. (1992) Hybrids and phylogenetic systematics. 2. The impact of hybrids on cladistic analysis. *Evolution* 6:1329–1346.

McGowan, C. and Davidson, W.S. (1992) Unidirectional natural hybridization between brown trout (*Salmo trutta*) and Atlantic salmon (*S. Salar*) in Newfoundland. *Can. J. Fish. Aquat. Sci* 49:1953–1958.

Moore, W.S. (1977) An evaluation of narrow hybrid zones in vertebrates. *Quart. Rev. Biol.* 52:263–277.

Moore, W.S. (1995) Inferring phylogenies from mtDNA variation: Mitochondrial-gene trees versus nuclear-gene trees. *Evolution* 49:718–726.

Moore, W.S. and Koenig, W.D. (1986) Comparative reproductive success of yellow-shafted, red-shafted and hybrid flickers across a hybrid zone. *Auk* 103:42–51.

Mort, M. (1990) Coexistence of *Daphnia* species and their hybrids: An experimental population genetics approach. *Arch. Hydrobiol.* 120:169–183.

Müller, J. (1993) *Räumliche und Zeitliche Variabilität der Genetischen Struktur Natürlicher Cladocerenpopulationen (Crustacea, Cladocera)*. Ph.D. thesis, Johannes Gutenberg-Universität, Mainz, Germany.

Müller, J. and Seitz, A. (1993) Habitat partitioning and differential vertical migration of some *Daphnia* genotypes in a lake. *Arch. Hydrobiol. Beih. (Ergebn. Limnol.)* 39:167–174.

Müller, J. and Seitz, A. (1994) Influence of differential natality and mortality on temporal fluctuations of *Daphnia* genotypes in natural populations. *In:* A.R. Beaumont (ed.): *Genetics and Evolution of Aquatic Organisms*. Chapman and Hall, London, pp 342–350.

Müller, J. and Seitz, A. (1995) Differences in genetic structure and ecological diversity between parental forms and hybrids in a *Daphnia* species complex. *Hydrobiologia* 307:25–32.

Patton, J.L. and Smith, M.F. (1993) Molecular evidence for mating asymmetry and female choice in a pocket gopher (*Thomomys*) hybrid zone. *Mol. Ecol.* 2:3–8.

Pijanowska, J. (1990) Cyclomorphosis in *Daphnia*: an adaptation to avoid invertebrate predation. *Hydrobiologia* 198:41–50.

Rand, D.M. and Harrison, R.G. (1989) Ecological genetics of a mosaic hybrid zone: Mitochondrial, nuclear, and reproductive differentiation of crickets by soil type. *Evolution* 43: 432–449.

Rieseberg, L.H. and Wendel, J.F. (1993) Introgression and its consequences in plants. *In:* R.G. Harrison (ed.): *Hybrid Zones and the Evolutionary Process.* Oxford University Press, New York, pp 70–109.

Schierwater, B., Ender, A., Schwenk, K., Spaak, P. and Streit, B. (1994) The evolutionary ecology of *Daphnia. In:* B. Schierwater, B. Streit, G.P. Wagner and R. DeSalle (eds): *Molecular Ecology and Evolution: Approaches and Applications.* Birkhäuser Verlag, Basel, pp 495–508.

Schwenk, K. (1993) Interspecific hybridization in *Daphnia:* Distinction and origin of hybrid matrilines. *Mol. Biol. Evol.* 10:1289–1302.

Schwenk, K., Ender, A. and Streit, B. (1995) What can molecular markers tell us about the evolutionary history of *Daphnia* species complexes? *Hydrobiologia* 307:1–7.

Scribner, K.T. (1993) Hybrid zone dynamics are influenced by genotype-specific variation in life-history traits: Experimental evidence from hybridizing *Gambusia* species. *Evolution* 47:632–646.

Scribner, K.T. and Avise, J.C. (1993) Cytonuclear genetic architecture in mosquitofish populations and the possible roles of introgressive hybridization. *Mol. Ecol.* 2:139–149.

Solignac, M. and Monnerot, M. (1986) Race formation, speciation, and introgression within *Drosophila simulans, D. mauritiana,* and *D. sechellia* inferred from mitochondrial DNA analysis. *Evolution* 40:531–539.

Spaak, P. (1994) *Genetical Ecology of a Coexisting Daphnia Hybrid Species Complex.* Ph.D. thesis, University of Utrecht, The Netherlands.

Spaak, P. (1995a) Cyclomorphosis as a factor explaining success of a *Daphnia* hybrid in Tjeukemeer. *Hydrobiologia* 307:283–289.

Spaak, P. (1995b) Sexual reproduction in *Daphnia:* Interspecific differences in a hybrid species complex. *Oecologia* 104:501–507.

Spaak, P. (1996) Temporal changes in the genetic structure of the *Daphnia* species complex in Tjeukemeer, with evidence for backcrossing. *Heredity* 76:539–548.

Spaak, P. and Hoekstra, J.R. (1993) Clonal structure of the *Daphnia* population in Lake Maarsseveen: Its implications for diel vertical migration. *Arch. Hydrobiol. Beih. Ergebn. Limnol.* 39:157–165.

Spaak, P. and Hoekstra, J.R. (1995) Life history variation and the coexistence of a *Daphnia* hybrid and its parental species. *Ecology* 76:553–564.

Spaak, P. and Hoekstra, J.R. (1997) Fish predation on a *Daphnia* hybrid species complex: A factor explaining species coexistence? *Limnol. Oceanogr.; in press.*

Sperling, F.A.H. (1987) Evolution of the *Papilio machaon* species group in western Canada (Lepidoptera, Papilionidae). *Can. Quaest. Entomol.* 23:198–315.

Sperling, F.A.H. and Spence, J.R. (1991) Structure of an asymmetric hybrid zone between two water strider species (Hemiptera: Gerridae: *Limnoporus). Evolution* 45:1370–1383.

Spitze, K. (1993) Population structure in *Daphnia obtusa* – quantitative genetic and allozymic variation. *Genetics* 135:367–374.

Stebbins, G.L. (1950) *Variation and Evolution in Plants.* Columbia University Press, New York.

Stebbins, G.L. and Daly, K. (1961) Changes in the variation pattern of a hybrid population of *Helianthus* over an eight-year period. *Evolution* 15:60–71.

Szymura, J.M. and Barton, N.H. (1986) Genetic analysis of a hybrid zone between the fire-bellied toads, *Bombina bombina* and *B. variegata,* near Cracow in southern Poland. *Evolution* 40:1141–1159.

Taylor, D.J. and Hebert, P.D.N. (1992) *Daphnia galeata mendotae* as a cryptic species complex with interspecific hybrids. *Limnol. Oceanogr.* 37:658–665.

Taylor, D.J. and Hebert, P.D.N. (1993a) Habitat-dependent hybrid parentage and differential introgression between neighboringly sympatric *Daphnia* species. *Proc. Natl. Acad. Sci. USA* 90:7079–7083.

Taylor, D.J. and Hebert, P.D.N. (1993b) A reappraisal of phenotypic variation in *Daphnia galeata mendotae* – the role of interspecific hybridization. *Can. J. Fish. Aquat. Sci.* 50:2137–2146.

Taylor, D.J. and Hebert, P.D.N. (1993c) Cryptic intercontinental hybridization in *Daphnia* (Crustacea) – the ghost of introductions past. *Proc. Roy. Soc. Lond.* B 254:163–168.

Taylor, D.J. and Hebert, P.D.N. (1994) Genetic assessment of species boundaries in the North American *Daphnia longispina* complex (Crustacea, Daphniidae). *Zool. J. Linn. Soc.* 110:27–40.

Taylor, D.J., Hebert, P.D.N. and Colbourne, J.K. (1996) Phylogenetics and evolution of the *Daphnia longispina* group (Crustacea) based on 12S rDNA sequence and allozyme variation. *Molec. Phylogenet. Evol.* 5:495–510.

Van Raay, T.J. and Crease, T.J. (1995) Mitochondrial DNA diversity in an apomictic *Daphnia* complex from the Canadian high arctic. *Mol. Ecol.* 4:149–161.

Vijverberg, J. and Richter, A.F. (1982) Population dynamics and production of *Daphnia hyalina* LEYDIG and *Daphnia cucullata* SARS in Tjeukemeer. *Hydrobiology* 95:235–259.

Vijverberg, J., Boersma, M., Van Densen, W.L.T., Hoogenboezem, W., Lammens, E.H.R.R. and Mooij, W.M. (1990) Seasonal variation in the interactions between piscivorous fish, planktivorous fish and zooplankton in a shallow eutrophic lake. *Hydrobiologia* 207:279–286.

Weider, L.J. (1993) Niche breadth and life history variation in a hybrid *Daphnia* complex. *Ecology* 74:935–943.

Weider, L.J. and Stich, H.B. (1992) Spatial and temporal heterogeneity of *Daphnia* in Lake Constance – intraspecific and interspecific comparisons. *Limnol. Oceanogr.* 37:1327–1334.

Weider, L.J. and Wolf, H.G. (1991) Life-history variation in a hybrid species complex of *Daphnia. Oecologia* 87:506–513.

Wolf, H.G. (1987) Interspecific hybridization between *Daphnia hyalina, D. galeata,* and *D. cucullata* and seasonal abundances of these species and their hybrids. *Hydrobiologia* 145:213–217.

Wolf, H.G. and Mort, M.A. (1986) Inter-specific hybridization underlies phenotypic variability in *Daphnia* populations. *Oecologia* 68:507–511.

Wolf, H.G. and Weider, L.J. (1991) Do life-history parameters of *Daphnia* as determined in the laboratory correctly predict species successions in the field? *Verh. internat. Verein. Limnol.* 24:2799–2801.

Zaret, T.M. (1975) Strategies for existence of zooplankton prey in homogeneous environments. *Verh. internat. Verein. Limnol.* 19:1484–1489.

Evolutionary Ecology of Freshwater Animals
ed. by B. Streit, T. Städler and C.M. Lively

Population biology, genetic structure, and mating system parameters in freshwater snails

T. Städler[1] and P. Jarne[2]

[1] Abteilung Ökologie und Evolution, Fachbereich Biologie, J.W. Goethe-Universität, Siesmayerstrasse 70, D-60054 Frankfurt, Germany
[2] Génétique et Environnement, Institut des Sciences de l'Evolution, Université Montpellier II, Place E. Bataillon, F-34095 Montpellier Cedex 5, France

Summary. Freshwater gastropods can reproduce by both uniparental and biparental means. In particular, self-fertilization in the hermaphrodite pulmonates (Basommatophora) and apomictic parthenogenesis in prosobranchs are viable alternatives to biparental sexuality in several species. The coexistence of different mating systems within and among extant populations provides opportunities to examine the forces directing their evolution. We review the models predicting genetic structure in subdivided populations, with an emphasis on the effects of inbreeding. Empirical population genetic data on freshwater pulmonates suggest a marked loss of genetic variability under selfing. We also consider the genetic and demographic factors thought to influence mating system evolution, and highlight approaches that should be used to estimate mating system parameters and inbreeding depression. We mainly draw empirical examples from our population biological studies on tropical species in the planorbid *Bulinus* and the European stream limpet, *Ancylus fluviatilis*. These ongoing studies contribute to a deeper understanding of the evolution of selfing versus outcrossing.

Introduction

A main concern of population genetics is the distribution of genetic variability within and among populations. This distribution is determined by the joint influence of many factors, e.g., mutation, selection, migration, and genetic drift. Another important factor is the mating system, because it mediates the transmission of genes across generations. Effects of the mating system were considered quite early by botanists, probably because many plants of agricultural interest are potential selfers. Since the widespread use of protein electrophoresis in population genetics, many studies have estimated mating system parameters in seed plants and have analyzed the influence of the mating system on the distribution of genetic variation (e.g., Hamrick and Godt, 1990; Schoen and Brown, 1991). In marked contrast, few such studies have been performed in hermaphroditic animals, despite substantial interest among population biologists in estimates derived from organisms other than seed plants (e.g., Bell, 1982; Charnov, 1982).

Population genetics is also concerned with the distribution of individuals over time and space. Real organisms may be spatially distributed in a more-or-less continuous fashion or, alternatively, as small subdivided populations. Analyzing the consequences of alternative mating systems for finite, subdivided populations is more challenging than for homogeneous, infinite populations. However, a recent revival of interest in both the genetics and dynamics of subdivided populations has greatly enhanced our knowledge of the fate of genes under these more complex circumstances (review in Slatkin, 1985, 1987; McCauley, 1991; Barton and Whitlock, 1996).

Why ought these issues be studied in freshwater snails? First, freshwater snails, and in particular the hermaphroditic basommatophorans, often occupy both patchily distributed and transient habitats that are subject to droughts (causing population crashes and even local extinctions) and floods, which may act as important agents of unidirectional dispersal (Russell-Hunter, 1978; Woolhouse and Chandiwana, 1989; Brown, 1994). These particular features of their population dynamics make freshwater snails an appropriate group to investigate genetic variability and population genetic structure in subdivided populations.

Second, freshwater snails exhibit a variety of mating systems, including self-fertilization in basommatophorans and parthenogenesis in proso-branchs (Hughes, 1989; Heller, 1993; Jarne and Städler, 1995). The co-occurrence of alternative reproductive modes within extant taxa presents exceptional opportunities for causal analyses of the factors involved in mating system evolution (for general overviews, see Bell, 1982; Jarne and Charlesworth, 1993, 1996). Until recently, the connection between the mating system and population genetic structure in freshwater snails has been largely neglected, despite thorough early work on the mating system considered both at the individual and the population level in *Bulinus truncatus* (Larambergue, 1939), and the extensive work of Selander and co-workers in the 1970s on terrestrial gastropods (for review, see Selander and Ochman, 1983). However, freshwater pulmonate snails have been largely studied in a biomedical context, mainly because some tropical species serve as intermediate hosts of human schistosome parasites (review in Brown, 1994). Historically, most of the interest has focused on protein electrophoretic studies performed to discriminate susceptible host (snail) populations (e.g., Jelnes, 1986).

In several prosobranch taxa, apomictic parthenogenesis rather than selfing is an alternative uniparental mating system, with major consequences for the distribution and maintenance of genetic variability. Although the genetics of clonal populations has received considerable attention in both plants and animals, sound population genetic data on clonal prosobranchs to date are limited. Therefore, we will restrict most considerations to an analysis of selfing versus outcrossing.

We first consider the models analyzing and predicting the distribution of within- and among-population genetic variability under inbreeding, and contrast allozyme and microsatellite markers as empirical tools in this regard. We then focus on the estimation of mating system parameters in pulmonates, and discuss inbreeding depression and other factors relevant to mating system evolution; particularly in this section, we draw heavily on our population biological studies in freshwater pulmonates. Finally, we consider the relationship between uniparental reproduction and population dynamics, particularly colonization potential.

Population genetic structure in subdivided populations

Snail dispersal and the nature of freshwater habitats

Freshwater snails occupy every kind of freshwater habitat. These habitats are often patchily distributed and may be viewed as islands of freshwater distributed in a "terrestrial sea"; they may be ponds, lakes, rivers, ditches, or artificial water bodies such as irrigation systems and reservoirs. Even when some of these habitats, such as large lakes, appear to be large and continuous to humans, discontinuity is likely to be imposed on snail populations by environmental factors such as water currents, wave action or food availability, as shown by numerous ecological studies (e.g., Russell-Hunter, 1978; Betterton, 1984; Woolhouse and Chandiwana, 1989).

　　The consequences of habitat patchiness for population genetic structure are expected to mainly depend on dispersal ability (Slatkin, 1985). Most snail species generally disperse little by active means, but many can disperse passively via animal agents such as birds or insects (e.g., Rees, 1965; Boag, 1986), or else during periods of flooding (Woolhouse, 1988; Madsen and Frandsen, 1989). Egg capsules and juvenile individuals likely can be dispersed when attached to drifting macrophytes or pieces of wood. Tropical and mediterranean freshwater habitats are often subject to annual periods of droughts (e.g., Vera et al., 1995); this may also hold for artificial water bodies regulated for agricultural purposes. It does not follow that affected populations completely disappear during droughts, at least in environments where snails can bury themselves; in some species, aestivation can last up to 11 months (Betterton et al., 1988; Vera et al., 1995). Habitats then fill up during the rainy season, and populations usually recover. This may have some significance for the genetics of populations since effective size is expected to fluctuate widely; it may also generate some within-habitat gene flow, such as within large ponds, assuming that individuals move such as to maintain themselves in the receding water body (Viard et al., 1996, 1997b).

Models of subdivided populations

Based on the biology of freshwater pulmonates and the nature of their habitats, it seems reasonable to consider the distribution of genetic variability by using models of subdivided populations (for reviews, see Slatkin, 1985; Barton and Whitlock, 1996). We will mainly consider the island model and its modified versions, because most predictions on the distribution of genetic variability in subdivided populations have been derived using this model (Slatkin, 1985, 1987; Wade and McCauley, 1988). The original island model (Wright, 1931) assumes an infinite number of populations exchanging migrants drawn from a unique pool, to which each population contributes equally at a proportion m. Finite versions of the model have since been derived, in which migrants may originate from only one source population. An alternative model of subdivided populations is the stepping-stone model (Kimura, 1953). It its simplest unidimensional version, populations are regularly distributed along a line or transect; each population exchanges a fraction $m/2$ of migrants with its two nearest neighbors. The two types of models sometimes behave similarly (Slatkin and Barton, 1989), and we will allude to the stepping-stone model only when necessary.

These models are not intended to describe a geographic reality. Rather, they represent a convenient framework for the distribution of variability, with population size and gene flow among populations being the important parameters (Slatkin, 1985, 1987). For example, it could be more realistic to describe the genetic structure of snails occupying a river using a unidimensional stepping-stone model with primarily downstream migration (due to passive dispersal). However it is possible to use the island model and the concept of "isolation by distance" (Wright, 1943), treating an upstream site as a reference population. Recent theoretical studies have considered the roles of local extinction, recolonization and the mating system in the island model. This is particularly relevant to the study of freshwater snail populations because of the transient nature of their habitats and their reproductive options of selfing or – in some species – parthenogenesis as uniparental alternatives to outcrossing.

A further aspect that must be taken into account is mutation. The vast majority of data describing the distribution of genetic variability in natural populations has been obtained with allozymes. These data can therefore be analyzed assuming the infinite alleles model (IAM) under which each mutation introduces a distinct, new allele. However, more and more microsatellite data will become available soon, for which the stepwise mutation model (SMM) may be more appropriate (for further distinctions and references, see Jarne and Lagoda, 1996). Under the SMM, each mutation introduces an allele differing from the progenitor allele by one "step", i.e., repeat unit of the microsatellite core sequence. In practice, this means that new microsatellite alleles are one repeat unit longer or shorter than the progenitor alleles. Microsatellites also have much higher mutation rates than allozymes, thus increasing the available variability (Charlesworth et al., 1994a; Jarne and Lagoda, 1996). We mention below where the distinction between the IAM and SMM seems important.

Reduction of within-population variability

Genetic drift and effective population size
Genetic drift results from the random sampling of gametes for zygote formation in finite populations (Hartl and Clark, 1989, chapter 2). Given enough time, genetic drift results in the fixation or loss of alleles at particular loci and thus in a decay of gene diversity. Its facility in eroding variability critically depends on effective population size, which may be extremely low at times of population crashes or colonization events (population bottlenecks). When the size of populations varies over generations, the magnitude of genetic drift is determined by the harmonic mean of population size (see Hartl and Clark, 1989). The magnitude of the loss of both heterozygosity and number of alleles further depends on levels of genetic variability before the bottleneck and the subsequent rate of population growth (Nei et al., 1975). These considerations are especially relevant for studies of tropical freshwater snails.

Rare alleles are particularly sensitive to the effects of bottlenecks. The restoration of variability occurs through mutation and/or gene flow, and at equilibrium between genetic drift and mutation under the IAM, the expected heterozygosity H is

$$H = \frac{4\,N\,u}{4\,N\,u + 1},$$

with N the effective population size and u the mutation rate to neutral alleles (Hartl and Clark, 1989, p. 124). Formulas under the SMM yield lower H values for similar values of $4\,N\,u$. However, as mentioned above, the mutation rate of microsatellites is much higher than that of allozyme loci. Regardless of the mutation model, little variability is maintained when effective population size is small.

Influence of the mating system
In infinite populations, selfing leads to the loss of heterozygous genotypes, compared to Hardy-Weinberg expectations (Fig. 1). In other words, the inbreeding coefficient of the population increases ($F_{IS} = F$, the probability that the two alleles at a given locus are identical by descent). Under the mixed-mating model, allowing a combination of random outcrossing and selfing, and assuming no inbreeding depression, at equilibrium

$$F = \frac{S}{2 - S},$$

with S the proportion of selfed offspring (i.e., selfing rate); this holds whatever the mutation model (Rousset, 1996).

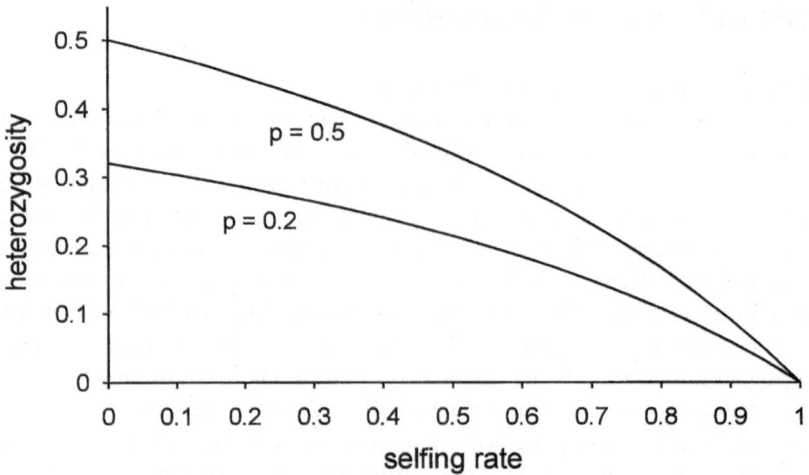

Figure 1. Observed heterozygosity at a polymorphic locus under inbreeding equilibrium. The curves correspond to allelic frequencies of 0.5 + 0.5 ($p = 0.5$) and 0.2 + 0.8 ($p = 0.2$), respectively. Hardy-Weinberg proportions, i.e., heterozygosity = $2 p q$, obtain under a selfing rate of zero. Under inbreeding equilibrium and partial selfing, heterozygosity = $2 p q (1 - F)$, with F the inbreeding coefficient. Thus, F measures the proportional reduction of observed heterozygosity.

Another important consequence of partial selfing is the maintenance and slow erosion of multilocus genotypic associations, i.e., gametic disequilibria. In the extreme situation of complete genomic homozygosity attained through selfing over many generations, genotypes are transmitted intact across generations when there is no external source of variability (e.g., mutation or gene flow). At that point, selfing and apomictic parthenogenesis have identical genetic consequences. A main difference between the two modes of uniparental reproduction is that heterozygosity, regardless how it was generated initially, can be maintained and replicated across generations via apomictic parthenogenesis, whereas selfing invariably erodes it (for examples, see Hughes, 1989).

Loss of variability in finite inbreeding populations
As mentioned above, populations of finite size are expected to gradually lose genetic variability. This effect of finite size is even more pronounced in clonal or selfing populations for at least four reasons.

First, the effective size N_{in} of an inbreeding population is

$$N_{in} = \frac{N_{out}}{1 + F},$$

with N_{out} the effective size of the corresponding random-mating population (Pollak, 1987). The genetic variability is therefore expected to be halved in populations with very high frequencies of self-fertilization. Orive (1993) showed theoretically that the ratio of the effective population size to the census population size is much lower in clonal organisms than in random-mating populations. Whether this will actually promote the loss of variability under natural conditions is not obvious, because census population sizes are generally not known for clonal populations.

A second reason is that selectively favored alleles can drive linked neutral alleles to very high frequencies in selfing populations, a process that has been referred to as "genetic hitchhiking" (Hedrick, 1980). Variability at the presumably neutral marker loci used to analyze genetic variability can therefore decrease in selfers because of genetic hitchhiking. This might also happen in multiclonal apomictic populations, after a selectively favored mutant is introduced. The third reason for the loss of variation in finite inbreeding populations is a form of generalized hitchhiking. Assuming that some loci are subject to mutation-selection balance and other loci are neutral, Charlesworth et al. (1993) showed that neutral alleles only go to fixation when occurring in gametes bearing the lowest number of deleterious mutations at the fitness-determining loci. As it does for hitchhiking, this "background selection" works at high selfing rates because genotypes are not recombined. Background selection results in the loss of both heterozygosity and number of alleles (Charlesworth et al., 1993); this result also holds for clonal populations. These ideas were initially developed with marker loci evolving under the IAM in mind. However, they can be extended to the SMM (P. Jarne and F. Viard, unpublished data), although a thorough quantitative analysis has yet to be done.

Finally, small populations are more likely to experience founder effects (Barrett and Kohn, 1991), and selfing populations may be more prone to bottlenecks than outcrossing populations, since they frequently seem to be involved in colonizing situations (see below). The loss of variability in inbreeding populations submitted to bottlenecks has yet to be appropriately analyzed theoretically. For populations of a given size, the absolute loss of alleles is stronger in inbreeding than in outbreeding populations, although the reverse holds for the relative loss of alleles; this is due to the lower effective size of inbreeding populations (Jarne, 1995).

Distinguishing between these factors is not straightforward, especially between the effects of genetic hitchhiking and background selection, and no tests have been proposed. Comparisons including species with various selfing rates may be useful to test whether the observed variability is diminished below the level expected due to reduced effective population size alone (Charlesworth et al., 1993, 1995; and see below). All the forces mentioned above are expected to reduce variability, whatever the genetic markers. How fast variability will recover after a marked reduction clearly depends on the mutation rate, and it seems likely that variability at microsatellite loci can recover very quickly (see Jarne and Lagoda, 1996).

Distribution of genetic variability among populations

The distribution of neutral alleles in subdivided sexual populations has been studied by Wright (1931), using the infinite island model. Wright demonstrated that no significant differentiation is to be expected when $N m > 1$, with N the effective population size and m the migration rate. When m decreases toward zero, a fraction p of the subpopulations are fixed for the allele of original frequency p; these results also hold for stepping-stone models. The genetic differentiation of subpopulations, as estimated by F_{ST}, is related to gene flow according to

$$F_{ST} = \frac{1}{1 + 4\, N m},$$

in the case of two alleles and with $u \ll m$. F_{ST} is the correlation of two randomly chosen alleles in a subpopulation relative to alleles in the whole population (Wright, 1951). Slatkin (1977, 1985) and others extended the analysis to the finite island model, and showed that a slightly modified version of the above formula can be used (Slatkin, 1985, p. 398). To account for sampling problems, F_{ST} must be estimated with the procedure developed by Weir and Cockerham (1984).

Models that allow migrants to be recruited at random from only one population, and that also consider local extinction and recolonization, have subsequently been analyzed by Slatkin (1977, 1985) and Wade and McCauley (1988). The results are still couched in terms of identity of alleles, and can be compared with the situation under the classical island model. Wade and McCauley (1988) found that F_{ST} may increase or decrease, depending on 1) how groups of colonists are formed, and 2) the numbers of individuals colonizing vacant habitats relative to twice the number moving into extant populations ($2\, N m$). The previous models were basically derived for random-mating populations and loci evolving under the infinite alleles model. More recently, Slatkin (1995) and Rousset (1996) showed that $N m$ cannot be estimated from F_{ST} under the stepwise mutation model, and introduced statistics more appropriate for microsatellite loci. The discrepancy is mainly due to the nature of mutation processes at these loci (see above), not to their higher mutation rate *per se*.

The influence of inbreeding on genetic differentiation has been analyzed by Whitlock and McCauley (1990), Maruyama and Tachida (1992), and Jarne (1995). These authors showed that inbreeding increases the genetic variance among populations and decreases the within-population genetic variance. In some way, inbreeding reduces the effective number of colonists. In clonal populations, we may expect an interplay of opposing forces: the establishment of new clones by way of migration, mutation, or "recruitment" from syntopic sexual genotypes (e.g., Dybdahl and Lively, 1995) being countered by genetic drift, i.e., clone extinction. A theoretical study by Maruyama and Kimura (1980) showed that population subdivision and recurrent episodes of extinction and colonization decrease both effective population size and the effective number of alleles (= "clones").

Genetic variability in freshwater pulmonates

Most data on genetic variability in freshwater pulmonates have been obtained using starch gel protein electrophoresis. Among the important genetic parameters discussed above, we focus on migration, genetic drift, and the mating system. Selection is not considered here, because allozymes are generally believed to be more or less neutral variants (Skibinski et al., 1993; but see Gillespie, 1991; Karl and Avise, 1992), and mutation rates are too low to have significant short-term effects. The null model is the island model at genetic equilibrium between migration and drift, an assumption that should be cautiously considered (see Viard et al., 1997b).

We have previously conducted an interspecific comparison of allozyme variability in freshwater pulmonates, based on published studies (for details and references, see Jarne and Städler, 1995). Unfortunately, sampling has rarely been designed to accommodate thorough population genetic analyses, and it is still virtually impossible to test for the main factors affecting genetic variability. The roles of genetic drift and bottlenecks may be inferred from a comparison of tropical and temperate genera, the former generally occupying less stable habitats in terms of water availability across seasons (e.g., Vera et al., 1995). This may partly explain the higher variability exhibited by the temperate genera *Lymnaea* and *Physa* when compared to the two tropical genera, *Biomphalaria* and *Bulinus* (see Jarne, 1995). As no estimate of effective population size is available, the role of genetic drift in freshwater pulmonates usually can only be suggested based on their fluctuating distribution in time and space. A recent study on *Bulinus globosus* provides an exception. Njiokou et al. (1994) showed that under some conditions, genetic drift and/or sampling effects can explain the observed variation in allelic frequencies over generations. No allozyme data pertaining to local extinctions and recolonizations are available.

More convincing is the comparison of selfing and outcrossing species. Of the 24 studies (16 species) evaluated by Jarne and Städler (1995), the inferred mating system is predominant selfing in eight studies (four species). As a group, selfing species have a lower proportion of polymorphic loci, a lower number of alleles per locus and lower expected heterozygosity (gene diversity) than outcrossing species, as expected on theoretical grounds (Tab. 1). However, the reduced effective population size of selfers alone cannot explain this loss of variability (see above). Rather, environmental effects, background selection or genetic hitchhiking must be invoked to account for the low variability in selfers, which, with the exception of *Ancylus fluviatilis* (Städler et al., 1995), are tropical

Table 1. Parameters describing the allozyme variability – a comparison of primarily selfing and primarily outcrossing freshwater pulmonate snails

Parameter	Selfers (nine studies)		Outcrossers (16 studies)		p
prop. of polymorphic loci	0.237	(0.307)	0.536	(0.271)	0.006
number of alleles per locus	1.056	(0.091)	1.410	(0.351)	<0.001
observed heterozygosity	0.001	(0.002)	0.081	(0.068)	<0.001
gene diversity	0.009	(0.012)	0.094	(0.081)	<0.001
G_{ST} (F_{ST}-equivalent)	0.634	(0.359)	0.302	(0.194)	0.057

All parameter estimates are averages across studies, with standard deviations in parentheses; p is the significance of a Mann-Whitney U-test. For references and other details, see Jarne and Städler (1995); data from *Ancylus fluviatilis* (T. Städler, unpublished data) have been added to the "selfer" column.

species. As expected, population differentiation as measured by an equivalent of F_{ST} is larger in selfers, even when the area sampled per study is taken into account (Tab. 1; Jarne, 1995).

Finally, it should be noted that allozyme variability in some tropical species is so low that inferences about the forces acting within populations and between populations separated by short geographic distances are impossible. However, this situation provided an impetus to characterize microsatellite loci (Jarne et al., 1994; Viard et al., 1996). These markers turned out to be highly polymorphic, and provided the first robust estimates of high selfing rates and levels of gene flow in *Bulinus truncatus* (Tab. 2). Based on temporal surveys, microsatellites also allowed to show that some populations of *B. truncatus* exhibit marked variation in allelic frequencies over a few generations, while others are more stable. The first scenario has been found for temporary ponds, while the second is more characteristic of permanent ponds. This has been interpreted as resulting from intra-pond migration in temporary ponds following water recession or run-off (Viard et al., 1997b). Nevertheless, the observed microsatellite polymorphism in *B. truncatus* is lower than in outcrossing species with large population size (references in Viard et al., 1996). Clearly, comparisons among populations or sister taxa, differing only in their selfing rate, would provide much-needed empirical data on the influence of the mating system on genetic variability.

Estimation of mating system parameters in freshwater pulmonates

Basommatophoran snails possess a single gonad, the ovotestis. A potential consequence is the occurrence of self-fertilization, since male and female gametes mature simultaneously and in close proximity (for reviews, see Duncan, 1975; Geraerts and Joosse, 1984; Jarne et al., 1993). Despite some cases of transient protandry among basommatophorans (e.g., Wethington and Dillon, 1993), once sexually mature, freshwater pulmonates are simultaneous hermaphrodites. Careful laboratory analyses have shown that

Table 2. Comparison of genetic variability in *Bulinus truncatus* based on allozymes versus microsatellites

Genetic marker	Populations studied	Number of loci (polym. loci)	# Alleles/locus (s.d.)	Obs. hetero-zygosity (s.d.)	Gene diversity (s.d.)	Inferred selfing rate (F_{IS}-based; s.d.)	Reference
allozymes	9	14 (3)	1.07 (0.27)	0.000 (0.000)	<0.001 (—)	—	Jelnes (1986)
allozymes	13	24 (2)	1.00 (0.00)	0.000 (0.000)	0.000 (0.000)	—	Mimpfoundi and Greer (1990)
allozymes	18	29 (10)	1.01 (0.02)	0.000 (0.000)	0.005 (0.010)	1.000 (0.000)	Njiokou et al. (1993a)
microsatellites	14	4 (4)	4.73 (2.15)	0.071 (0.065)	0.508 (0.203)	0.929 (0.059)	Viard et al. (1996)

The number of polymorphic loci given in parentheses refers to the entire geographic scale of each study; # alleles/locus refers to the mean number of alleles per locus, including monomorphic loci; – denotes missing data, or inference not possible. All standard deviations (s.d.) were calculated over populations.

selfing is possible in all species studied. Unfortunately, these studies have been limited to very few genera, namely *Ancylus, Biomphalaria, Bulinus, Lymnaea* and *Physa*. Without additional data on natural populations and further taxa, it remains unclear how general these observations are, and whether species with the potential to self-fertilize do so with appreciable frequency under natural conditions. In this section, we focus on the estimation of mating system parameters, in particular the selfing rate, using genetic markers.

Genetic markers

The selfing rate must be estimated using genetic markers. Behavioral observations and experiments generally provide few pertinent data, because copulation *per se* is a necessary, but by no means sufficient, event for outcrossing to occur (Larambergue, 1939; Njiokou et al., 1993b; Städler et al., 1993; Doums et al., 1996). Freshwater snails are prone to copulate in normally outcrossing species, particularly after periods of isolation, whereas copulations seem to occur less frequently in selfers (Doums et al., 1996). Nevertheless, copulation may serve functions other than providing allosperm, such as stimulation of ovulation and/or egg-laying (see Geraerts and Joosse, 1984). Another complication occurs under laboratory conditions because the presence of multiple partners may lead to lower selfing rates than is typical of natural populations. Moreover, a more or less continuous production of eggs over the whole life cycle, once maturity has been attained, introduces the possibility of temporal variation in the selfing rate.

Ideal genetic markers would be neutral variants with codominant, Mendelian inheritance. Further, they ought to be usable at all stages of the life cycle, provide data for both laboratory and field populations, and allow analyses of large numbers of individuals. Predictably, no single available marker satisfies all these requirements. Pigment markers have been used in a few outcrossing species, e.g., *Biomphalaria glabrata, Bulinus* spp., and *Physa heterostropha*. They have been very useful to show that outcrossing actually occurs, and that allosperm can be stored for several months (for review, see Jarne et al., 1993). However, as phenotypic characters with a genetic basis, pigment markers are likely to be under selective pressure. For instance, normally pigmented and albino laboratory strains of *B. glabrata*, when paired, differ in their propensity to outcross as males (Vianey-Liaud, 1995), which can be interpreted as the result of sperm competition. The utility of pigment markers is further hampered by the recessivity of albinism (e.g., Paraense, 1955), which severely limits their applicability in natural settings.

Both allozymes and DNA markers can be used to estimate the selfing rate. While DNA markers other than microsatellites might be useful to some extent, we anticipate that microsatellites will be the markers of choice in species with limited allozyme polymorphism, primarily due to their high variability and codominant inheritance, facilitating a range of population genetic analyses (Viard et al., 1996; Jarne and Lagoda, 1996). Two logically and experimentally differing approaches can be followed to estimate the selfing rate.

Indirect estimation procedures

As already mentioned, the selfing rate, S, can be inferred from population genetic structure, using the equation $F_{IS} = S/(2 - S)$ (see Fig. 1). This approach assumes mixed-mating and genetic equilibrium. Positive F_{IS} values, i.e., deficiencies of heterozygotes, have been the main – and often the only – argument in favor of predominant selfing in some species of freshwater pulmonates (see Jarne and Städler, 1995). However, positive values of F_{IS} may reflect factors other than selfing such as temporal or spatial variance in allele frequencies (Wahlund effect) or biparental inbreeding. Nevertheless, these other factors are less likely to confound estimation of the mating system as the selfing rate increases. Moreover, data on population genetic structure may have the advantage of reflecting selfing rates averaged over a longer timespan, unless S fluctuates widely across generations.

Importantly, this indirect approach will underestimate the true selfing rate (primary mating system) if mean survival of selfed offspring is lower than that of outcrossed offspring, i.e., with inbreeding depression > 0 (see below). Similarly, this approach necessarily "ignores" that selfing is not a feature of entire populations, but rather an attribute of individuals. Variation in the selfing rate may be expected among individuals and, by extension, among populations and species, but data on population genetic structure only allow inferences of the population-level mating system. Direct approaches are therefore required to uncover biologically important within-population variation in the selfing rate.

Direct estimates via progeny arrays

A partial solution in cases of low within-population allozyme polymorphism (preventing F_{IS} estimates) has been to cross individuals drawn from populations fixed for different alleles and then scoring the resulting progeny at the marker locus or loci. Using this kind of crossing study, *Bulinus truncatus* (Njiokou et al., 1993b) and *Ancylus fluviatilis* (Städler et al., 1993) were estimated to self-fertilize in excess of 90%. However, in the absence of additional data pertaining to the actual selfing rate of source populations, it is difficult to dismiss ecological differences among populations or partial reproductive isolation as factors artificially lowering outcrossing rates in a crossing experiment, especially when high proportions of selfed offspring are obtained. Clearly, these ambiguities call for additional within-population estimates of the mating system. Subsequently, pairs of individuals drawn from single populations were shown to exhibit high selfing rates when crossed in the laboratory (*A. fluviatilis*: Loew, 1992, allozymes; *B. truncatus*: Doums et al., 1996, microsatellites). Such crossing experiments must be performed using individuals sampled very recently from natural populations to minimize laboratory-induced variation in the selfing rate.

The most relevant direct approach involves progeny-array analyses, also known as parent–offspring comparisons (Brown, 1990). Progeny-array analysis rests on comparisons of maternal snails ("mothers") and samples of their offspring, each scored for several polymorphic marker loci; thus, the genotypic data are family-structured. The basic model assumes mixed mating and yields maximum-likelihood (ML) estimates of the selfing rate (Ritland and Jain, 1981; Ritland, 1986). Extensions of this model allow the estimation of additional parameters and consideration of situations more realistic than mixed mating (see Brown, 1990). In contrast to the voluminous work on seed plants utilizing progeny-array data to estimate selfing rates, such analyses have generally not been used in freshwater snails. Although family-structured data were gathered in several earlier studies, authors only tested for pure selfing versus pure outcrossing (e.g., Mulvey and Vrijenhoek, 1981; Vrijenhoek and Graven, 1992).

However, a recent study on a highly polymorphic population of *Ancylus fluviatilis* has provided the first progeny-array based ML estimates of mating system parameters in animal hermaphrodites (Städler et al., 1995; see Tab. 3). Based on 42 maternal families and 848 progeny scored for five polymorphic allozyme loci, the population selfing rate was found to be in the range 85–87%. Moreover, the small difference between multilocus estimates and those derived from the mean of single-locus values suggested a limited role of biparental inbreeding. These progeny-array data

Table 3. ML estimates of population-level selfing rates (S) and correlated-mating parameters based on progeny-array data in *Ancylus fluviatilis* and *Bulinus truncatus*

Parameter	Ancylus fluviatilis		Bulinus truncatus	
multilocus S (S_m)	0.865	(0.056)	0.845	(0.076)
single-locus S (S_s)	0.873	(0.061)	0.871	(0.072)
correlation of selfing (r_s)	0.661	(0.139)	0.580	(0.195)
correlation of paternity (r_p)	0.810	(0.111)	0.817	(0.291)

Data on *A. fluviatilis* (one population, 42 families with 848 progeny, allozymes) are from Städler et al. (1995), those on *B. truncatus* (five populations, 57 families with 447 progeny, microsatellites) are from Viard et al. (1997a). See text for an explanation of correlated-mating parameters r_s and r_p. All estimates were obtained with the program MLTR (K. Ritland, unpublished data). Standard deviations (in parentheses) are based on bootstrapping the data 300 times (*A. fluviatilis*); estimates for *B. truncatus* are presented as mean values (s.d.) for the five populations.

should closely portray the natural population selfing rate, as maternal snails were sampled at the height of the reproductive season, after copulations and transfer of allosperm had occurred in the field. These data thus represent the strongest evidence yet for high selfing rates in *A. fluviatilis*, substantiating the earlier inter-population crossing experiment (Städler et al., 1993). Also, partial selfing was confirmed at the individual level, as all outcrossing snails also produced some selfed progeny.

Importantly, these progeny-array data illustrate the detectability of among-individual heterogeneity in the mating system, as only 38% of the maternal cohort contributed to the population outcrossing rate, with individual outcrossing rates ranging from zero to 0.89 (Städler et al., 1995). This kind of information is unavailable, in principle, from data on population genetic structure. Moreover, parameters of Ritland's (1989) "sibling-pair" model entail information on the extent of inter-individual differences in the mating system and the extent of multiple paternity among outcrossed sibs. If all families were to exhibit the same (population) selfing rate, r_s, the correlation of selfing within progeny arrays would equal zero, whereas $r_s = 1$ obtains when families are either purely selfed or purely outcrossed. Similarly, r_p, the correlation of outcrossed paternity within sibships, equals zero when all outcrosses are half-sibs, whereas $r_p = 1$ indicates that outcrossed sibs were sired by the same father (Ritland, 1989). Clearly, the progeny-array data in *A. fluviatilis* show strong differences in individual selfing rates and suggest that the extent of multiple paternity is limited (Tab. 3; Städler et al., 1995). While it remains unknown whether these differences have a genetic basis, nonrandom patterns of genetic relatedness due to correlated matings in partially self-fertilizing populations could have biological implications in situations such as resource competition among sibs or selective embryo abortion in brooders (Lively and Johnson, 1994).

Very recent progeny-array data on five populations of *Bulinus truncatus*, using highly polymorphic microsatellites as genetic markers, likewise demonstrate high selfing rates with considerable inter-individual variability (Viard et al., 1997a); the degree of concordance with data in *A. fluviatilis* is indeed striking (Tab. 3). Although the number of families per population and family size were fairly small, the use of several populations spanning a range of aphally ratios allowed to test for an association between the selfing rate and the sexual morph (aphallic versus euphallic; see Larambergue, 1939). High selfing rates overall were found for both sexual morphs (Viard et al., 1997a), and population structure data on 14 populations of *B. truncatus* likewise fail to support a correlation between aphally ratio and the population selfing rate (Tab. 2; Viard et al., 1996). Hence, at present there seems to be no empirical basis for the assumption of higher selfing rates in populations with higher proportions of aphallic individuals (Schrag et al., 1994a, b; see Johnson et al., this volume).

While these recent progeny-array data on two predominantly selfing species underscore the existence of variation in the selfing rate, this variation calls for an explanation of the circumstances under which it may evolve and be maintained. Although the potential rewards from progeny-array data seem obvious, large sample sizes and several polymorphic loci are required to obtain robust estimates, especially for correlated mating parameters and individual outcrossing rates (Ritland, 1989). Therefore, we expect that microsatellite markers will be used for this purpose in largely selfing species, but largely outcrossing species in the genera *Lymnaea* and *Physa* exhibit enough allozyme polymorphism for these estimates to be feasible.

Inbreeding depression and other forces

Inbreeding depression is thought to be of general importance in the evolution of selfing versus outcrossing. Here, we deliberately emphasize a population genetic perspective and approaches to parameter estimation, keeping in mind that ecological factors surely deserve equal attention in explaining the evolutionary trajectories of mating systems (see Jarne and Charlesworth, 1993; Johnson et al., this volume). More than half a century ago, Fisher (1941) pointed out that, all else being equal, a selfing mutant occurring in an outcrossing population would have an immediate fitness advantage because it can transmit genes by selfing but also through male outcrossing. Disregarding male gamete discounting, it thus has a 50% genetic advantage known as the cost of outcrossing (Charlesworth, 1980). Naturally, this advantage may disappear when selfed offspring have lower mean fitness than outcrossed progeny. If we operationally define inbreeding depression as

$$\delta = 1 - \frac{w_s}{w_o}$$

(Charlesworth and Charlesworth, 1987), with w_s and w_o denoting the mean fitness of selfed and outcrossed offspring, respectively, the cost of outcrossing is exactly compensated when $\delta = 0.5$. This simple approach assumes that both the selfing rate and inbreeding depression are fixed entities. However, the mating system can be expected to evolve with inbreeding history and population structure due to associations between mating system loci and fitness-determining loci (see Uyenoyama et al., 1993; Waller, 1993), requiring the genetic basis of inbreeding depression to be specified. This more appropriate dynamic perspective has shown that a threshold value of $\delta = 0.5$ cannot be taken as an absolute rule in predicting the evolution of selfing rates.

Most recent theory has assumed that inbreeding depression is caused by partially recessive deleterious alleles at fitness-determining loci, maintained by mutation-selection balance (Lande and Schemske, 1985; Charlesworth and Charlesworth, 1987; Charlesworth et al., 1990b; Lande et al.,

1994). Inbreeding causes these deleterious alleles to be exposed to selection because of increased homozygosity at all loci. Consequently, a "purging" of the genetic load under continued inbreeding is expected, and historically inbreeding populations ought to exhibit less inbreeding depression than habitually outcrossing populations.

However, it does not follow that predominately selfing populations are expected to show no fitness reduction of selfed progeny; recurrent mutations to mildly deleterious alleles should be able to maintain substantial inbreeding depression even in highly selfing populations (Charlesworth et al., 1990a; Lande et al., 1994). This theoretical proposition has found some empirical support from recent studies in angiosperms (Charlesworth et al., 1994b; Johnston and Schoen, 1995). Based on a large interspecific comparison of data in seed plants, Husband and Schemske (1996) found that primarily selfing taxa exhibit inbreeding depression mainly for later stages in the life cycle whereas outcrossers show strong components of inbreeding depression at all stages. This suggests that early-acting recessive lethals have been purged in selfers, while mutations with small effects are still maintained. The magnitude of inbreeding depression in highly selfing populations is of particular theoretical interest, as such data may allow estimates of the genomic mutation rate to deleterious alleles (Kondrashov, 1985; Charlesworth et al., 1990a; Johnston and Schoen, 1995).

The experimental approach in freshwater pulmonates

Estimating the magnitude of inbreeding depression in freshwater pulmonates is not a trivial task. Like in the extensive studies on seed plants, the traditional experimental approach involves fitness comparisons of selfed progeny and presumably outcrossed progeny (e.g., Jarne and Delay, 1990; Jarne et al., 1991). There are at least two potential problems with this approach. First, unlike in plants, outcrossing cannot be enforced in partially self-fertilizing snails. Thus, paired or grouped snails will produce a mixture of selfed and outcrossed progeny in partial selfers such as *Bulinus truncatus* and *Ancylus fluviatilis*. Although genetic markers can, in principle, be used to verify the origin of each offspring (see Doums et al., 1996), the sheer number of individuals in such experiments limits this possibility in practice. Consequently, there is no unambiguous fitness comparison of progeny generated by selfing and outcrossing, respectively. However, even in the absence of data on the selfing rate of paired individuals, a maximum value of inbreeding depression for juvenile survival can be estimated. If x is the proportion of surviving selfed offspring (sired by isolated individuals), then $1-x$ represents the maximum inbreeding depression. Such reasoning has been used by Doums et al. (1996) to infer very limited inbreeding depression in three highly selfing populations of *B. truncatus* and *Biomphalaria straminea*.

Second, poorly-understood interactions among conspecifics and chemical stimuli may be responsible for "grouping effects", resulting in lower fecundity of grouped individuals, irrespective of density, when the mating system is controlled for (Doums et al., 1994). Moreover, Vernon (1995) recently claimed that an absence of "social facilitation", rather than inbreeding depression, might be responsible for the low fecundity and poor hatching success in isolated individuals of primarily outcrossing freshwater pulmonates. She found that individuals of *Biomphalaria glabrata* prevented from outcrossing, but in the presence of a conspecific, had higher fecundity and hatching rates than isolated, selfing snails. While previous studies have indeed confounded effects of the mating system (selfing or outcrossing) and experimental condition (isolation or pairing), Vernon's (1995) study seems far from definitive because of low sample size, unusually low hatching success, and the use of strains that had been maintained in the laboratory for numerous generations. Thus, the generality of her results is difficult to evaluate without additional studies using appropriate designs.

In snails as in seed plants, the experimental approach usually yields estimates of inbreeding depression based only on fitness *components*, not lifetime fitness (e.g., Jarne and Delay, 1990; Doums et al., 1996). In snails, these components usually are fecundity, hatching success of eggs, and posthatching juvenile survival. Furthermore, it is well known from studies in plants that factors such as competition, predation or parasitism may affect estimates of inbreeding depression. Hence, greenhouse studies under fairly benign environmental conditions usually yield lower inbreeding depression than studies conducted in the field, although this does not always hold (references in Charlesworth et al., 1994 b). Because all estimates in freshwater pulmonates have been laboratory-based, one may expect that the magnitude of inbreeding depression has been systematically underestimated.

The temporal pattern of inbreeding depression uncovered by Husband and Schemske (1996; see above) in seed plants implies that comparisons of juvenile survival up to maturity and reproduction might be required in order to arrive at meaningful estimates of inbreeding depression in highly selfing snails. Such an experimental regime is feasible in species of the tropical genera *Biomphalaria* and *Bulinus* that are characterized by short generation times and easy laboratory maintenance (see Doums et al., 1996), but is clearly unattainable in annual species such as *Ancylus fluviatilis*. At any rate, the plant data suggest that robust estimates of inbreeding depression will be hard to obtain in highly selfing snails, using the experimental approach.

Assuming that inbreeding depression is caused by partially recessive deleterious alleles, limited depression is expected in highly selfing populations, and large depression is expected in normally outcrossing populations (Lande and Schemske, 1985; Charlesworth et al., 1990b). Despite the caveats discussed above, the limited data available in freshwater pulmonates qualitatively agree with theoretical expectations. While severe in-breeding depression was found in largely outcrossing freshwater pulmo-nates (e.g., Jarne and Delay, 1990; Jarne et al., 1991), limited inbreeding depression and low neutral polymorphism, high selfing rates, and limited propensity to copulate are associated in what has been referred to as the "selfing syndrome" (e.g., *Bulinus truncatus*; Doums et al., 1996). Inter-estingly, among-population differences in the magnitude of inbreeding depression have been shown for *Bulinus globosus* (Njiokou et al., 1992), in parallel with evidence for high versus low selfing rates in these populations derived from population genetic structure. However, no definitive con-clusions are warranted as yet, given the limited extent of robust data in freshwater pulmonates.

The marker-based approach

Ritland (1990) introduced a non-experimental population genetic approach to estimating lifetime inbreeding depression with genetic markers. In partially self-fertilizing populations, the inbreeding coefficient, F, usually increases from adults to zygotes due to partial selfing. However, F should then decrease among the juvenile cohort if there is selection against products of selfing, i.e., inbreeding depression. The magnitude of this de-cline in F from zygotes to adults reflects the intensity of inbreeding depres-sion. With one exception (Dole and Ritland, 1993), only Ritland's (1990) equilibrium estimator has been used to date in studies on angiosperms:

$$\delta = 1 - \frac{2F(1-S)}{S(1-F)},$$

with F the inbreeding coefficient of the adult cohort (maternal parents) and S the population-level selfing rate of these adults, measured directly via progeny arrays (see above). This estimator assumes constant F values across generations, and thus suffers from large statistical variance (Rit-land, 1990).

Using progeny-array data from two predominantly selfing populations of *Ancylus fluviatilis*, this approach yields estimates of inbreeding depression in excess of 0.5 (T. Städler, unpublished data). Taken at face value, this would indicate substantial inbreeding depression in a species characterized by selfing rates >0.85 (Tab. 3; Städler et al., 1995). However, standard errors of these inbreeding depression estimates are large, and the approach may not be suitable for populations with very high selfing rates (Ritland, 1990). Interestingly, plant studies where estimates of inbreeding depression are available from both the experimental and marker-based approach have found higher values of inbreeding depression based on genetic markers (e.g., Dole and Ritland, 1993; Eckert and Barrett, 1994). Whether these

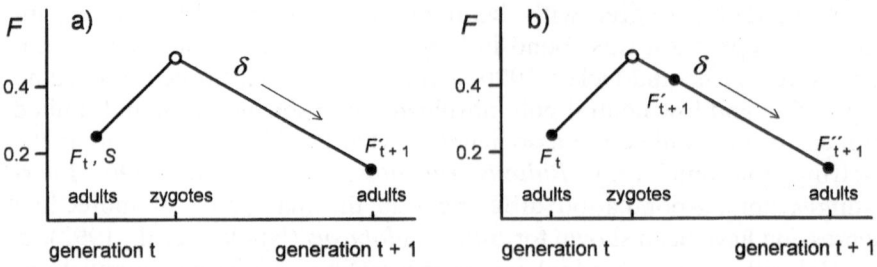

Figure 2. Expected change in the inbreeding coefficient F over two generations in partially selfing populations. Two marker-based approaches may be used to estimate the magnitude of inbreeding depression (δ): (a) Determination of adult F and adult selfing rate S via progeny arrays in generation t, plus adult F in generation t + 1; (b) Determination of adult F in both generations and of juvenile F in generation t + 1, all from population data. The primes denote F after selection against selfed progeny in generation t + 1, and filled circles indicate the time of sampling with the estimated parameters. The scale of the y-axis is arbitrary. Modified after Ritland (1990).

discrepancies are due to inherent biases of the marker approach (e.g., due to associations between marker loci and fitness-determining loci in partially selfing populations) or alternatively, due to incomplete "lifetime" fitness comparisons obtained experimentally, is presently unknown.

At any rate, multigenerational methods, rather than the equilibrium estimator, should be used to obtain robust estimates, as originally advocated by Ritland (1990). The two possible approaches are illustrated in Figure 2. Essentially, combining progeny-array data in generation t (simultaneous estimation of adult F and S) with population data in generation t + 1 (F of the next adult cohort) substitutes data for the equilibrium assumption (Fig. 2(a)). Alternatively, three estimates of F from population data can yield information on the extent of inbreeding depression between juvenile and adult life-cycle stages in generation t + 1 (Fig. 2(b)). The marker-based approach seems particularly suitable for annual species with non-overlapping generations and difficult laboratory maintenance for which experimental studies are not feasible; however, it should also be valuable in other partially self-fertilizing species.

Other factors in mating system evolution

The cost of outcrossing and inbreeding depression should not be portrayed as the only relevant forces at work (see Tab. 4). For instance, a potential role for reproductive assurance is suggested by the indirect evidence for founder effects, from data on population genetic structure (Viard et al., 1997b). Further, selfing could be favored under strong selection for local adaptation, because selfing reproduces particular multilocus genotypes more faithfully. Extensive studies in angiosperms have found instances of local adap-

Table 4. Possible advantages and disadvantages of self-fertilization (see text for discussion)

Type of hypotheses	Advantages	Disadvantages
Genetical	Avoidance of cost of outcrossing (see text)	Inbreeding depression (see text)
		Fertility usually lowered in self-fertilizing individuals
		Male gamete discounting = the reduced production of male gametes with increasing mean selfing rate
		Biparental inbreeding = inbreeding resulting not from selfing but from non-random outcrossing among relatives
Ecological	Local adaptation = the aptitude of reproducing locally adapted genotypes *Alternatively:* Maintenance of "general-purpose genotypes" across generations due to restricted recombination; colonization potential	Variable physical environments: the inaptitude to promote variability or generate diverse genotypes across generations and among offspring to cope with spatial and temporal variability
	Cost of copulation = the many costs involved in finding a partner, courting, copulating; exposure to predators	Variable biotic environments: the inaptitude to cope with coevolving antagonists, in particular parasites, which are thought to exert frequency-dependent selection (Red Queen hypothesis)
	Reproductive assurance = the assurance of reproducing even when isolated	

Male gamete discounting and biparental inbreeding cannot easily be classified as either advantageous or disadvantageous.

tation at very small spatial scales (references in Waser, 1993). Naturally, movement of individuals, gametes, or dispersive stages such as seeds (i.e., gene flow) would tend to counteract selection for local adaptation. As snails are not strictly sessile organisms, the spatial scale at which local adaptation – if any – may be anticipated is larger than in angiosperms. The only relevant work in snails is Dillon's elegant study in the gonochoric prosobranch *Goniobasis proxima* (Dillon, 1988). Artificial introductions with genetically marked individuals failed to indicate any local adaptation in these highly subdivided populations, and even suggested an advantage for introduced snails, which partly displaced the resident population.

Biparental inbreeding, and hence population structure, likely plays a role in unstable and small populations. Under some circumstances that have been studied theoretically, biparental inbreeding can generate evolutionarily stable intermediate selfing rates (Ronfort and Couvet, 1995). This is a satisfying result because most previous genetic models predict that only pure selfing and pure outcrossing are stable – predictions that do not fully account for the observed distribution of selfing rates among and within species. Similarly, recent strategy models considering the joint effects of parasites and inbreeding depression have shown that these forces could interact to select for evolutionarily stable intermediate selfing rates (Lively and Howard, 1994). Parasites and metapopulation structure as factors in mating system evolution are explored in more depth by Johnson et al. (this volume); Table 4 summarizes the main arguments.

Mating systems and population dynamics

Population genetics and population dynamics follow the fate of genes and individuals, respectively, over space and time. Mainly for historical reasons, the two fields are rarely considered jointly. However, knowledge about population dynamics is a prerequisite for genetic analyses of subdivided populations. Some parameters, such as extinction rates or the rates of increase of populations must be derived from empirical investigations of population dynamics (e.g., Betterton et al., 1988; Woolhouse, 1992). Although the processes involved in population dynamics and genetics do not act at the same time scale, we call for studies incorporating both approaches; artificial systems may prove to be more appropriate for small-scale studies. Population dynamics is most obviously associated with the mating system through density-dependent effects and colonization potential.

Density-dependent effects

Only a few studies on the population dynamics of natural populations of freshwater snails have been performed using appropriate techniques (e.g.,

Woolhouse and Chandiwana, 1989; Woolhouse, 1992). These studies indicate large spatial and temporal variation in density. The connection between density, mating system, and population variability is even less-well understood. However, density may modify the distribution of genetic variability through its action on the mating system for the following reasons:

1) Density modifies the opportunity to mate and therefore to outcross. 2) Inbreeding depression may vary with population density; this has been shown empirically in plants (e.g., Charlesworth et al., 1994b). 3) Density may alter certain life-history traits. For instance, the reproductive output of *Bulinus truncatus* under laboratory conditions is reduced at high densities (Doums et al., 1994). Whether reduced reproductive output results in a modification of the mating system, e.g., through an effect on inbreeding depression, has not been analyzed. 4) Parasite load may vary with the local density of intermediate hosts (see Lively, 1992). Such a pattern may reflect greater opportunities for parasite transmission in high-density populations of hosts, and may also affect genetic variability of host populations through changes in the mating system or differential attack of particular genotypes/clones.

The colonization potential

A critical feature of metapopulation models is the (re)colonization rate of vacant habitats. Some freshwater snails have a high potential for colonization, as exemplified by the rapid spread of planorbids in irrigation systems (Betterton, 1984; Madsen and Frandsen, 1989), the invasion of the French Antilles by the parthenogenetic prosobranch *Melanoides tuberculata* (Pointier et al., 1992, 1993) and the rapid colonization of Europe by the New Zealand prosobranch *Potamopyrgus antipodarum* (Ponder, 1988; Hughes, 1989). Although much remains to be learned about the role of reprodutive systems in colonizing species, circumstantial evidence suggests that selfing or parthenogenetic lineages can be particularly successful colonizers.

Weedy species and so-called "general-purpose genotypes" have long been viewed as being better at colonizing empty or marginal environments (Baker, 1965). General-purpose genotypes are characterized by small temporal variance in fitness in fluctuating environments, compared to more specialized genotypes (see Lynch, 1984). Because selection can operate on entire multilocus associations of alleles, apomictic parthenogens and highly inbred lineages that persist for considerable time are expected to be tolerant of diverse ecological conditions. On the other hand, an array of clones that can continuously be spun-off from sexual genotypes, and where individual clones may be short-lived, is more akin to an assemblage of "frozen" sexual genotypes and is not necessarily expected to possess generalist qualities (Vrijenhoek, 1984; Dybdahl and Lively, 1995).

A general problem for an empirical assessment of the relationship between colonizing potential and the mating system in hermaphrodites is the potential circularity. It is difficult to discriminate whether high selfing rates are just a consequence of colonization (i.e., due to genetic bottlenecks), rather than one of the factors causally related to its initial success. The limited dataset in freshwater pulmonates prevents a rigorous test of these alternative hypotheses. We note that the highly selfing *Bulinus truncatus* has a very large geographic range, but other outcrossing *Bulinus* species are also widely distributed. Two very successful colonizers have opposite mating systems. *Physa heterostropha* is most probably an outcrosser in North America (Dillon and Wethington, 1995). *Physa acuta* is likely conspecific with *P. heterostropha* (R.T. Dillon, personal communication), and recently invaded Europe where it is now widespread. Preliminary work indicates that *P. acuta* also is a preferential outcrosser (P. Jarne, unpublished data). On the other hand, *Lymnaea truncatula*, most likely a selfer, very recently invaded South America from the Old World (Jabbour-Zahab et al., 1997). At least at this crude level of analysis, the empirical evidence therefore seems equivocal. By and large, selfing is probably not a *necessary* prerequisite for being a good colonizer.

The clonal prosobranchs *Melanoides tuberculata* and *Potamopyrgus antipodarum* have recently extended their geographic range, most likely mediated by human activities. Based on morphology and life-history comparisons, Winterbourn (1972) and Ponder (1988) argued convincingly for the New Zealand origin of European *P. antipodarum (= P. jenkinsi)*, the latter having been introduced in the mid-19th century. Most European populations appear to be composed of parthenogenetic females (Wallace, 1992). In contrast, many New Zealand populations harbor mixtures of sexual and clonal individuals (Lively, 1987, 1992), and clones are now known to be triploid (Wallace, 1992; Dybdahl and Lively, 1995). The high number and spatial structuring of allozyme-detected clones in some New Zealand populations suggests frequent "freezing" of sexual genotypes into new clonal lineages (Dybdahl and Lively, 1995; Fox et al., 1996).

Given the absence of diploid sexuals in Europe (preventing new clones being spun-off from a local sexual population) and the unstable population dynamics with frequent population crashes (references in Hughes, 1989), we may expect European populations of *P. antipodarum* to be dominated by a limited number of widespread clones. In accord with this prediction, a DNA fingerprinting study of British populations documented three widely dispersed, genetically differentiated lineages, attributed to different waves of colonization (Hauser et al., 1992). Two of these clonal lineages were subsequently found to comprise Danish populations of *P. antipodarum* (Jacobsen et al., 1996), with no additional clones detected. However, mitochondrial haplotypes derived from DNA sequence analysis, and representing more extensive geographic sampling, suggest that the minimum number of colonizing lineages must be greater than previously thought; these mitochondrial DNA data clearly point to multiple introductions derived from New Zealand source populations (M. Frye and T. Städler, unpublished data).

Melanoides tuberculata has invaded much of the intertropics, most probably from Africa. However, the history of invasion has been followed in detail only in the French Antilles, where *M. tuberculata* has enjoyed considerable success (Pointier et al., 1992, 1993). Here, it may have contributed to the extinction of indigenous populations of *Biomphalaria glabrata* and *B. straminea*. Moreover, morphologically distinguishable lineages of colonizing *M. tuberculata* differ significantly in life-history features (Pointier et al., 1992). These patterns are consistent with a model of persistent interclonal selection and the geographic spread of a few generalist genotypes. Recent genetic evidence suggests that at least some clones of *M. tuberculata* are polyploid, and there may be variation in ploidy among clones (B. Delay, unpublished data). Although it still seems premature to causally link clonality, polyploidy, and colonization potential, these two apomictic prosobranchs appear to be better colonizers, at least in the recent past, than the average prosobranch species.

Conclusions and prospects

While the study of both freshwater pulmonates and prosobranchs has much to offer, this potential is only beginning to be appreciated. For example, the vigorous research program to elucidate the evolutionary significance of alternative reproductive modes in hermaphrodite vascular plants has not yet been matched by similar efforts in hermaphrodite animals (see Jarne and Charlesworth, 1996). Conversely, the genetics and ecology of clonal animals has been thoroughly investigated, particularly among vertebrates and aquatic arthropods; however, relevant studies in prosobranchs have only recently begun. The conceptual and theoretical tools as well as suitable molecular techniques are now available to address fundamental issues in population biology, taking advantage of the diversity of reproductive strategies realized among freshwater snails.

Population genetic structure

The potential of protein electrophoretic data to unravel the relationship between the selfing rate and population genetic structure in species exhibiting a reasonable amount of genetic variability (e.g., *Lymnaea peregra*) has not been fully exploited. Most allozyme studies on freshwater pulmonates have not even attempted to estimate the inbreeding coefficient, instead drawing tentative conclusions from qualitative agreement (or otherwise) with Hardy-Weinberg equilibrium at polymorphic loci, and/or from overall levels of genetic variability. Regardless, caution should be exercised when selfing rates are inferred solely from observed heterozygote deficits, because such deficits might also be generated by biparental inbreeding due to unrecognized population substructure. However, under significant levels of early-acting inbreeding depression, studies of population genetic structure might completely "miss" a selfing component in the mating system, i.e., if selfed zygotes do not survive to the time of sampling. This pervasive possibility argues strongly for direct estimates of the selfing rate via progeny arrays (see below). In populations with limited allozyme polymorphism, microsatellites are the markers of choice (e.g., Viard et al., 1996, 1997a). In either case, robust datasets on the genetic structure of freshwater pulmonates have the potential to test predictions of models concerned with inbreeding species and those with limited powers of dispersal (Viard et al., 1997b). Studies of population dynamics and rates of local extinction and recolonization are needed to empirically assess the suitability and assumptions of metapopulation models (e.g., Dybdahl, 1994).

Evolution of mating systems

Jointly with data on population structure, independently derived evidence on mating system parameters in self-fertile hermaphrodite snails will help to explore the evolutionary significance of selfing versus outcrossing. In particular, direct estimates of the selfing rate via progeny arrays are a powerful approach to uncover inter-individual differences in the mating system (Städler et al., 1995; Viard et al., 1997a) and to estimate male outcrossing paternity. Progeny-array data thus allow a much finer dissection of reproductive phenomena within populations, inaccessible to indirect estimates of the population-level mating system. Moreover, estimates of the level of inbreeding depression among populations with different selfing rates may be used to assess the genetic basis of inbreeding depression; such data also promise to clarify the relationship between selfing rates and population genetic structure.

    In addition to population genetic work, life-history studies appear to be promising, especially in tropical planorbids with easy laboratory maintenance and rapid generation turnover, such as some species in the genus *Bulinus*. For example, it seems feasible to monitor the course of inbreeding depression over several generations in an experimental setting, thus directly addressing the purging of deleterious alleles. Compared to seed plants, however, highly selfing pulmonates present some additional, but not insurmountable, problems to precise estimation of inbreeding depression. Populations with different selfing rates should lend themselves to test for differential allocation of resources to male versus female function. Similarly, we expect that both genetic data (allozymes, nuclear and mitochondrial DNA markers) and life-history studies will significantly improve our understanding of clonal origins, clonal richness, and clonal turnover in parthenogenetic prosobranchs, parameters with important implications for theories on the maintenance of sex (see Dybdahl and Lively, 1995; Fox et al., 1996).

Acknowledgements
Our understanding of mixed mating systems and clonal prosobranchs owes much to numerous discussions with Curt Lively and Jukka Jokela. Rob Dillon, Claudie Doums, Steve Johnson, Curt Lively and Frédérique Viard offered helpful comments on the manuscript, for which we are grateful. We also thank Nick Barton, Bernard Delay, Rob Dillon, Curt Lively and Marc Vianey-Liaud for access to unpublished manuscripts and/or information. T. S. was supported by the ESF *Programme in Population Biology* during a 6-month-stay in Montpellier. P. J. is supported by grants from CNRS to UMR 5554 (Institut des Sciences de l'Evolution) and by the Ministère de l'Environnement (EGPN 94019). This is contribution No. 97-055 from Institut des Sciences de l'Evolution.

References

Baker, H.G. (1965) Characteristics and mode of origin of weeds. *In:* H.G. Baker and G.L. Stebbins (eds): *The Genetics of Colonizing Species.* Academic Press, New York, pp 147–172.

Barrett, S.C.H. and Kohn, J.R. (1991) Genetic and evolutionary consequences of small population size in plants: Implications for conservation. *In:* D.A. Falk and K.E. Holsinger (eds): *Genetics and Conservation of Rare Plants.* Oxford University Press, New York, pp 3–30.

Barton, N.H. and Whitlock, M.C. (1996) The evolution of metapopulations. *In:* I.A. Hanski and M.E. Gilpin (eds): *Metapopulation Biology: Ecology, Genetics, and Evolution.* Academic Press, London, pp 183–210.

Bell, G. (1982) *The Masterpiece of Nature – The Evolution and Genetics of Sexuality.* University of California Press, Berkeley.

Betterton, C. (1984) Spatiotemporal distributional patterns of *Bulinus rohlfsi* (Clessin), *Bulinus forskalii* (Ehrenberg) and *Bulinus senegalensis* (Müller) in newly-irrigated areas in northern Nigeria. *J. Mollusc. Stud.* 50:137–152.

Betterton, C., Ndifon, G.T. and Tan, R.M. (1988) Schistosomiasis in Kano State, Nigeria. II. Field studies on aestivation in *Bulinus rohlfsi* (Clessin) and *Bulinus globosus* (Morelet) and their susceptibility to local strains of *Schistosoma haematobium* (Bilharz). *Ann. Trop. Med. Parasitol.* 82:571–579.

Boag, D.A. (1986) Dispersal in pond snails: Potential role of waterfowl. *Can. J. Zool.* 64:904–909.

Brown, A.H.D. (1990) Genetic characterization of plant mating systems. *In:* A.H.D. Brown, M.T. Clegg, A.L. Kahler and B.S. Weird (eds): *Plant Population Genetics, Breeding, and Genetic Resources.* Sinauer Associates, Sunderland, pp 145–162.

Brown, D.S. (1994) *Freshwater Snails of Africa and Their Medical Importance*, Second Edition. Taylor and Francis Ltd., London.

Charlesworth, B. (1980) The cost of sex in relation to mating system. *J. Theor. Biol.* 84:655–671.

Charlesworth, B., Charlesworth, D. and Morgan, M.T. (1990a) Genetic loads and estimates of mutation rates in highly inbred plant populations. *Nature* 347:380–382.

Charlesworth, B., Morgan, M.T. and Charlesworth, D. (1993) The effect of deleterious mutations on neutral molecular variation. *Genetics* 134:1289–1303.

Charlesworth, B., Sniegowski, P. and Stephan, W. (1994a) The evolutionary dynamics of repetitive DNA in eukaryotes. *Nature* 371:215–220.

Charlesworth, D. and Charlesworth, B. (1987) Inbreeding depression and its evolutionary consequences. *Annu. Rev. Ecol. Syst.* 18:237–268.

Charlesworth, D., Morgan, M.T. and Charlesworth, B. (1990b) Inbreeding depression, genetic load, and the evolution of outcrossing rates in a multilocus system with no linkage. *Evolution* 44:1469–1489.

Charlesworth, D., Lyons, E.E. and Litchfield, L.B. (1994b) Inbreeding depression in two highly inbreeding populations of *Leavenworthia. Proc. Roy. Soc. Lond.* B 258:209–214.

Charlesworth, D., Charlesworth, B. and Morgan, M.T. (1995) The pattern of neutral molecular variation under the background selection model. *Genetics* 141:1619–1632.

Charnov, E.L. (1982) *The Theory of Sex Allocation.* Princeton University Press, Princeton.

Dillon, R.T., Jr. (1988) Evolution from transplants between genetically distinct populations of freshwater snails. *Genetica* 76:111–119.

Dillon, R.T., Jr. and Wethington, A.R. (1995) The biogeography of sea islands: Clues from the population genetics of the freshwater snail *Physa heterostropha. Syst. Biol.* 44:400–408.

Dole, J. and Ritland, K. (1993) Inbreeding depression in two *Mimulus* taxa measured by multi-generational changes in the inbreeding coefficient. *Evolution* 47:361–373.

Doums, C., Delay, B. and Jarne, P. (1994) A problem with the estimate of self-fertilization depression in the hermaphrodite freshwater snail *Bulinus truncatus*: The effect of grouping. *Evolution* 48:498–504.

Doums, C., Viard, F., Pernot, A.-F., Delay, B. and Jarne, P. (1996) Inbreeding depression, neutral polymorphism, and copulatory behavior in freshwater snails: A self-fertilization syndrome. *Evolution* 50:1908–1918.

Duncan, C.J. (1975) Reproduction. *In:* V. Fretter and J. Peake (eds): *Pulmonates,* vol. 1, *Functional Anatomy and Physiology.* Academic Press, London, pp 309–365.

Dybdahl, M.F. (1994) Extinction, recolonization, and the genetic structure of tidepool copepod populations. *Evol. Ecol.* 8:113–124.

Dybdahl, M.F. and Lively, C.M. (1995) Diverse, endemic and polyphyletic clones in mixed populations of a freshwater snail (*Potamopyrgus antipodarum*). *J. Evol. Biol.* 8:385–398.

Eckert, C.G. and Barrett, S.C.H. (1994) An analysis of inbreeding depression in partially self-fertilizing *Decodon verticillatus* (Lythraceae): Experimental and population genetic approaches. *Evolution* 48:952–964.

Fisher, R.A. (1941) Average excess and average effect of a gene substitution. *Ann. Eugen.* 11:53–63.

Fox, J.A., Dybdahl, M.F., Jokela, J. and Lively, C.M. (1996) Genetic structure of coexisting sexual and clonal subpopulations in a freshwater snail (*Potamopyrgus antipodarum*). *Evolution* 50:1541–1548.

Geraerts, W.P.M. and Joosse, J. (1984) Freshwater snails (Basommatophora). *In:* A.S. Tompa, N.H. Verdonk and J.A.M. van den Biggelaar (eds): *The Mollusca,* vol. 7, *Reproduction.* Academic Press, Orlando, pp 141–207.

Gillespie, J.H. (1991) *The Causes of Molecular Evolution.* Oxford University Press, New York.

Hamrick, J.L. and Godt, M.J.W. (1990) Allozyme diversity in plant species. *In:* A.H.D. Brown, M.T. Clegg, A.L. Kahler and B.S. Weird (eds): *Plant Population Genetics, Breeding, and Genetic Resources.* Sinauer Associates, Sunderland, pp 43–63.

Hartl, D.L. and Clark, A.G., (1989) *Principles of Population Genetics.* Second Edition. Sinauer Associates, Sunderland.

Hauser, L., Carvalho, G.R., Hughes, R.N. and Carter, R.E. (1992) Clonal structure of the introduced freshwater snail *Potamopyrgus antipodarum* (Prosobranchia: Hydrobiidae), as revealed by DNA fingerprinting. *Proc. Roy. Soc. Lond.* B 249:19–25.

Hedrick, P.W. (1980) Hitchhiking: A comparison of linkage and partial selfing. *Genetics* 94:791–808.

Heller, J. (1993) Hermaphroditism in molluscs. *Biol. J. Linn. Soc.* 48:19–42.

Hughes, R.N. (1989) *A Functional Biology of Clonal Animals.* Chapman and Hall, London.

Husband, B.C. and Schemske, D.W. (1996) Evolution of the magnitude and timing of inbreeding depression in plants. *Evolution* 50:54–70.

Jabbour-Zahab, R., Pointier, J.-P., Jourdane, J., Jarne, P., Oviedo, J.A., Bargues, M.D., Mas-Coma, S., Anglès, S., Perera, G., Balzan, C., Khallaayoune, K., Renaud, F. (1997) Phylogeography and genetic divergence of some lymnaeid snails, intermediate hosts of human and animal fascioliasis with special reference to lymnaeids from the Bolivian Altiplano. *Acta Tropica* 64:191–203.

Jacobsen, R., Forbes, V.E. and Skovgaard, O. (1996) Genetic population structure of the prosobranch snail *Potamopyrgus antipodarum* (Gray) in Denmark using PCR-RAPD fingerprints. *Proc. Roy. Soc. Lond.* B 263:1065–1070.

Jarne, P. (1995) Mating system, bottlenecks and genetic polymorphism in hermaphroditic animals. *Genet. Res.* 65:193–207.

Jarne, P. and Charlesworth, D. (1993) The evolution of the selfing rate in functionally hermaphrodite plants and animals. *Annu. Rev. Ecol. Syst.* 24:441–466.

Jarne, P. and Charlesworth, D. (1996) Hermes meets Aphrodite: An animal perspective. *Trends Ecol. Evol.* 11:105–107.

Jarne, P. and Delay, B. (1990) Inbreeding depression and self-fertilization in *Lymnaea peregra* (Gastropoda: Pulmonata). *Heredity* 64:169–175.

Jarne, P. and Lagoda, P.J.L. (1996) Microsatellites, from molecules to populations and back. *Trends Ecol. Evol.* 11:424–429.

Jarne, P. and Städler, T. (1995) Population genetic structure and mating system evolution in freshwater pulmonates. *Experientia* 51:482–497.

Jarne, P., Finot, L., Delay, B. and Thaler, L. (1991) Self-fertilization versus cross-fertilization in the hermaphroditic freshwater snail *Bulinus globosus*. *Evolution* 45:1136–1146.

Jarne, P., Vianey-Liaud, M. and Delay, B. (1993) Selfing and outcrossing in hermaphrodite freshwater gastropods (Basommatophora): Where, when and why. *Biol. J. Linn. Soc.* 49:99–125.

Jarne, P., Viard, F., Delay, B. and Cuny, G. (1994) Variable microsatellites in the highly selfing snail *Bulinus truncatus* (Basommatophora: Planorbidae). *Mol. Ecol.* 3:527–528.

Jelnes, J.E. (1986) Experimental taxonomy of *Bulinus* (Gastropoda: Planorbidae): The West and North African species reconsidered, based upon an electrophoretic study of several enzymes per individual. *Zool. J. Linn. Soc.* 87:1–26.

Johnston, M.O. and Schoen, D.J. (1995) Mutation rates and dominance levels of genes affecting total fitness in two angiosperm species. *Science* 267:226–229.

Karl, S.A. and Avise, J.C. (1992) Balancing selection at allozyme loci in oysters: Implications form nuclear RFLPs. *Science* 256:100–102.

Kimura, M. (1953) "Stepping stone" model of population. *Annu. Rept. Natl. Inst. Genet. Japan* 3:62–63.

Kondrashov, A.S. (1985) Deleterious mutation as an evolutionary factor. II. Facultative apomixis and selfing. *Genetics* 111:635–653.

Lande, R. and Schemske, D.W. (1985) The evolution of self-fertilization and inbreeding depression in plants. I. Genetic models. *Evolution* 39:24–40.

Lande, R., Schemske, D.W. and Schultz, S.T. (1994) High inbreeding depression, selective interference among loci, and the threshold selfing rate for purging recessive lethal mutations. *Evolution* 48:965–978.

Larambergue, M. de (1939) Étude de l'autofécondation chez les gastéropodes pulmonés. Recherches sur l'aphallie et la fécondation chez *Bulinus (Isidora) contortus* Michaud. *Bull. Biol. France Belg.* 73:19–231.

Lively, C.M. (1987) Evidence from a New Zealand snail for the maintenance of sex by parasitism. *Nature* 328:519–521.

Lively, C.M. (1992) Parthenogenesis in a freshwater snail: Reproductive assurance versus parasitic release. *Evolution* 46:907–913.

Lively, C.M. and Howard, R.S. (1994) Selection by parasites for clonal diversity and mixed mating. *Phil. Trans. Roy. Soc. Lond.* B 346:271–281.

Lively, C.M. and Johnson, S.G. (1994) Brooding and the evolution of parthenogenesis: Strategy models and evidence from aquatic invertebrates. *Proc. Roy. Soc. Lond.* B 256:89–95.

Loew, M. (1992) *Genetische Untersuchungen zum Reproduktionssystem von* Ancylus fluviatilis *(Gastropoda: Basommatophora)*. Diploma thesis. University of Frankfurt, Frankfurt am Main, Germany.

Lynch, M. (1984) Destabilizing hybridization, general-purpose genotypes and geographic parthenogenesis. *Quart. Rev. Biol.* 59:257–290.

Madsen, H. and Frandsen, F. (1989) The spread of freshwater snails including those of medical and veterinary importance. *Acta Tropica* 46:139–146.

Maruyama, K. and Tachida, H. (1992) Genetic variability and geographical structure in partially selfing populations. *Jpn. J. Genet.* 67:39–51.

Maruyama, T. and Kimura, M. (1980) Genetic variability and effective population size when local extinction and recolonization of subpopulations are frequent. *Proc. Natl. Acad. Sci. USA* 77:6710–6714.

McCauley, D.E. (1991) Genetic consequences of local population extinction and recolonization. *Trends Ecol. Evol.* 6:5–8.

Mimpfoundi, R. and Greer, G.J. (1990) Allozyme comparisons and ploidy levels among species of the *Bulinus truncatus/tropicus* complex (Gastropoda: Planorbidae) in Cameroon. *J. Mollusc. Stud.* 56:63–68.

Mulvey, M. and Vrijenhoek, R.C. (1981) Multiple paternity in the hermaphroditic snail, *Biomphalaria obstructa. J. Hered.* 72:308–312.

Nei, M., Maruyama, T. and Chakraborty, R. (1975) The bottleneck effect and genetic variability in populations. *Evolution* 29:1–10.

Njiokou, F., Bellec, C., N'Goran, E.K., Yapi Yapi, G., Delay, B. and Jarne, P. (1992) Comparative fitness and reproductive isolation between two *Bulinus globosus* (Gastropoda: Planorbidae) populations. *J. Mollusc. Stud.* 58:367–376.

Njiokou, F., Bellec, C., Berrebi, P., Delay, B. and Jarne, P. (1993a) Do self-fertilization and genetic drift promote a very low genetic variability in the allotetraploid *Bulinus truncatus* (Gastropoda: Planorbidae) populations? *Genet. Res.* 62:89–100.

Njiokou, F., Bellec, C., Jarne, P., Finot, L. and Delay, B. (1993b) Mating system analysis using protein electrophoresis in the self-fertile hermaphrodite species *Bulinus truncatus* (Gastropoda: Planorbidae). *J. Mollusc. Stud.* 59:125–133.

Njiokou, F., Delay, B., Bellec, C., N'Goran, E.K., Yapi Yapi, G. and Jarne, P. (1994) Population genetic structure of the schistosome-vector snail *Bulinus globosus*: Examining the role of genetic drift, migration and human activities. *Heredity* 72:488–497.

Orive, M.E. (1993) Effective population size in organisms with complex life-histories. *Theor. Popul. Biol.* 44:316–340.

Paraense, W.L. (1955) Self and cross-fertilization in *Australorbis glabratus. Mems Inst. Oswaldo Cruz* 53:285–291.

Pointier, J.-P., Delay, B., Toffart, J.L., Lefèvre, M. and Romero-Alvarez, R. (1992) Life history traits of three morphs of *Melanoides tuberculata* (Gastropoda: Thiaridae), an invading snail in the French West Indies. *J. Mollusc. Stud.* 58:415–423.

Pointier, J.-P., Thaler, L., Pernot, A.-F. and Delay, B. (1993) Invasion of the Martinique island by the parthenogenetic snail *Melanoides tuberculata* and the succession of morphs. *Acta Oecologica* 14:33–42.

Pollak, E. (1987) On the theory of partially inbreeding finite populations. I. Partial selfing. *Genetics* 117:353–360.

Ponder, W.F. (1988) *Potamopyrgus antipodarum* – a molluscan coloniser of Europe and Australia. *J. Mollusc. Stud.* 54:271–285.

Rees, W.J. (1965) The aerial dispersal of Mollusca. *Proc. Malac. Soc. Lond.* 36:269–282.

Ritland, K. (1986) Joint maximum likelihood estimation of genetic and mating structure using open-pollinated progenies. *Biometrics* 47:35–43.

Ritland, K. (1989) Correlated matings in the partial selfer *Mimulus guttatus. Evolution* 43:848–859.

Ritland, K. (1990) Inferences about inbreeding depression based upon changes in the inbreeding coefficient. *Evolution* 44:1230–1241.

Ritland, K. and Jain, S. (1981) A model for the estimation of outcrossing rate and gene frequencies using *n* independent loci. *Heredity* 47:35–52.

Ronfort, J. and Couvet, D. (1995) A stochastic model of selection on selfing rates in structured populations. *Genet. Res.* 65:209–222.

Rousset, F. (1996) Equilibrium values of measures of population subdivision for stepwise mutation processes. *Genetics* 142:1357–1362.

Russell-Hunter, W.D. (1978) Ecology of freshwater pulmonates. *In:* V. Fretter and J. Peake (eds): *Pulmonates,* vol. 2A, *Systematics, Evolution and Ecology.* Academic Press, London, pp 335–383.

Schoen, D.J. and Brown, A.H.D. (1991) Intraspecific variation in population gene diversity and effective population size correlates with the mating system in plants. *Proc. Natl. Acad. Sci. USA* 88:4494–4497.

Schrag, S.J., Mooers, A.Ø., Ndifon, G.T. and Read, A.F. (1994a) Ecological correlates of male outcrossing ability in a simultaneous hermaphrodite snail. *Am. Nat.* 143:636–655.

Schrag, S.J., Ndifon, G.T. and Read, A.F. (1994b) Temperature-determined outcrossing ability in wild populations of a simultaneous hermaphrodite snail. *Ecology* 75:2066–2077.

Selander, R.K. and Ochman, H. (1983) The genetic structure of populations as illustrated by molluscs. *Isozymes* 10:93–123.

Skibinski, D.O.F., Woodmark, M. and Ward, R.D. (1993) A quantitative test of the neutral theory using pooled allozyme data. *Genetics* 135:233–248.

Slatkin, M. (1977) Gene flow and genetic drift in a species subject to frequent local extinctions. *Theor. Popul. Biol.* 12:253–262.

Slatkin, M. (1985) Gene flow in natural populations. *Annu. Rev. Ecol. Syst.* 16:393–430.

Slatkin, M. (1987) Gene flow and the geographic structure of natural populations. *Science* 236: 787–792.

Slatkin, M. (1995) A measure of population subdivision based on microsatellite allele frequencies. *Genetics* 139:457–462.

Slatkin, M. and Barton, N.H. (1989) A comparison of three indirect methods for estimating average levels of gene flow. *Evolution* 43:1349–1368.

Städler, T., Loew, M. and Streit, B. (1993) Genetic evidence for low outcrossing rates in polyploid freshwater snails (*Ancylus fluviatilis*). *Proc. Roy. Soc. Lond.* B 251:207–213.

Städler, T., Weisner, S. and Streit, B. (1995) Outcrossing rates and correlated matings in a predominantly selfing freshwater snail. *Proc. Roy. Soc. Lond.* B 262:119–125.

Uyenoyama, M.K., Holsinger, K.E. and Waller, D.M. (1993) Ecological and genetic factors directing the evolution of self-fertilization. *Oxford Surv. Evol. Biol.* 9:327–381.

Vera, C., Bremond, P., Labbo, R., Mouchet, F., Sellin, E., Boulanger, D., Pointier, J.P., Delay, B. and Sellin, B. (1995) Seasonal fluctuations in population densities of *Bulinus senegalensis* and *B. truncatus* (Planorbidae) in temporary pools in a focus of *Schistosoma haematobium* in Niger: Implications for control. *J. Mollusc. Stud.* 61:79–88.

Vernon, J.G. (1995) Low reproductive output of isolated, self-fertilizing snails: Inbreeding depression or absence of social facilitation? *Proc. Roy. Soc. Lond.* B 259:131–136.

Vianey-Liaud, M. (1995) Bias in the production of heterozygous pigmented embryos from successively mated *Biomphalaria glabrata* (Gastropoda: Planorbidae) albino snails. *Malacol. Rev.* 28:97–106.

Viard, F., Bremond, P., Labbo, R., Justy, F., Delay, B. and Jarne, P. (1996) Microsatellites and the genetics of highly selfing populations in the freshwater snail *Bulinus truncatus*. *Genetics* 142:1237–1247.

Viard, F., Doums, C. and Jarne, P. (1997a) Selfing, sexual polymorphism and microsatellites in the hermaphroditic freshwater snail *Bulinus truncatus*. *Proc. Roy. Soc. Lond.* B 264: 39–44.

Viard, F., Justy, F. and Jarne, P. (1997b) Metapopulation dynamics from temporal variations at microsatellite loci in the selfing snail *Bulinus truncatus*. *Genetics; in press*.

Vrijenhoek, R.C. (1984) Ecological differentiation among clones: The frozen niche-variation model. *In:* K. Wöhrmann and V. Loeschcke (eds): *Population Biology and Evolution*. Springer-Verlag, Berlin, pp 217–231.

Vrijenhoek, R.C. and Graven, M.A. (1992) Population genetics of Egyptian *Biomphalaria alexandrina* (Gastropoda, Planorbidae). *J. Hered.* 83:255–261.

Wade, M.J. and McCauley, D.E. (1988) Extinction and recolonization: Their effects on the genetic differentiation of local populations. *Evolution* 42:995–1005.

Wallace, C. (1992) Parthenogenesis, sex and chromosomes in *Potamopyrgus*. *J. Mollusc. Stud.* 58:93–107.

Waller, D.M. (1993) The statics and dynamics of mating system evolution. *In:* N.W. Thornhill (ed.): *The Natural History of Inbreeding and Outbreeding*. University of Chicago Press, Chicago and London, pp 97–117.

Waser, N.M. (1993) Population structure, optimal outbreeding, and assortative mating in angiosperms. *In:* N.W. Thornhill (ed.): *The Natural History of Inbreeding and Outbreeding*. University of Chicago Press, Chicago and London, pp 173–199.

Weird, B.S. and Coqauvin, C.C. (1984) Estimating *F*-statistics for the analysis of population structure. *Evolution* 38:1358–1370.

Wethington, A.R. and Dillon, R.T., Jr. (1993) Reproductive development in the hermaphroditic freshwater snail *Physa* monitored with complementing albino lines. *Proc. Roy. Soc. Lond.* B 252:109–114.

Whitlock, M.C. and McCauley, D.E. (1990) Some population genetic consequences of colony formation and extinction: Genetic correlations within founding groups. *Evolution* 44: 1717–1724.

Winterbourn, M.J. (1972) Morphological variation of *Potamopyrgus jenkinsi* (Smith) from England and a comparison with the New Zealand species, *Potamopyrgus antipodarum* (Gray). *Proc. Malac. Soc. Lond.* 40:133–145.

Woolhouse, M.E.J. (1988) Passive dispersal of *Bulinus globosus*. *Ann. Trop. Med. Parasitol.* 82:315–317.

Woolhouse, M.E.J. (1992) Population biology of the freshwater snail *Biomphalaria pfeifferi* in the Zimbabwe highveld. *J. Appl. Ecol.* 29:687–694.

Woolhouse, M.E.J. and Chandiwana, S.K. (1989) Spatial and temporal heterogeneity in the population dynamics of *Bulinus globosus* and *Biomphalaria pfeifferi* and in the epidemiology of their infection with schistosomes. *Parasitology* 98:21–34.

Wright, S. (1931) Evolution in Mendelian populations. *Genetics* 16:97–159.

Wright, S. (1943) Isolation by distance. *Genetics* 28:114–138.

Wright, S. (1951) The genetical structure of populations. *Ann. Eugen.* 15:323–354.

Evolutionary Ecology of Freshwater Animals
ed. by B. Streit, T. Städler and C. M. Lively
© 1997 Birkhäuser Verlag Basel/Switzerland

Evolution and ecological correlates of uniparental and biparental reproduction in freshwater snails

S. G. Johnson[1], C. M. Lively[2], S. J. Schrag[3]

[1] Department of Biological Sciences, University of New Orleans, New Orleans, LA 70148, USA
[2] Department of Biology, Indiana University, Bloomington, IN 47505, USA
[3] Department of Biology, Emory University, Atlanta, GA 30322, USA

Summary. We review the spatial and temporal correlates of uniparental and biparental reproduction in three species of freshwater snails as they pertain to the ecological hypotheses for the
maintenance of biparental sex. The biogeographic evidence from two species (*Potamopyrgus
antipodarum* and *Bulinus truncatus*) presently supports the Red Queen hypothesis that biparental reproduction is selected as a way to reduce the risk to progeny of parasite attack. Uniparental reproduction in these species is associated with low levels of infection by parasites
(castrating digenetic trematodes), suggesting that parthenogenesis or self-fertilization can replace
cross-fertilization when the risk of infection is low. In addition, in *B. truncatus,* the timing
of cross-fertilization coincides with the season in which parasite attack is highest. In a third
species (*Campeloma decisum*), parthenogenetic reproduction is correlated with latitude and the
presence of a non-castrating trematode that may prevent cross-fertilization; these patterns
suggest that parthenogenesis has been selected as a mechanism to assure reproduction. We also
discuss the taxonomic distribution of parthenogenesis in aquatic invertebrates, and suggest
that brooding may be an exaptation for the evolution of parthenogenetic reproduction in these
animals.

Introduction

The apparent evolutionary stability of biparental sex in many eukaryotes
remains a paradox in evolutionary biology (Williams, 1971; Maynard
Smith, 1978; Bell, 1982). If sexual and parthenogenetic females have
similar fecundity, parthenogenetic females should rapidly replace sexual
females because sexual females invest in sons which do not directly bear
any progeny, while parthenogens invest only in daughters (Maynard Smith,
1971, 1978). Similarly, an allele for self-fertilization that does not also
affect reproductive success through male function should rapidly spread in
outcrossing hermaphroditic populations (Williams, 1971, 1975; Nagylaki,
1976; Lloyd, 1979). Offspring produced by selfing, just like offspring
produced asexually, receive all their genes from only one parent. This can
translate into a 3/2 advantage for a mutation to pure self-fertilization in a
non-selfing randomly mating population, assuming no inbreeding depression and no loss of ability to contribute male gametes via outcrossing
(Charlesworth, 1980). Hence, cross-fertilization is subject to invasion and
replacement by uniparental forms of reproduction in both dioecious and
hermaphroditic populations, and would seem to require a general explanation for its maintenance. Such theories should also explain the predomi-

nance of asexual lineages in novel or disturbed habitats at high altitudes and latitudes (review in Bell, 1982).

In this paper, we first briefly describe the major ecological hypotheses for the maintenance of biparental sex. These ideas are mainly concerned with the adaptive significance of producing variable progeny in variable environments. We then present a review of our studies on three freshwater gastropods, which were designed to discriminate among these alternative hypotheses. These studies involve investigations of the spatial and temporal distribution of uniparental and biparental reproduction within and among populations of a New Zealand prosobranch (*Potamopyrgus antipodarum*), an African pulmonate (*Bulinus truncatus*), and a North American prosobranch (*Campeloma decisum*). Finally, we consider the relationship between brooding and the phylogenetic distribution of parthenogenesis in aquatic invertebrates.

The ecological hypotheses for maintenance of biparental sex

The adaptive variation hypotheses

Three hypotheses for the maintenance of cross-fertilization in natural populations postulate that there is an advantage to producing variable progeny in variable environments. They differ primarily in whether or not the advantage stems from frequency-dependent selection and, if so, whether the frequency dependence comes from intraspecific or interspecific interactions.

Lottery hypothesis

The lottery hypothesis predicts a selective advantage for outcrossing in environments that vary unpredictably over time (Fisher, 1930; Williams, 1975). When offspring experience environments different from their parents, the cost of outcrossing could be offset by the benefit of producing genetically variable offspring, thereby increasing the likelihood that some survive the new conditions. The advantage to outcrossing under this view is frequency independent, and operates by increasing the expected geometric mean fitness of outcrossing individuals. The lottery hypothesis predicts that biparental reproduction will be more common in unstable environments where conditions are likely to vary unpredictably between generations.

The tangled bank hypothesis

The tangled bank hypothesis predicts a selective advantage of outcrossing in heterogeneous environments when there is high intraspecific competition for resources (Ghiselin, 1974; Bell, 1982). If parents and offspring experience similar environments and resources are limited, offspring which are able to utilize new niches may face less competition. Hence there may be a density-dependent rare advantage that selects for cross-fertilization. The tangled bank hypothesis predicts that biparental reproduction will be more common in stable environments where there is high competition for resources. The biogeographic distribution of uniparental reproduction is consistent with the tangled bank hypothesis in that selfing and partheno-genesis are more common in more homogeneous, less stable habitats at high altitudes or latitudes (Bell, 1982).

These predictions are based on the assumption that biparental repro-duction leads to higher levels of genetic variability among offspring (Mitchell-Olds and Waller, 1985). This assumption does not always hold for the case of outcrossing versus selfing (Schmitt and Ehrhardt, 1987). If selfed sibships were more genetically diverse than outcrossed sibships, then patterns opposite to our predictions might be found.

The Red Queen hypothesis

Fluctuations in the biotic environment, in particular some forms of antag-onistic coevolution, can also favor outcrossing (Levin, 1975; Jaenike, 1978; Bremermann, 1980; Hamilton, 1980, 1993; Lloyd, 1980; Bell, 1982; reviews in Ebert and Hamilton, 1996; Hurst and Peck, 1996; Clay and Kover, 1996). The most common version of this hypothesis, which we call the Red Queen hypothesis after Bell (1982), focuses on the interaction between parasites and their hosts. When parasites substantially decrease host fitness, coevolution between parasites and hosts can result in fre-quency-dependent selection favoring rare host genotypes (Hamilton, 1980). A central prediction of the Red Queen hypothesis is that selection for rare genotypes should be stronger in populations where parasites exert a greater selective pressure on their hosts. Uniparental reproduction should be favored in habitats where parasite pressure is weak.

The risk of parasitism may be expected to vary among locations in a host's geographic range. For example, in parasites that cycle between two or more host species (e.g., digenetic trematodes), the absence of one host species in some locations will eliminate the parasite. In parasites such as microsporidians that directly transmit infections between individuals of the same host species, infection may not be sustained in low-density host popu-lations. Hence, parthenogens may replace sexuals in some areas due to relaxation of negative frequency-dependent selection for the production of rare genotypes by sex. Levin (1975) argued that the association between uniparental reproduction in plants and marginal habitats may be due to the absence of parasites. Glesener and Tilman (1978) made a similar case for geographical parthenogenesis in animals. However, as Jaenike (1978) first noted, the advantage to parthenogenesis may be short-lived in evolutionary time because the original parasites and sexual individuals may eventually colonize the area. Due to the inability to produce genetically variable progeny, the asexual hosts may be devastated when recolonized by parasites, and there would be a strong advantage to colonizing sexual individuals in the presence of parasites.

In a similar fashion, outcrossing simultaneous hermaphrodites will be more successful at generating novel genotypes than those which reproduce by selfing, and this could lead to a stable mixture of outcrossed and selfed progeny within a single parent (Lively and Howard, 1994). Parasites should also select against common clones in gonochoric populations; parasites may eventually eliminate such clones in the presence of either soft-truncation selection resulting from intraspecific competition for resources (Hamilton et al., 1990), or if they aid the accumulation of mutations through the action of Muller's ratchet by periodically depressing the numbers of clonal individuals (Howard and Lively, 1994; Lively and Howard, 1994). Like the tangled bank model, the biogeographic distribution of uniparental reproduction can be accounted for by the Red Queen hypothesis: marginal, disturbed habitats at high altitude or latitude may have weaker parasite pressure (Bell, 1982).

The reproductive assurance hypothesis

Hypotheses based on reproductive assurance argue that uniparental forms of reproduction will evolve as mechanisms to assure reproductive success in population where access to mates is severely limited (Gerritsen, 1980; Lloyd, 1980), or where male gametes are prevented from fertilizing eggs (Johnson, 1994). These hypotheses have limited explanatory power because they do not address why cross-fertilization is advantageous when access to male gametes in not limited. Reproductive assurance is nonetheless likely to explain the occurrence of uniparental reproduction in some species. In what follows, we review our studies of freshwater snails, which were designed in part to contrast the adaptive variation hypotheses with the reproductive assurance hypothesis. The results show that reproductive assurance is not sufficient to explain the distribution of uniparental reproduction.

Ecological correlates of sex in *Potamopyrgus antipodarum*

Potamopyrgus antipodarum is a small (< 10 mm) prosobranch snail, native to New Zealand. Some populations consist entirely of parthenogenetic females, while other populations have sexual females and males. Parthenogenesis in this species appears to be apomictic (Phillips and Lambert, 1989), and parthenogenetic individuals have recently been shown to be triploid (Wallace, 1992). Moreover, based on allozyme studies, sexual and parthenogenetic individuals are now known to coexist within the same lake, and the parthenogens are almost certainly derived from the local sexual population (Dybdahl and Lively, 1995a). There is therefore no indication that the clones are either migrants or of hybrid origin, and they are therefore likely to be ecologically similar to the local sexual population from which they are derived. Consistent with this idea, Jokela et al. (1997) found that clonal *P. antipodarum* mature at the same size and produce the same number of embryos as sexual individuals living in the same habitats. In addition, the diversity of the locally derived clones is very high, with over 150 different clones in some locations (Fox et al., 1996). This coexistence of sexual and locally derived parthenogenetic individuals makes *P. antipodarum* ideal for comparative studies contrasting the different ecological theories of sex.

The snail is very common and widely distributed in lakes and streams throughout New Zealand. Some simple predictions can be made based on the working hypothesis that streams are more variable in time than lakes, and, as a consequence, competition for resources is more intense in lake populations. New Zealand streams are very prone to rearrangement by flooding, as well as changes in pH (Winterbourn et al., 1981), so the working assumption has some empirical support.

If sex is an adaptation to uncertain physical conditions, as suggested by the lottery model, then males and sexual females would be expected to be more common in the highly unpredictable stream habitats. If, on the other hand, sex is an adaptation to produce variable progeny in highly competitive, but physically stable habitats (the tangled bank hypothesis), then sexual populations should be more commonly found in the large stable lakes. Finally, if sexual reproduction is an adaptive defense against parasites, then sexual females should predominate in populations where there is a high risk of attack by parasites, independent of habitat type. Hence, two of the predictions are habitat specific; the lottery model predicts more sex in streams, while the tangled bank predicts more sex in lakes. The Red Queen hypothesis makes no specific prediction with respect to habitat, unless it is known in advance that one type of habitat is associated with a higher risk of parasite attack.

Field studies

In order to contrast these hypotheses, lake and stream populations were sampled from the glaciated regions of the South Island of New Zealand (Lively, 1987). Snails were sampled from 21 lakes and 29 streams, and the gender and state of infection by digenetic trematodes were determined for 40 to several hundred individuals (usually 100). Prevalence of infection was determined for each of a dozen trematode species at each location, and the frequency of males in the population was calculated. Sexual females could not be visually discriminated from parthenogens, but males could easily be distinguished from either. Because males are produced only by sexual females, they were used as an indicator of the frequency of sexual females in the population (Lively, 1987). A more recent genetic analysis by Fox et al. (1996) has shown that the correlation between male frequency and the frequency of diploid sexual females is positive and highly significant ($r^2 = 0.54$; df $= 13$; $p = 0.002$).

In contrast to the prediction of the lottery model, more males were found in lakes than in streams. Thus, if streams are more variable in time, as assumed, the lottery model can be rejected. The lottery model also fails to explain the geographic distribution of parthenogenesis across many plant and animal taxa (Bell, 1982).

The result of more males in lake populations is, however, consistent with the tangled bank model, which predicts that sex should be favored in the stable lake habitats where competition is more likely to occur. But, there were also more parasites in lake populations, which is consistent with the Red Queen hypothesis. In order to contrast these two alternatives, a stepwise multiple regression was used to determine whether habitat (lake versus stream) or parasites explained more of the variation in male frequency. Parasites explained more of the variation. The mean frequency of males was indeed greater in lakes, but there was a substantial amount of variation for male frequency within both lakes and streams. In fact, within lakes, male frequency varied from 0 to 40 percent; and it varied from 0 to 20 percent in streams (Lively, 1987). Parasites (castrating digenetic trematodes; see Tab. 1) were correlated with this variation in both habitats. Within lakes, males were positively and significantly correlated with the bird-final-host parasite, *Microphallus* sp. Within streams, males were similarly correlated with the eel-final-host parasite, *Stegodexamene anguilli*. The sum of these two parasites explained 34% of the variation in both lakes and streams. Hence, the difference between lakes and streams, which originally favored the tangled bank hypothesis, appears to have been driven by

Table 1. Host species, parasite species, life cycle of each parasite and effects of infection on host fitness

Host species	Parasite taxa	Other hosts	Effect on host fitness
P. antipodarum	*Microphallus* sp.	duck	castrator
	Stegodexamene sp.	eel	castrator
B. truncatus	*Xiphidiocercariae*	unknown	castrator (?)
C. decisum	*Leucochloridiomorpha constantiae*	duck	slight on female fitness
	Sellacotyle mustelae	fish, mustelid	castrator
	Linstowiella szidati	fish, bird	castrator
	Cercariae leptacantha	unknown	castrator

parasites. Thus the evidence falls in favor of the Red Queen hypothesis (Lively, 1987). These results are consistent with the findings of Schrag et al. (1994a), reviewed herein.

This kind of strong inference approach is very useful as a way of rejecting some of the alternative hypotheses (Platt, 1964). It is, however, less than conclusive evidence for any particular hypothesis, because alternative explanations for the same pattern might also exist. Nonetheless, as far as this study was concerned, any alternative must be regarded as post hoc. One especially appealing, but somewhat complicated, alternative involves reproductive assurance. What if, as discussed in the introduction, cross-fertilization has little to do with either competition or parasitism. Assume, for the sake of argument, that sex is maintained by recombinational repair (Bernstein et al., 1985) or by some kind of exogenous repair mechanism (e.g., Muller's ratchet (Muller, 1964) or deterministic mutation accumulation (Kondrashov, 1988)). Then sex should be lost only when populations become so sparse that mates become difficult to find. Moreover, it might be that in such sparse populations, the density of snails is too low to support a stable population of parasites (Anderson and May, 1979a; May and Anderson, 1979). Hence, under this two-part hypothesis, dense populations would be sexual and have parasites, while sparse populations would be expected to be parthenogenetic and have low parasitism, which are the results seen in Lively (1987).

Fortunately, any hypothesis involving reproductive assurance is testable, and can be falsified by data which show that density is unrelated to reproductive mode. In a survey of 65 lake populations over a 5-year period, Lively (1992) found no support for the reproductive assurance idea. The results of this more thorough survey, however, were consistent with expectations under the Red Queen hypothesis. Specifically, there were no populations with high male frequency where parasites were rare or absent, but sexual populations were common and correlated with the percent of individuals castrated by digenetic trematodes (Lively, 1992). Taken together, the results of the field surveys among populations of *P. antipodarum* are inconsistent with the expectations of the lottery model and the reproductive assurance hypothesis. The tangled bank hypothesis cannot be rejected; but, at present, the data provide stronger support for the Red Queen hypothesis.

In a very recent study, the distribution of males was examined within a single lake population (Lake Alexandrina) on the South Island of New Zealand. Surprisingly, the distribution of males was parallel to that observed among populations in that it was correlated with the presence of castrating trematodes (Jokela and Lively, 1995a). In general, the size-specific prevalence of infection was higher in shallow waters where ducks (one of the final avian hosts; see Tab. 1) tend to feed, and it is in this habitat that sexual individuals are most common (Jokela and Lively, 1995a, b; Fox et al. 1996). Hence it would appear that parthenogenetic females have been more successful at replacing sexual individuals in deeper water where the risk of infection is lower. This hypothesis would only make sense if individuals are not moving between shallow and deep areas, which seems to be the case based on the results of recent electrophoretic survey (Fox et al., 1996). In addition, snails in the shallow-water habitats tend to begin reproduction at a smaller size (Jokela and Lively, 1995 a), which is consistent with optimization theory for age/size-related mortality (Gadgil and Bossert, 1970; Law et al., 1979; Michod, 1979) and with the empirical results of Lafferty (1993) for a marine snail. Taken together, these results suggest that the New Zealand snails reproduce earlier and are more likely to reproduce sexually in the shallow-water habitats where the risk of infection is highest.

Reciprocal cross-infection experiments

A direct test of the Red Queen hypothesis requires experimental evidence that parasites can rapidly track genotypes as they become common. Such tracking is essential, otherwise the descendants of clonal mutants would rapidly replace their sexual ancestors (Lively, 1996). One form of evidence for this kind of tracking is local adaptation. If parasites are locally adapted to their "home" host population, it is likely that they are tracking genotypes in that population as they become common. One powerful way to test the idea that parasites are locally adapted is to conduct reciprocal cross-infection experiments.

Two reciprocal cross-infection experiments were conducted to determine whether *Microphallus* is locally adapted to its host populations (Lively, 1989). In the first experiment, involving two lakes on opposite sides of the southern Alps of New Zealand (Lakes Mapourika and Alexandrina), Lake Mapourika parasites were significantly more infective to snails from the same lake. Similarly, Lake Alexandrina parasites were significantly more infective to snails collected from Lake Alexandrina. Similar results were observed in the second experiment, which involved three lakes on the west side of the southern Alps (Mapourika, Wahapo, and Paringa). Parasites from all three lakes were better at infecting snails drawn from their local host populations. Hence, *Microphallus* is adapted to its local host populations, and likely, therefore, to be engaged in local coevolutionary interactions with its host. If this were not the case, the Red Queen hypothesis would clearly be false.

In addition, the most common trematode (*Microphallus*) in Lake Alexandrina is also strongly adapted to shallow-water snails in the same lake. In a reciprocal cross-infection experiment, parasites drawn from shallow water were more infective to shallow-water snails than snails from deep water; but the converse was not true: parasites drawn from deep water were not more infective to deep-water snails (Lively and Jokela, 1996). Hence, the parasite seems to be coevolving with its snail host in shallow water where it is easily recycled by the final hosts (ducks) that forage in shallow water. This result may help to explain the predominance of sexual individuals in the shallow-water habitats of Lake Alexandrina: the risk of infection is higher in shallow water, and the host-parasite interaction is more likely to be coevolutionary.

Maintenance of euphally and aphally in *Bulinus truncatus*

Bulinus truncatus (Audouin), one of the few well-studied pulmonates because of its role as an intermediate host for the trematode *Schistosoma haematobium*, is a tetraploid snail which lives in a range of freshwater habitats from lakes to temporary pools in northern and western Africa and the Middle East. All *B. truncatus* are self-compatible simultaneous hermaphrodites. Individuals within this species, however, develop into one of two sexual morphs referred to as phally. Euphallics develop an ovotestis and fully functional male and female tracts (Fig. 1(A), (B)). Aphallics, in contrast, do not develop the distal portions of the male tract although functional sperm are still produced by the ovotestis (Geraerts and Joosse, 1984). Thus, both euphallics and aphallics can self-fertilize, but aphallics cannot donate sperm. Laboratory studies indicate that there can be a cost to the growth and maintenance of a fully developed male tract (Jarne et al., 1992; Schrag and Rollinson, 1994; Doums and Jarne, 1996). Nonetheless, the proportion of euphallics in natural populations varies from zero to one (Larambergue, 1939; Brown and Wright, 1972; Schrag et al., 1994a, b). Additionally, *B. truncatus* has a short generation time of 4 weeks from the egg to egg stage (Schrag and Read, 1992); phally can be scored without dissection in lightly pigmented strains; and over small geographic distances (e.g., 10 km along the same road (Schrag et al., 1994a, b)) the proportion of euphallics has been found to very by as much as 81%, allowing for direct tests of factors maintaining this variability.

Factors maintaining outcrossing have been investigated in a number of theoretical and empirical studies of plant populations (Jain, 1976; Clay, 1983; Waller, 1984; Clay and Antonovics, 1985; Mitchell-Olds and Waller, 1985; Holtsford and Ellstrand, 1990; Schmitt and Gamble, 1990; Jarne and Charlesworth, 1993; Uyenoyama et al., 1993). Phally dimorphisms, which have arisen at least 14 independent times in pulmonate snails (Schrag and Read, 1996), present a unique opportunity to examine factors maintaining outcrossing ability within an animal species. An important assumption behind such investigations, however, is that the level of euphally within a population correlates with the level of outcrossing. If euphallics never donated sperm, selection would be expected to remove euphally because of the growth and maintenance costs associated with developing a full male tract. Clearly, in the extreme case, 0% euphally correlates with 0% outcrossing. Nonetheless, levels of outcrossing associated with intermediate levels of euphally remain to be determined. Recent progress in population genetics studies based on the analysis of microsatellite loci as opposed to

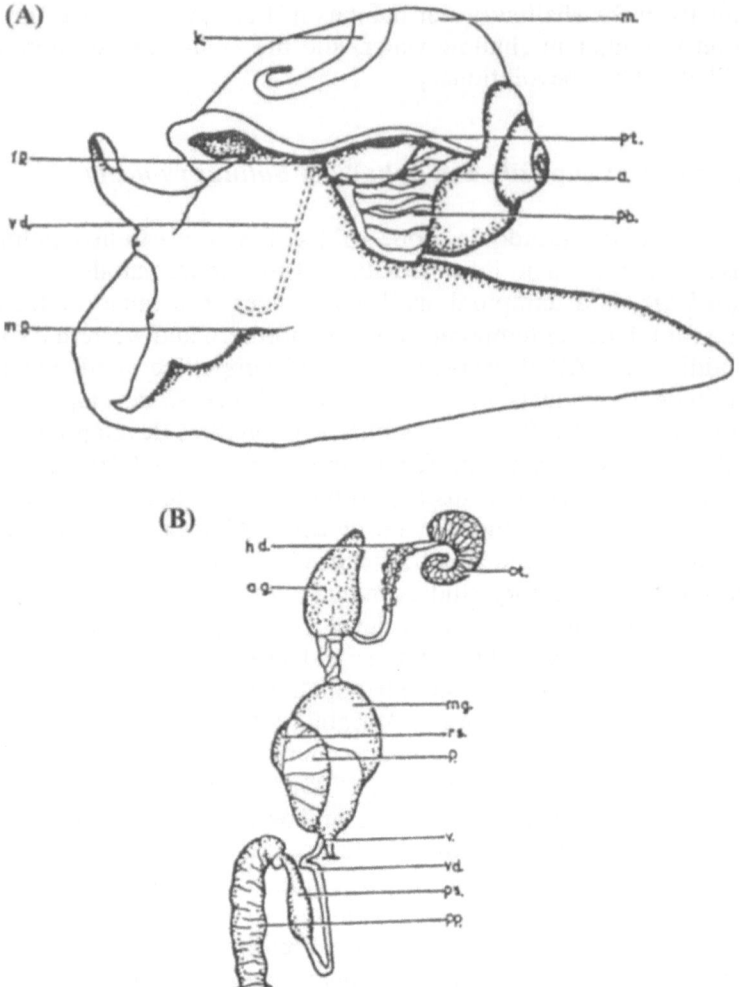

Figure 1. Schematic diagram of a euphallic *B. truncatus*. (A) Shell removed from one side: the vas deferens (vd) is clearly visible through the body wall behind the left tentacle. (B) Genital tract; the distal portions (penis (pp) and prostate (ps)) are not present in aphallics and the vas deferens (vd) does not fully develop. From Wright (1957).

allozyme variability (Jarne et al., 1994; Viard et al., 1996) promises to allow direct empirical investigations of the relationship between euphally and outcrossing in laboratory and field populations of *B. truncatus*. Initial studies (Jarne et al., 1994; Viard et al., 1996; Doums et al., 1996) suggest that, overall, levels of outcrossing in natural populations may be low. Assessing the relationship between euphally and outcrossing in natural populations (e.g., Viard et al., 1996) will remain difficult, however, because seasonal fluctuations in euphally within populations can be extreme (see below).

Mechanism of phally determination

Phally in *B. truncatus* develops during the first weeks after oviposition, prior to sexual maturity, and does not change once it develops (Larambergue, 1939). Laboratory studies of *B. truncatus* have demonstrated that phally may have a strong heritable component in some populations (Larambergue, 1939) while in others it is determined by temperature that juveniles experience prior to maturity, with colder temperatures favoring the development of euphally (Schrag and Read, 1992; Schrag et al., 1992, 1994b). There is suggestive evidence that other pulmonate species may also have environmentally-determined phally (Watson, 1934; Nicklas and Hoffmann, 1981). Field observations of *B. truncatus* in northern Nigeria have demonstrated that colder temperatures during phally development similarly correlate with high levels of euphally in natural populations (Schrag et al., 1994b). A small number of field populations in this region, however, maintained extremely low levels of euphally despite decreasing temperatures, suggesting that populations may have differed in their temperature-sensitivities. Laboratory observations confirmed this between-population variation and further suggested that individuals within some populations varied in their temperature-sensitivities (Schrag et al., 1994b).

Investigating the adaptive significance of euphally

The following investigations of hypotheses for the maintenance of euphally in *B. truncatus* assume that levels of euphally within populations correlate with levels of outcrossing. To test among competing hypotheses for the maintenance of outcrossing and selfing in *B. truncatus*, a number of ecological variables including water temperature and proportion of euphallics were measured in 49 *B. truncatus* populations in northern Nigeria (Schrag et al., 1994a). Habitats ranged from man-made dams and lakes to small irrigation channels and temporary pools. Proportion of euphallics (based on an average of 50 snails per population) in this region ranged from 0 to 81%, and even sites within close proximity often differed dramatically in levels of euphally.

Across populations there was no association between population density and euphally, inconsistent with standard interpretations of the reproductive assurance hypothesis for the maintenance of euphally. This result is similar to observations of *P. antipodarum* described above. Furthermore, there was no evidence of a link between euphally and resource availability, contrary to predictions of the tangled bank hypothesis. Levels of euphally were also not associated with habitat instability (estimated by habitat type, human activity, rate of desiccation, and changes in water chemistry), contrary to predictions of the lottery hypothesis. This lack of association is also inconsistent with the tangled bank hypothesis which predicts that more stable habitats will have higher proportions of euphallics.

Conductivity, in contrast, explained a small but significant percentage of variation in proportion of euphallics, with lower ion concentrations correlated with higher levels of euphally (Schrag et al., 1994a). Conductivity in freshwater is a summary measure reflecting the total concentration of major ions (Beadle, 1974). A similar association between conductivity (in particular, concentrations of Mg^{++}, Ca^{++}, Na^+ and Cl^-) and prevalence of males was found across bisexual and parthenogenetic populations of the freshwater snail, *Melanoides tuberculata*, in Israel (Heller and Farstey, 1990). However, the relationships between water chemistry, snail biology and habitat ecology are not well understood (Brown, 1980; Jordan and Webbe, 1982), although calcium is one of the essential elements for snail growth (Brown, 1980). Middle ranges of conductivity are often optimal for African freshwater snails, while both high and low extremes result in increased hatching time, delayed egg production and decreased fertility (Brown, 1980).

Overall prevalence of trematode infection was not significantly correlated with the proportion of euphallics, when the effects of time of year (a good indicator of temperature) and mean snail age were controlled for (depending on the length of the pre-patent period, there is a minimum age below which patent infections cannot be scored). However, prevalence of the most abundant trematode taxa, Xiphidiocercariae (which in Nigerian *Bulinus* may consist of two trematode species based on morphology (Ndifon and Umar-Yahaya, 1988–1990)), correlated positively and significantly with proportion of euphallics ($r^2 = 10\%$; $n = 49$ when the effects of snail size and time of year were statistically removed), in support of the Red Queen hypothesis (Schrag et al, 1994a). Indices of trematode diversity which incorporated both prevalence and richness, also correlated significantly with the proportion of euphallics. It may be that parasite diversity *per se* is an important source of parasite-mediated selection for outcrossing; however, because the diversity indices correlated strongly with prevalence of *Xiphidiocercariae* infection, this data set did not allow these measures to be distinguished. There was no association between conductivity and trematode prevalence (Schrag et al., 1994a).

Of the site-specific variables considered, the two that correlated significantly with proportion of euphallics have been found to correlate similarly in other freshwater snails (trematode prevalence in *P. antipodarum* (Lively, 1987, 1992) and conductivity in *M. tuberculata* (Heller and Farstey, 1990)), despite the fact that *B. truncatus* belongs to a different sub-class and has a different breeding system from both *Potamopyrgus* and *Melanoides*. The correlation most easily explained by competing hypotheses for the maintenance of outcrossing suggests that euphally is maintained by parasite-mediated selection for genetically variable offspring.

Why might phally be temperature-sensitive?

Adaptive arguments successfully explain the maintenance of environmental sex determination (ESD) in a wide range of taxa (e.g., Tingley and Anderson, 1986; Conover and Heins, 1987; Naylor et al., 1988; but see Bull and Charnov (1989) for a counter-example in reptiles). An analogous argument may explain environmental phally determination: temperature-sensitive phally determination will be favored by natural selection when 1) factors determining the relative fitness of selfed and outcrossed offspring vary, 2) these factors are correlated with temperature, and 3) parents cannot predict and offspring cannot control the conditions offspring experience.

Are these three conditions plausible for the case of *B. truncatus*? The across-population correlation between trematode infection and proportion of euphallics described above suggests that the first condition may hold: male outcrossing ability appears to have a selective advantage when parasite pressure is high. In support of the second and third conditions, field observations in natural populations of *Bulinus* snails suggest that water temperature is a good predictor of future levels of parasitism (Woolhouse and Chandiwana, 1989; Schrag, 1993). In contrast, parasite prevalence within sites can fluctuate wildly over short periods of time so that current parasite prevalence is a poor predictor of future parasite prevalence (Woolhouse and Chandiwana, 1989; Schrag, 1993). Furthermore, low snail mobility suggests that snails themselves cannot choose among habitats.

In northern Nigeria, prevalence of *Xiphidiocercariae* reaches peak levels in snail populations between March and July (Betterton, 1984). If outcrossed offspring are favored when parasite pressure is high, and this generates a selective advantage to outcrossing and hence euphally, then euphally should be more common among the parents of snails facing this seasonal rise in parasite pressure. Temperature-sensitive phally determination during the juvenile stage would ensure this: juveniles hatching during the months of December and January are more likely to develop into euphallics because this is when temperatures are coldest. These juveniles will reach maturity and start producing offspring just when parasite pressures peak.

Thus, several lines of evidence support the idea that seasonal variation in parasite pressure maintains temperature-sensitive phally determination. A central prediction based on the patterns of variation observed in the field is that populations with low levels of parasitism and/or low levels of fluctuation in parasite pressure should have lower temperature-sensitivities. Long-term field observations of seasonal fluctuations in water temperature, trematode prevalence and proportion of euphallics across and within populations would be necessary to test this prediction.

If temperature is an indirect cue for parasite pressure, this might also explain why temperature-sensitivity varies on such a small geographic scale. Thus, while sites in close proximity may not experience large differences in temperature conditions, if levels of parasite pressure differ between sites on a small scale, small scale variation in levels of temperature-sensitive phally determination would be expected. In the Kano City region, only three out of 22 populations where within-site variation in proportion of euphallics was monitored showed little change in proportion of euphallics, suggesting that weak temperature-sensitivity is relatively rare.

Ecological correlates of parthenogenetic and sexual reproduction in *Campeloma decisum*

Campeloma decisum is an ovoviviparous, dioecious prosobranch that is widely distributed and locally common in streams and lakes throughout eastern North America. The biogeographic distribution of parthenogenetic and sexual reproduction in *Campeloma* was considered a classic case of geographical parthenogenesis: parthenogenetic reproduction in northern, glaciated habitats and sexual reproduction in southern, unglaciated habitats (Bell, 1982). Parthenogenesis could be favored in glaciated regions to assure reproduction where male density is low. Alternatively, the Red Queen hypothesis would predict that parthenogenesis may be favored in these glaciated habitats because parasites are rare in these marginal habitats or, under the tangled bank hypothesis, there is relaxed selection for the production of outcrossed progeny in spatially homogeneous environments in which competition is limited. Given that the original description of the biogeographic distribution of reproductive mode in *Campeloma* is consistent with the reproductive assurance hypothesis and these two adaptive variation hypotheses, Johnson (1992 a, b, 1994) explored the biogeographic distribution of reproductive mode in a more thorough fashion, and also examined the correlation between reproductive mode and risk of parasitism. He also explored the possibility that a digenetic trematode may severely limit sperm availability, and that parthenogenesis is favored to assure reproduction.

Johnson (1994) sampled three geographic regions: northern glaciated regions ranging from Indiana to northern Michigan and Wisconsin; unglaciated habitats in the southeastern United States (Virginia and North Carolina) and unglaciated habitats in the south-central United States (Louisiana, Mississippi, Arkansas, and Missouri). Parthenogenetic populations contained no males, while other populations contained >40% males, and were categorized as "sexual", although some sexual populations may contain mixtures of sexual and parthenogenetic females. Thus, the continuous range in functional gender that is seen in *P. anti-*

podarum or *B. truncatus* is not observed in *Campeloma*. The most common parasite of *C. decisum* is the digenetic trematode, *Leucochloridiomorpha constantiae*. This parasite lives as unencysted metacercaria in the female brood chamber. Prevalence and intensity of infection (mean number of metacercariae per host) were determined. The consequences of *L. constantiae* infection on female fitness are very weak, although male fitness may be severely limited by the presence of metacercariae (see below; Johnson, 1992 b). Prevalence of infection by three other trematodes (*Sellacotyle mustelae*, *Cercariae leptacantha*, and *Linstowiella szidati*) was also recorded. These trematodes produce larval stages in the snail reproductive system and cause strong effects on fitness (Tab. 1 and Johnson, 1994).

Parasite prevalence was significantly higher in parthenogenetic populations relative to sexual populations (Johnson, 1994). All 26 parthenogenetic populations were infected by metacercariao of *L. constantiae* and all 14 sexual populations were uninfected. Similarly, the prevalence of the three castrating trematodes in parthenogenetic populations (12%) was three times higher than the prevalence of these trematodes in sexual populations (4%). Parthenogenesis is not maintained in areas where parasites are absent or rare.

One potentially confounding aspect of these results is that the degree of geographical separation is much greater between sexual and parthenogenetic *Campeloma* populations than that observed in *Potamopyrgus* and *Bulimus*. In addition, sexual and parthenogenetic individuals apparently do not coexist within populations as seen in *Potamopyrgus*. Hence, there is a potential effect of latitude on the correlation between parasite load and reproductive mode, so this study does not represent a very strong test of the Red Queen hypothesis. As outlined below, experimental tests of the Red Queen in this system should alleviate this problem.

Reproductive assurance and parthenogenesis

The biogeographic distribution of sexual and parthenogenetic populations is, for the most part, a typical pattern of parthenogenetic populations in glaciated northern regions and sexual populations in unglaciated southern regions (Johnson, 1994). However, parthenogenesis has arisen in the southeastern United States where there was no Pleistocene glaciation. The presence of parthenogens in North Carolina and Virginia and the regional coexistence of sexuals and parthenogens in the coastal plain of North Carolina suggest that colonization of glaciated regions is not necessary for the establishment of parthenogenesis. However, the predominance of parthenogens in the northern U.S. suggests that parthenogens may have had an advantage in colonizing northern temperate regions after the Pleistocene glaciers receded.

The absolute congruence of parthenogenetic reproduction in the snail and the presence of the free-living stage (metacercaria) of *Leucochloridiomorpha constantiae* in the brood chamber of female snails led to the hypothesis that this parasite can severely limit sperm availability and that parthenogenesis assures reproduction (Johnson, 1992b). The sperm-limitation hypothesis requires that the prevalence and intensity of infection is high and that infections persist throughout an individual's lifetime. Strong selection against male function is required because sexuals would potentially invade these areas where parasites are common. Infection levels are high and persistent in many parthenogenetic populations (Johnson, 1992b). Given the persistent, high level of infection in parthenogenetic populations, what are the mechanisms by which this parasite limited sperm availability when the parasite was introduced historically into a sexual population?

The introduction of *L. constantiae* into sexual populations may have led indirectly to the elimination of males via sperm ingestion or sperm blockage, thus leading to strong selection for those rare females capable of parthenogenetic reproduction. Sperm are stored in an undifferentiated, open seminal receptacle (see Fig. 2; adapted from Vail, 1977), and metacercariae would have easy access to the stored sperm and would be able to

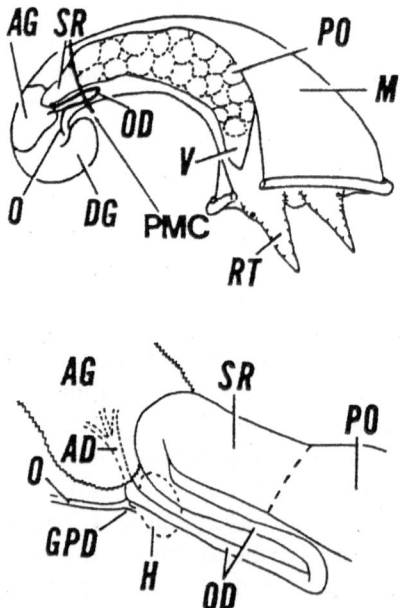

Figure 2. Reproductive anatomy of female *Campeloma*. AD, albumen gland duct; AG, albumen gland; DG, digestive gland; H, heart; M, mantle; O, ovary; OD, oviduct; PMC, posterior end of mantle cavity; PO, pallial oviduct (brood chamber); RT, right tentacle; SR, seminal receptacle; V, vagina. From Vail (1977).

ingest or displace them. Severe sperm limitation in infected populations could result, leading to strong selection for females capable of parthenogenetic reproduction. This hypothesis requires high infectivity of *L. constantiae* when introduced into sexual snail populations and little genetic variation for resistance.

An alternative, perhaps more parsimonious, hypothesis exists to account for the concordance between the presence of *L. constantiae* and the biogeographic distribution of parthenogenesis. There may have been strong selection for females capable of parthenogenetic reproduction to assure reproduction during colonization of marginal habitats, and the high prevalence and intensity of infection by *L. constantiae* may be derived from the ability of parasites to exploit the absence of genetic variation in non-recombining parthenogens. The association between parthenogenesis and parasitism may result from the ability of parasites to rapidly track and infect locally common clonal genotypes. Sexual populations may be exposed to the parasite, but high levels of infection might not occur because there is genetic variation for resistance in sexual host populations. If infection levels remain low, then sexual females could still reproduce and would probably have higher fecundity than parthenogens, because those females capable of parthenogenetic reproduction may have reduced fecundity after switching to meiotic parthenogenesis (Templeton, 1982). This hypothesis is distinct from the sperm limitation hypothesis in that parthenogenesis is selected to assure reproduction in colonizing females and that high infectivity results from parasite exploitation of common host resistance genotypes. These two hypotheses offer contrary predictions regarding the infectivity of the parasite in sexual and parthenogenetic populations: the sperm limitation hypothesis predicts that sexual and parthenogenetic individuals should experience similar, high levels of infection, while the second hypothesis predicts that there is local parasite adaptation to the host and that parthenogens should be more susceptible to infection by this parasite than sexual individuals. Under the Red Queen hypothesis, there should be greater variability in the prevalence and intensity of infection among sexual individuals. Future experimental work will address these two hypotheses.

Spontaneous and hybrid origins of parthenogenesis

Most parthenogens that originate by hybridization between genetically divergent sexual ancestors have elevated levels of heterozygosity, and the alternate alleles at these heterozygous loci are found in the two putative sexual ancestors. Parthenogens that arise by spontaneous mutation show similar or lower levels of genetic variation relative to their sexual ancestor. There are two classes of parthenogens in *C. decisum*: homozygous parthenogens from North Carolina and Wisconsin and heterozygous parthenogens from Indiana, Michigan and one population from Wisconsin

(Johnson, 1992a). Parthenogenetic and sexual individuals from North Carolina populations were genetically identical at the 19 enzyme loci scored. They were fixed for the same allele at those nine loci that vary across the entire geographic range. The homozygous clones probably reproduce by automictic parthenogenesis. The genetic consequences of automixis over many generations are analogous to complete selfing, in that recombination cannot generate rare genotypes, even though there may be multiple clonal lineages derived from a heterozygous founder. Sexual populations from Mississippi, Louisiana, and Arkansas are nearly fixed for an alternate allele at these nine loci. All parthenogens from Indiana and Michigan are heterozygous at six or seven enzyme loci, and share the alleles which are fixed alternately in the North Carolina and south-central sexual populations. These heterozygous clones are probably apomictic parthenogens.

The presence of two modes of parthenogenesis potentially offers some insight into the advantages and disadvantages of heterozygous or homozygous parthenogens. From the perspective of the Red Queen hypothesis, the consequences of automictic and apomictic parthenogenesis in the host are similar: neither parthenogen can generate rare resistance genotypes, assuming that automictic parthenogens have undergone many generations of automictic reproduction. Some evidence suggests that hybrid parthenogens in poeciliid fishes (Lively et al., 1990) and Australian geckos (Moritz et al., 1991) are more susceptible to parasites than coexisting sexuals, but whether hybrid parthenogens are more susceptible to parasitism will depend on the level of recombination and the dysgenic consequences of hybrid genomes. Possibly, the presence of two non-recombining parental genomes in apomictic parthenogens increases the number of resistance genes, thereby increasing the probability of detecting parasite antigens. In this case, hybrids may be less susceptible to parasites as may be the case in certain geckos from the Pacific Islands (Hanley et al., 1995; Brown et al., 1995). Because parasites are under such strong selection to evade host immune responses, mutations to novel virulence genes would be favored and hybrid parthenogens may be driven to low frequency by debilitating parasites. In all of the above cases, however, it is unclear that these macroparasites can generate strong selection against host genotypes to drive the dynamic changes in host genotype frequencies envisioned by the Red Queen hypothesis. It is also important to remember that one assumption of the Red Queen hypothesis is that parthenogenetic individuals are similar genotypically or phenotypically (albeit less variable) compared to their sexual ancestor (Dybdahl and Lively, 1995a). Hybrid parthenogens may have lesser or greater than a two-fold advantage depending upon their competitive abilities and fecundity relative to their sexual ancestors. It is surprising that very few studies have addressed whether asexual descendants and their sexual ancestor have equal fecundity. Lamb and Willey (1979) suggest that certain parthenogenetic insects often suffer

33% reduction in hatching success relative to their sexual ancestor. If these insects are automictic parthenogens, the high rate of embryonic death may be the result of expression of recessive lethals during the origin of parthenogenesis. In this case, parthenogens do not have a two-fold advantage.

Another interesting pattern is the biogeographic distribution and ploidy levels of these two *Campeloma* parthenogens. Heterozygous clones, which are probably polyploid, predominate in the northern region, whereas homozygous (diploid?) clones are common in the southeastern region (Johnson, 1992a), although recent work has discovered hybrid parthenogens throughout the panhandle of western Florida (Johnson, unpublished data). Whether the selective advantage of polyploidy and heterozygosity in northern temperate regions is due to some general-purpose genotype (Lynch, 1984) in these environments deserves careful scrutiny.

Brooding and the phylogenetic distribution of parthenogenesis

Whereas the ecological correlates of cross-fertilization have been addressed by these and other studies, the mechanisms responsible for the spotty taxonomic distribution of parthenogenesis have received little empirical attention. Parthenogenesis is phylogenetically widespread, but many clades contain no known parthenogenetic varieties (Bell, 1982; Lynch, 1984). This pattern may result from two mechanisms. Members of some clades may never be released from the selective forces that act to maintain cross-fertilization (i.e., coevolved biological enemies) or parthenogenesis cannot spread when rare in some groups, because of developmental defects associated with its early evolution (Templeton, 1982; Uyenoyama, 1984). These developmental abnormalities are frequent enough during the early evolution of parthenogenesis that the evolution of parthenogenesis may be more paradoxical than the maintenance of outcrossing (Uyenoyama, 1984). If so, brooding lineages in which selective abortion of defective offspring is possible may be expected to be more susceptible to the initial establishment of parthenogenetic mutants (Lively and Johnson, 1994). In addition, if zygotes are overproduced, competition among embryos in the confines of the brood chamber may also favor the initial establishment of parthenogens if defective embryos are outcompeted.

Stearns (1987) argued that, if zygotes are relatively cheap, overproduction of zygotes would lead to competition within the brood chamber (the "selective arena" in Stearns' terminology), which would increase the mean fitness of the brood. The model of Lively and Johnson (1994) is a special case of this idea: when the early evolution of parthenogenesis is burdened by developmental abnormalities, the selective arenas of brooders could favor the spread of rare parthenogenetic mutants because defective embryos could be culled without much cost to the parent. Zygote overproduction greatly increases the probability that a rare parthenogenetic mutant will spread to fixation compared to non-brooding parthenogens. Brooding may aid in the early evolution of parthenogenesis, but ecological forces, such as escape from parasites, will ultimately determine if parthenogenesis is maintained within a lineage.

An alternative to the selective arena formulation is that inferior offspring are selectively aborted and replaced. In this situation, there would be some cost accrued by a parthenogenetic mutant to replacing embryos. A strategy model of selective abortion indicates that the range of values that favors the spread of parthenogenesis increases with the number of times aborted eggs are replaced (Lively and Johnson, 1994). Although selective arenas give the greatest advantage to parthenogenesis in these strategy models, selective abortion and replacement of embryos also increase the probability that parthenogenesis will spread.

A basic prediction from these strategy models is that parthenogenesis would be more likely to arise in brooding lineages. Because brooding and parthenogenesis are rare traits in most taxa, recent comparative analyses have focused on taxa in which parthenogenesis has arisen independently in various species and for which variation in development exists. There was a strong association between brooding and parthenogenesis in various lineages (Lively and Johnson, 1994). In the Cnidaria and Mollusca, parthenogenesis evolved significantly more often in brooding lineages than in non-brooding lineages. Many taxa show consistent relationships between brooding and the evolution of parthenogenesis, although parthenogenesis and brooding are not associated in sipunculids, gastrotrichs, and tardigrades. All parthenogenetic species in these taxa oviposit uncleaved eggs. It would be valuable to determine whether selective abortion or selective arenas operate in brooding species, particularly in sexual species that are ancestral to brooding parthenogenetic groups.

Developmental defects in parthenogenetic eggs are not strictly required for the hypothesis. Rapid mutation accumulation could also result in a bias toward the evolution of parthenogenesis in brooding species. Suppose, as suggested by Kondrashov (1988), that mutations are on the order of at least one per genome per generation, and that all individuals having more than a threshold number of mutations (k) die, but individuals with fewer mutations suffer no loss in fitness (i.e., the fitness function is truncated at k mutations). The equilibrium mean number of mutations in parthenogens after selection is $k - 1$, with very little variance. Then, assuming a Poisson distribution of mutations with a mean of one, approximately $2/3$ of the parthenogenetic offspring will have k or more mutations and they will die. But if zygotes are sufficiently overproduced, or there is selective abortion with replacement of offspring, the parthenogenetic female might recover a sufficient number of offspring to have a reproductive advantage over sexual females. This model assumes that deleterious mutations at different loci act synergistically such that the rate of decline in fitness against number of deleterious mutations is steeper than a linear function (Kondrashov, 1988).

The few empirical estimates of synergistic epistasis suggest it is weak or non-existent in *Drosophila* (Mukai, 1969) and in partially selfing populations of the monkey flower, *Mimulus* (Willis, 1993).

Alternative hypotheses can also predict an association between brooding and uniparental reproduction. The evolution of brooding may be coupled with low offspring dispersal, and there may be subsequent selection against the production of variable progeny in order to preserve locally adapted genotypes (Mitter et al., 1979). Another hypothesis for the evolution of uniparental reproduction by self-fertilization has been suggested by Strathmann et al. (1984) and Eernisse (1988). Their idea is that the limited dispersal of brooded offspring leads to an inbred population structure, which in turn may increase the likelihood that alleles for self-fertilization can spread when rare. These ideas are not mutually exclusive, and further investigations of the association between brooding and uniparental reproduction are warranted.

Finally, if brooding does serve as an exaptation for the evolution of parthenogenesis, then it might also lead to greater extinction rates in certain ecological situations. This increase would be expected if the parthenogens rapidly replace their sexual ancestors, and then later become extinct due to an inability to track climatic changes, or due to the stochastic accumulation of mutations through Muller's ratchet (Muller, 1964). In such a situation, we would expect species-level selection against the evolution of brooding (Nunney, 1989).

Synthesis and conclusions

There are two major hurdles for the successful invasion and spread of uniparental reproduction. The first and foremost of these is overcoming the genetic disadvantages of uniparental reproduction. For selfing the primary problem is inbreeding depression. Reductions in fitness due to inbreeding depression, however, must be severe (greater than 50%) to prevent the spread of selfing within populations (Charlesworth, 1980; Maynard Smith, 1989). Although there are no direct measures of the magnitude of inbreeding depression in populations of *B. truncatus*, theoretical and empirical analyses of population structure suggest it is unlikely to be this severe (Jarne et al., 1992; Njiokou et al., 1993a; Doums et al., 1996). Furthermore, if a population passes through the initial negative effects of inbreeding, it can evolve into a viable selfing population and is unlikely to revert back to outcrossing (Maynard Smith, 1989). Observations of *B. truncatus* populations which are 100% aphallic in the field and which do not show intra-population variation in proportion of euphallics due to temperature (Schrag et al., 1994b) suggest that inbreeding depression is not an insurmountable obstacle to the establishment of 100% selfing populations in *B. truncatus*.

A similar hurdle exists for the evolution of parthenogenesis. The subversion of meiosis can result in developmental defects, which can have similar effects to those resulting from inbreeding depression (Templeton, 1982; Uyenoyama, 1984). If these effects are severe enough (i.e., greater than a 50% reduction in fitness), then parthenogenesis cannot spread when rare. Furthermore, lineages may vary in the fitness effects of mutation to parthenogenesis, and this could potentially explain the "patchy" distribution of parthenogenetic reproduction in eukaryotes (Bell, 1982; Lynch, 1984). However, the patchy distribution of apomictic parthenogenesis might also be explained by differences in life-history strategies among lineages. For example, lineages that brood their offspring could be especially vulnerable to invasion by parthenogenesis, because brooding could allow for the culling of defective embryos through competition among brood mates or through selective abortion (Lively and Johnson, 1994); the two parthenogenetic prosobranchs reviewed here are brooders. For similar reasons, selective abortion (or arenas) might allow for the initial spread of selfing even under situations where inbreeding depression is greater than 50%.

If these initial genetic hurdles are overcome, then uniparental reproduction would be expected to spread when rare. The question then becomes whether it would become fixed within populations. Uniparental reproduction may fix if advantages of reproductive assurance outweigh the advantages to cross-fertilization, as appears to be the case for *Campeloma decisum*. In this unique system, sperm limitation by a digenetic trematode may favor females capable of parthenogenetic reproduction and select against sexual females in some populations, even though there would be selection for recombinant progeny. Over time, parasites should disproportionally infect these common, invariant host genotypes. Hence, there is support for the exploitation of common host genotypes as predicted by the Red Queen hypothesis. Future effort should be directed to testing the mechanisms by which sexual reproduction is maintained in southern populations.

However, if access to mates is not limited, then uniparental reproduction might not be expected to fix in local populations. The results of field studies of *Potamopyrgus antipodarum* and *Bulinus truncatus* are both consistent with the idea that parasites select for at least partial cross-fertilization. In *Potamopyrgus*, sexual populations are associated with higher parasite loads, which suggests that parthenogens have been able to replace sexual individuals only where parasites are rare or absent (Lively, 1987, 1992). There was no indication that parthenogenetic reproduction is associated with populations of low density (Lively, 1992), which would have been consistent with the reproductive assurance hypothesis.

Similarly, across *Bulinus* populations, higher levels of euphally were associated with higher prevalence of trematode infection. Furthermore, seasonal increases in euphally, due to colder temperatures during juvenile development, coincide with increases in parasite pressure. To further test whether the Red Queen hypothesis can explain the distribution of aphallics and euphallics, it will be important to determine whether key assumptions of this hypothesis hold for the case of *B. truncatus*. Fitness consequences of *Xiphidiocercariae* infection (the most common trematode observed in the field remains undescribed and cannot yet be propagated in the laboratory) on host snails are not known, although laboratory infections of *B. truncatus* with other trematode species have shown significant fitness reductions including parasitic castration in some cases (Anderson and May, 1979 b; Fryer et al., 1990; Schrag and Rollinson, 1994). Furthermore, estimates of genetic variability and outcrossing rate within populations will be important to directly assess the relationship between levels of euphally and outcrossing rates, and the potential for outcrossing to generate genetic variability. Electrophoretic analyses of *B. truncatus* population structure have been of limited utility due to high levels of fixed (non-segregating) heterozygosity patterns typical of tetraploids (Wright and Rollinson, 1981; Jelnes, 1986; Njiokou et al., 1993 a), although available evidence points to high levels of selfing in natural populations (Njiokou et al., 1993 a, b). Nonetheless, observations of genetic variability in temperature-sensitivity within and between populations (Schrag et al., 1994 b) suggest that genetic variation is present within populations. Advances in methods of genetic analysis (Jarne et al., 1994; Jones et al., 1994; Viard et al., 1996; see also Städler et al., 1995 for the first full application of progeny-array analysis to an animal hermaphrodite) may soon make it possible to determine whether individuals are the product of selfing or outcrossing, facilitating tests of both the Red Queen hypothesis and the role of inbreeding depression in maintaining euphally and aphally.

At first glance, the Red Queen hypothesis appears unfalsifiable because it can predict that parthenogens suffer either lower or greater parasitism relative to their sexual ancestors. The prediction of a positive relation between levels of recombination and parasite pressure stems from the idea that sex should be favored in populations with a high risk of parasitism and asex should be favored in marginal habitats with no or low risk of parasitism. A negative relationship between levels of recombination and parasite pressure can result if parthenogenesis is favored by other ecological mechanisms (reproductive assurance) and parasites recolonize hosts. If sexuals cannot invade these marginal habitats or reversion to sex is constrained, parthenogens should be overparasitized. In a similar fashion, if sexuals and a limited number of clonal genotypes have an equal likelihood of exposure to debilitating parasites over many generations, clonal individuals should be overparasitized. Clearly, any "snapshot" of a coevolutionary cycle between hosts and parasites could reveal either higher or lower parasite loads in parthenogens relative to sexuals (Dybdahl and Lively, 1995 b). If correlational evidence rejects the null hypothesis of equal parasite loads in sexuals and parthenogens, experimental exposure of sexual and clonal individuals to virulent parasites is needed to address whether parasites can overexploit clonal genotypes.

Studies of *Potamopyrgus* and *Bulinus* provide some of the strongest evidence for the maintenance of sex by parasites. Additional correlational support for the Red Queen hypothesis comes from a number of studies (comparative evidence: Bell, 1982; Burt and Bell, 1987; field evidence: Antonovics and Ellstrand, 1984; Schmitt and Antonovics, 1986; Lively et al., 1990; Burt and Bell, 1991; Moritz et al., 1991). A powerful adjunct to these correlational studies will be experimental manipulations of parasite pressure in mixed populations of outcrossing and selfing/partheno-genetic individuals. These studies should answer the fundamental question of whether parasites are the major selective mechanism by which rare genotypes generated by sex are favored over uniparental progeny.

Using individual-based simulation models, recent extensions of the Red Queen hypothesis have focused on the effect of host and parasite population structure on the coevolutionary dynamics between parasites, sexual hosts, and asexual hosts (Ladle et al., 1993; Judson, 1995; Keeling and Rand, 1995). Ladle et al. (1993) suggest that the relative host to parasite migration rate has a strong influence on the maintenance of sex by debilitating parasites: high host: parasite migration rate favors asexual hosts (escape from parasites); sexual hosts with an increase from zero migration can win even under moderate to high parasite migration; asexual hosts will win in cases with no host migration and parasites with moderate to high migration rates. This latter finding is especially intriguing in the context of why parthenogenesis and brooding are strongly coupled in gastropods. Brooding may lead to limited dispersal of snail hosts and parasite dispersal may be relatively higher especially in parasites with complex life cycles in which one of the hosts is very mobile (i.e., birds). We note, however, that *Potamopyrgus* and *Campeloma* are superb colonizers of marginal habitats.

Keeling and Rand (1995) also find interesting conditions under which sex or asex is stable to invasion. Under the realistic notion that host-parasite ecologies are spatially extended, they find that clumping of genetically identical asexual hosts are especially prone to local extinction by rapidly-evolving detrimental parasites. Clumps of heterogeneous sexual individuals cause stochastic fluctuations in spatial structure of host genotypes and therefore it is difficult for specialized parasites to cause local extinction. The patterns are most evident when parasites have very strong effects on host fitness and the transmission rate of parasites is high. These findings are contingent upon the average number of parasite mutations in a host life span: if the value is low, asexual hosts should win; and if the parasite mutation rate is high, sexual reproduction would be favored. Perhaps this latter finding illuminates the phylogenetic pattern of asexuality being common in short-lived small hosts and biparental reproduction more common in long-lived larger hosts. All of these recent extensions of the Red Queen suggest that future work concentrate on relative host and parasite migration rate, quantifying the strength of selection by parasites on host fitness, transmission rates of parasites and parasite mutation rate.

Acknowledgements
This paper is an updated version of our paper originally published in *Experientia* as part of a multi-author review organized by B. Streit and T. Städler. We thank Laurence Hurst, Philippe Jarne, Thomas Städler, and Steve Stearns for critical reviews of the manuscript and Laurence Hurst for the suggestion that brooding lineages may be more prone to extinction.

References

Anderson, R.M. and May, R.M. (1979a) Population biology of infectious diseases: Part I. *Nature* 280:361–366.

Anderson, R.M. and May, R.M. (1979b) Prevalence of schistosome infections within molluscan populations: Observed patterns and theoretical predictions. *Parasitology* 79:63–94.

Antonovics, J. and Ellstrand, N.C. (1984) Experimental studies of the evolutionary significance of sexual reproduction. I. A test of the frequency-dependent selection hypothesis. *Evolution* 38:103–115.

Beadle, L.C. (1974) *The Inland Waters of Tropical Africa: An Introduction To Tropical Limnology.* Longman, London.

Bell, G. (1982) *The Masterpiece of Nature: The Evolution and Genetics of Sexuality.* University of California Press, Berkeley.

Bernstein, H., Byerly, H.C., Hopf, F.A. and Michod, R.E. (1985) Genetic damage, mutation, and the evolution of sex. *Science* 229:1277–1281.

Betterton, C. (1984) Spatiotemporal distributional patterns of *Bulinus rohlfsi* (Clessin), *Bulinus forskalii* (Ehrenberg) and *Bulinus senegalensis* (Müller) in newly-irrigated areas in northern Nigeria. *J. Moll. Stud.* 50:137–152.

Bremermann, H.J. (1980) Sex and polymorphism as strategies in host-pathogen interactions. *J. Theor. Biol.* 87:671–702.

Brown, D.S. (1980) *Freshwater Snails of Africa and Their Medical Importance.* Taylor and Francis, London.

Brown, D.S. and Wright, C.A. (1972) On a polyploid complex of freshwater snails (Planorbidae: *Bulinus*) in Ethiopia. *J. Zool. (Lond.)* 167:97–132.

Brown, S.G., Kwan, S. and Shero, S. (1995) The parasitic theory of sexual reproduction: Parasitism in unisexual and bisexual geckos. *Proc. Roy. Soc. Lond.* B 260:317–320.

Bull, J.J. and Charnov, E.L. (1989) Enigmatic reptilian sex ratios. *Evolution* 43:1561–1566.

Burt, A. and Bell, G. (1987) Mammalian chiasma frequencies as a test of two theories of recombination. *Nature* 326:803–805.

Burt, A. and Bell, G. (1991) Seed reproduction is associated with a transient escape from parasite damage in American beech. *Oikos* 61:145–148.

Charlesworth, B. (1980) The cost of sex in relation to mating system. *J. Theor. Biol.* 84:655–671.

Charnov, E.L. and Bull, J.J. (1977) When is sex environmentally determined? *Nature* 266:828–830.

Clay, K. (1983) Differential establishment of seedlings from chasmogamous and cleistogamous flowers in the grass, *Danthonia spicata. Oecologia* 36:734–741.

Clay, K. and Antonovics, J. (1985) Demographic genetics of the grass *Danthonia spicata*: Success of progeny from chasmogamous and cleistogamous flowers. *Evolution* 39:205–210.

Clay, K. and Kover, P.X. (1996) The Red Queen hypothesis and plant/pathogen interactions. *Annu. Rev. Phytopathol.* 34:29–50.

Conover, D.O. and Heins, S.W. (1987) Adaptive variation in environmental and genetic sex determination in a fish. *Nature* 326:496–498.

Doums, C. and Jarne, P. (1996) The evolution of phally polymorphism in *Bulinus truncatus* (Gastropoda: Planorbidae): The cost of male function analysed through life-history traits and sex allocation. *Oecologia* 106:464–469.

Doums, C., Viard, F., Pernot, A.-F., Delay, B. and Jarne, P. (1996) Inbreeding depression, neutral polymorphism and copulatory behavior in freshwater snails: A self-fertilization syndrome. *Evolution* 50:1908–1918.

Dybdahl, M.F. and Lively, C.M. (1995a) Diverse, endemic and polyphyletic clones in mixed populations of a freshwater snail (*Potamopyrgus antipodarum*). *J. Evol. Biol.* 8:385–398.

Dybdahl, M.F. and Lively, C.M. (1995b) Host-parasite interactions: Infection of common clones in natural populations of a freshwater snail (*Potamopyrgus antipodarum*). *Proc. Roy. Soc. Lond.* B 260:99–103.

Ebert, D. and Hamilton, W.D. (1996) Sex against virulence: The coevolution of parasitic diseases. *Trends Ecol. Evol.* 11:79–82.

Eernisse, D.J. (1988) Reproductive patterns in six species of *Lepidochitona* (Mollusca: Polyplacophora) from the Pacific Coast of North America. *Biol. Bull.* 174:287–302.

Fisher, R.A. (1930) *The Genetical Theory of Natural Selection*. Oxford University Press, Oxford.

Fox, J.A., Dybdahl, M.F., Jokela, J. and Lively, C.M. (1996) Genetic structure of coexisting sexual and clonal subpopulations in a freshwater snail (*Potamopyrgus antipodarum*). *Evolution* 50:1541–1548.

Fryer, S.E., Oswald, R.C., Probert, A.J. and Runham, N.W. (1990) The effect of *Schistosoma haematobium* infection on the growth and fecundity of three sympatric species of bulinid snails. *J. Parasitol.* 76:557–563.

Gadgil, M. and Bossert, W. (1970) Life history consequences of natural selection. *Amer. Nat.* 104:1–24.

Geraerts, W.P.M. and Joosse, J. (1984) Freshwater snails (Basommatophora). *In:* A.S. Tompa, N.H. Verdonk and J.A.M. van den Biggelaar (eds): *The Mollusca*, vol. 7, *Reproductions*, Academic Press, Orlando, pp 141–207.

Gerritsen, J. (1980) Sex and parthenogenesis in sparse populations. *Am. Nat.* 115:718–742.

Ghiselin, M.T. (1974) *The Economy of Nature and the Evolution of Sex*. University of California Press, Berkeley.

Glesener, R.R. and Tilman, D. (1978) Sexuality and the components of environmental uncertainty: Clues from geographic parthenogenesis in terrestrial animals. *Amer. Nat.* 112:659–673.

Hamilton, W.D. (1980) Sex versus non-sex versus parasite. *Oikos* 35:282–290.

Hamilton, W.D. (1993) Haploid dynamic polymorphism in a host with matching parasites: Effects of mutations/subdivision, linkage, and patterns of selection. *J. Hered.* 84:328–338.

Hamilton, W.D., Axelrod, R. and Tanese, R. (1990) Sexual reproduction as an adaptation to resist parasites (A review). *Proc. Natl. Acad. Sci. USA* 87:3566–3573.

Hanley, K.A., Fisher, R.N. and Case, T.J. (1995) Lower mite infestation in an asexual gecko compared with its sexual ancestor. *Evolution* 49:418–426.

Heller, J. and Farstey, V. (1990) Sexual and parthenogenetic populations of the freshwater snail *Melanoides tuberculata* in Israel. *Israel J. Zool.* 37:75–87.

Holtsford, T.P. and Ellstrand, N.C. (1990) Inbreeding effects in *Clarkia tembloriensis* (Onagraceae) populations with different natural outcrossing rates. *Evolution* 44:2031–2046.

Howard, R.S. and Lively, C.M. (1994) Parasitism, mutation accumulation and the maintenance of sex. *Nature* 367:554–557.

Hurst, L.D. and Peck, J.R. (1996) Recent advances in understanding of the evolution and maintenance of sex. *Trends Ecol. Evol.* 11:46–52.

Jaenike, J. (1978) An hypothesis to account for the maintenance of sex within populations. *Evol. Theory* 3:191–194.

Jain, S.K. (1976) The evolution of inbreeding in plants. *Annu. Rev. Ecol. Syst.* 7:469–495.

Jarne, P. and Charlesworth, D. (1993) The evolution of the selfing rate in functionally hermaphrodite plants and animals. *Annu. Rev. Ecol. Syst.* 24:441–466.

Jarne, P., Finot, L., Bellec, C. and Delay, B. (1992) Aphally versus euphally in self-fertile hermaphrodite snails from the species *Bulinus truncatus* (Pulmonata: Planorbidae). *Amer. Nat.* 139:424–432.

Jarne, P., Vianey-Liaud, M. and Delay, B. (1993) Selfing and outcrossing in hermaphrodite freshwater gastropods (Basommatophora): Where, when and why. *Biol. J. Limn. Soc.* 49:99–125.

Jarne, P., Viard, F., Delay, B. and Cuny, G. (1994) Variable microsatellites in the highly selfing snail *Bulinus truncatus* (Basommatophora: Planorbidae). *Mol. Ecol.* 3:527–528.

Jelnes, J.E. (1986) Experimental taxonomy of *Bulinus* (Gastropoda: Planorbidae): The West and North African species reconsidered, based upon an electrophoretic study of several enzymes per individual. *Zool. J. Limn. Soc.* 87:1–26.

Johnson, S.G. (1992a) Spontaneous and hybrid origins of parthenogenesis in *Campeloma decisum* (freshwater prosobranch snail). *Heredity* 68:253–261.

Johnson, S.G. (1992b) Parasite-induced parthenogenesis in a freshwater snail: Stable, persistent patterns of parasitism. *Oecologia* 89:533–541.

Johnson, S.G. (1994) Parasitism, reproductive assurance, and the evolution of reproductive mode in a freshwater snail. *Proc. Roy. Soc. Lond.* B 255:209–213.

Jokela, J. and Lively, C.M. (1995a) Parasites, sex, and early reproduction in a mixed population of freshwater snails. *Evolution* 49:1268–1271.

Jokela, J. and Lively, C.M. (1995b) Spatial variation for infection by digenetic trematodes in a population of freshwater snails (*Potamopyrgus antipodarum*). *Oecologia* 103:509–517.

Jokela, J., Lively, C.M., Dybdahl, M.F. and Fox, J.A. (1977) Evidence for a cost of sex in the freshwater snail *Potamopyrgus antipodarum*. *Ecology* 78:452–460.

Jones, C.S., Okamura, B. and Noble, L.R. (1994) Parent and larval RAPD fingerprints reveal outcrossing in freshwater bryozoans. *Mol. Ecol.* 3:193–199.

Jordan, P. and Webbe, G. (1982) *Schistosomiasis: Epidemiology, Treatment and Control.* William Heinemann Medical Books Ltd., London.

Judson, O.P. (1995) Preserving genes: A model of the maintenance of genetic variation in a metapopulation under frequency-dependent selection. *Genet. Res.* 65:175–191.

Keeling, M.J. and Rand, D.A. (1995) A spatial mechanism for the evolution and maintenance of sexual reproduction. *Oikos* 74:414–424.

Kondrashov, A.S. (1988) Deleterious mutations and the evolution of sexual reproduction. *Nature* 336:435–440.

Ladle, R.J., Johnstone, R.A. and Judson, O.P. (1993) Coevolutionary dynamics of sex in a metapopulation: Escaping the Red Queen. *Proc. Roy. Soc. Lond.* B 253:155–160.

Lafferty, K.D. (1993) The marine snail, *Cerithidea californica*, matures at smaller sizes where parasitism is high. *Oikos* 68:3–11.

Lamb, R.Y. and Willey, R.B. (1979) Are parthenogenetic and related bisexual insects equal in fertility? *Evolution* 33:774–775.

Larambergue, M. de (1939) Étude de l'autofécondation chez les gastéropodes pulmonés: Recherches sur l'aphallie et la fécondation chez. *Bulinus (Isidora) contortus* Michaud. *Bull. Biol. France Belg.* 73:19–231.

Law, R., Bradshaw, A.D. and Putwain, P.D. (1979) Optimal life histories under age-specific predation. *Amer. Nat.* 114:399–417.

Levin, D.A. (1975) Pest pressure and recombination systems in plants. *Amer. Nat.* 109:437–451.

Lively, C.M. (1987) Evidence from a New Zealand snail for the maintenance of sex by parasitism. *Nature* 328:519–521.

Lively, C.M. (1989) Adaptation by a parasitic trematode to local populations of its snail host. *Evolution* 43:1663–1671.

Lively, C.M. (1992) Parthenogenesis in a freshwater snail: Reproductive assurance versus parasitic release. *Evolution* 46:907–913.

Lively, C.M. (1996) Host-parasite coevolution and sex. *Bio Science* 46:107–114.

Lively, C.M. and Howard, R.S. (1994) Selection by parasites for clonal diversity and mixed mating. *Phil. Trans. Roy. Soc. Lond.* B 346:271–281.

Lively, C.M. and Johnson, S.G. (1994) Brooding and the evolution of parthenogenesis: Strategy models and evidence from aquatic invertebrates. *Proc. Roy. Soc. Lond.* B 256:89–95.

Lively, C.M. and Jokela, J. (1996) Clinal variation for local adaptation in a host-parasite interaction. *Proc. Roy. Soc. Lond.* B 263:891–897.

Lively, C.M., Craddock, C. and Vrijenhoek, R.C. (1990) Red Queen hypothesis supported by parasitism in sexual and clonal fish. *Nature* 344:864–866.

Lloyd, D.G. (1979) Some reproductive factors affecting the selection of self-fertilization in plants. *Amer. Nat.* 113:67–79.

Lloyd, D.G. (1980) Benefits and handicaps of sexual reproduction. *Evol. Biol.* 13:69–111.

Lynch, M. (1984) Destabilizing hybridization, general-purpose genotypes and geographic parthenogenesis. *Quart. Rev. Biol.* 59:257–290.

May, R.M. and Anderson, R.M. (1979) Population biology of infectious diseases: Part II. *Nature* 280:455–460.

Maynard Smith, J. (1971) The origin and maintenance of sex. *In:* G.C. Williams (ed.): *Group Selection,* Aldine Atherton, Chicago, pp 163–175.

Maynard Smith, J. (1978) *The Evolution of Sex.* Cambridge University Press, Cambridge.

Maynard Smith, J. (1989) *Evolutionary Genetics.* Oxford University Press, Oxford.

Michod, R.E. (1979) Evolution of life histories in response to age-specific mortality factors. *Amer. Nat.* 113:531–550.

Mitchell-Olds, T. and Waller, D.M. (1985) Relative performance of selfed and outcrossed progeny in *Impatiens capensis. Evolution* 39:533–544.

Mitter, C., Futuyma, D.J., Schneider, J.C. and Hare, J.D. (1979) Genetic variation and host plant relations in a parthenogenetic moth. *Evolution* 33:777–790.

Moritz, C., McCallum, H., Donnellan, S. and Roberts, J.D. (1991) Parasite loads in parthenogenetic and sexual lizards (*Heteronotia binoei*): Support for the Red Queen hypothesis. *Proc. Roy. Soc. Lond.* B 244:145–149.

Mukai, T. (1969) The genetic structure of natural populations of *Drosophila melanogaster.* VII. Synergistic interactions of spontaneous mutant polygenes affecting viability. *Genetics* 61:749–761.

Muller, H.J. (1964) The relation of recombination to mutational advance. *Mutat. Res.* 1:2–9.

Nagylaki, T. (1976) A model for the evolution of self-fertilization and vegetative reproduction. *J. Theor. Biol.* 58:55–58.

Naylor, C., Adams, J. and Greenwood, P.J. (1988) Variation in sex determination in natural populations of a shrimp. *J. Evol. Biol.* 1:355–368.

Ndifon, G.T. and Umar-Yahaya, A. (1988–1990) Cercariae of freshwater snails in Kano, Nigeria. *Nigerian J. Parasit.* 9–11:69–75.

Nicklas, N.L. and Hoffmann, R.J. (1981) Apomictic parthenogenesis in a hermaphroditic terrestrial slug, *Deroceras laeve* (Müller). *Biol. Bull.* 160:123–135.

Njiokou, F., Bellec, C., Berrebi, P., Delay, B. and Jarne, P. (1993a) Do self-fertilization and genetic drift promote a very low genetic variability in the allotetraploid *Bulinus truncatus* (Gastropoda: Planoribidae) populations? *Genet. Res.* 62:89–100.

Njiokou, F., Bellec, C., Jarne, P., Finot, L. and Delay, B. (1993b) Mating system analysis using protein electrophoresis in the self-fertile hermaphrodite species *Bulinus truncatus* (Gastropoda: Planorbidae). *J. Moll. Stud.* 59:125–133.

Nunney, L. (1989) The maintenance of sex by group selection. *Evolution* 43:245–257.

Phillips, N.R. and Lambert, D.M. (1989) Genetics of *Potamopyrgus antipodarum* (Gastropoda: Prosobranchia): Evidence for reproductive modes. *N. Z. J. Zool.* 16:435–445.

Platt, J.R. (1964) Strong inference. *Science* 146:347–353.

Schmitt, J. and Antonovics, J. (1986) Experimental studies of the evolutionary significance of sexual reproduction. IV. Effect of neighbor relatedness and aphid infestation on seedling performance. *Evolution* 40:830–836.

Schmitt, J. and Ehrhardt, D.W. (1987) A test of the sib-competition hypothesis for outcrossing advantage in *Impatiens capensis. Evolution* 41:579–590.

Schmitt, J. and Gamble, S.E. (1990) The effect of distance from the parental site on offspring performance and inbreeding depression in *Impatiens capensis*: A test of the local adaptation hypothesis. *Evolution* 44:2022–2030.

Schrag, S.J. (1993) Factors influencing selfing and outcrossing in the hermaphrodite, *Bulinus truncatus. D. Phil. Thesis,* University of Oxford.

Schrag, S.J. and Read, A.F. (1992) Temperature determination of male outcrossing ability in a simultaneous hermaphrodite. *Evolution* 46:1698–1707.

Schrag, S.J. and Read, A.F. (1996) Loss of male outcrossing ability in simultaneous hermaphrodites: Phylogenetic analyses of pulmonate snails. *J. Zool (Lond.)* 238:287–299.

Schrag, S.J. and Rollinson, D. (1994) Effects of *Schistosoma haematobium* infection on reproductive success and male outcrossing ability in the simultaneous hermaphrodite, *Bulinus truncatus* (Gastropoda: Planorbidae). *Parasitology* 108:27–34.

Schrag, S.J., Rollinson, D., Keymer, A.E. and Read, A.F. (1992) Heritability of male outcrossing ability in the simultaneous hermaphrodite, *Bulinus truncatus* (Gastropoda: Planorbidae). *J. Zool. (Lond.)* 226:311–319.

Schrag, S.J., Mooers, A.Ø, Ndifon, G.T. and Read, A.F. (1994a) Ecological correlates of male outcrossing ability in a simultaneous hermaphrodite snail. *Am. Nat.* 143:636–655.

Schrag, S.J., Ndifon, G.T. and Read, A.F. (1994b) Temperature-determined outcrossing ability in wild populations of a simultaneous hermaphrodite snail. *Ecology* 75:2066–2077.

Städler, T., Weisner, S. and Streit, B. (1995) Outcrossing rates and correlated matings in a predominately selfing freshwater snail. *Proc. Roy. Soc. Lond.* B 262:119–125.

Stearns, S.C. (1987) The selection-arena hypothesis. *In:* S.C. Stearns (ed.): *The Evolution of Sex and Its Consequences*, Birkhäuser Verlag, Basel, pp 337–349.

Strathmann, R.R., Strathmann, M.R. and Emson, R.H. (1984) Does limited brood capacity link adult size, brooding, and simultaneous hermaphroditism? A test with the starfish *Asterina phylactica. Amer. Nat.* 123:796–818.

Templeton, A.R. (1982) The prophecies of parthenogenesis. *In:* H. Dingle and J.P. Hegmann (eds): *Evolution and Genetics of Life Histories*, Springer-Verlag, New York, pp 75–101.

Tingley, G.A. and Anderson, R.M. (1986) Environmental sex determination and density-dependent population regulation in the entomogenous nematode *Romanomermis culcivorax. Parasitology* 92:431–449.

Uyenoyama, M.K. (1984) On the evolution of parthenogenesis: A genetic representation of the "cost of meiosis". *Evolution* 38:87–102.

Uyenoyama, M.K., Holsinger, K.E. and Waller, D.M. (1993) Ecological and genetic factors directing the evolution of self-fertilization. *Oxford Surv. Evol. Biol.* 9:327–381.

Vail, V.A. (1977) Comparative reproductive anatomy of three viviparid gastropods. *Malacologia* 16:519–540.

Viard, F., Bremond, P., Labbo, R., Justy, F., Delay B. and Jarne, P. (1996) Microsatellites and the genetics of highly selfing populations in the freshwater snail *Bulinus truncatus. Genetics* 142:1237–1247.

Wallace, C. (1992) Parthenogenesis, sex and chromosomes in *Potomopyrgus. J. Moll. Stud.* 58:93–107.

Waller, D.M. (1984) Differences in fitness between seedlings derived from cleistogamous and chasmogamous flowers in *Impatiens capensis. Evolution* 38:427–440.

Watson, H. (1934) Genital dimorphism in *Zonitoides. J. Conch.* 20:33–42.

Williams, G.C. (1971) Introduction. *In:* G.C. Williams (ed.): *Group Selection,* Aldine Atherton, Chicago, pp 1–15.

Williams, G.C. (1975) *Sex and Evolution.* Princeton University Press, Princeton.

Willis, J.H. (1993) Effects of different levels of inbreeding on fitness components in *Mimulus guttatus. Evolution* 47:864–876.

Winterbourn, M., Rounick, J.S. and Cowie, B. (1981) Are New Zealand stream ecosystems really different? *N.Z.J. Mar. Freshw. Res.* 15:321–328.

Woolhouse, M.E.J. and Chandiwana, S.K. (1989) Spatial and temporal heterogeneity in the population dynamics of *Bulinus globosus* and *Biomphalaria pfeifferi* and in the epidemiology of their infection with schistosomes. *Parasitology* 98:21–34.

Wright, C.A. (1957) *A Guide to Molluscan Anatomy for Parasitologists in Africa.* British Museum of Natural History, London.

Wrigth, C.A. and Rollinson, D. (1981) Analysis of enzymes in the *Bulinus tropicus/truncatus* complex (Mollusca: Planorbidae). *J. Nat. Hist.* 15:873–885.

Evolutionary Ecology of Freshwater Animals
ed. by B. Streit, T. Städler and C. M. Lively
© 1997 Birkhäuser Verlag Basel/Switzerland

Genetic similarity, parasitism, and metapopulation structure in a freshwater bryozoan

B. Okamura

School of Animal and Microbial Sciences, The University of Reading, Whiteknights, PO Box 228, Reading RG6 6AJ, UK

Summary. Many freshwater organisms exhibit mixed life histories that include sexual and asexual phases of reproduction. Ecological and genetic studies have revealed insights about the consequences of mixed life histories for populations of planktonic cladocerans and rotifers and aquatic macrophytes. However, little is known of the consequences of mixed life histories for populations of benthic colonial freshwater invertebrates. A series of investigations of the ecology and genetics of local populations of the freshwater bryozoan, *Cristatella mucedo*, are reviewed in this context. Asexual reproduction achieved by colony growth, fission, and the production of numerous resistant statoblasts promotes the persistence and dispersal of clones and leads to extensive genetic similarity within and between populations in southern England. Genetic characterization of parent and larval colonies indicates that sexual reproduction results in inbreeding amongst genetically similar clonal stock within sites and so generates little genetic variation. Ecological sampling confirms that sexual reproduction is limited in duration and suggests that in some sites and/or years, a sexual phase is foregone entirely. Myxozoans parasitize these genetically similar host populations and apparently indiscriminately attack the highly-related bryozoan clones present within a site. As myxozoans adversely affect host fitness they may, in part, be responsible for the drastic reductions and occasional extinctions observed in *C. mucedo* populations. The presence of harmful parasites and the relative insignificance of sex suggest that host-parasite coevolution as predicted by the Red Queen does not apply in this system. Rather, evidence indicates that a metapopulation structure allows the persistence of genetically-similar sub-populations as long as rates of asexual replication and dispersal are sufficient to provide a means of escape from parasites and other adverse conditions. Thus metapopulation structure may provide at least a short-term alternative to the regular production of genetic novelty through sex. Perhaps due to features such as a sessile nature, asexual replication via vegetative growth, and indirect fertilization, the consequences of mixed life histories for bryozoans of southern England contrast with those in planktonic cladocerans and rotifers but show some similarities to those of aquatic macrophytes.

Introduction

The life histories of many higher organisms combine cyclical bouts of sexuality with extended periods of clonal reproduction that may be achieved by a variety of asexual mechanisms (Abrahamson, 1980; Bell, 1982; Hughes and Cancino, 1985). Since so many animals and plants utilize both modes of reproduction, it can be assumed that the relative balance of the two modes confers distinct advantages. However, despite the common incidence of mixed life histories, relatively little is known of the ecological conditions that promote phases of both sexual and asexual modes of reproduction within a given life history or of the associated genetic and ecological consequences (Jackson, 1985; Silander, 1985; Eckert and Barrett, 1993).

Benefits that can result from such life histories include escape from paying the costs of sex, the proliferation of highly adapted genotypes by clonal reproduction, and the introduction of genetic variation by sexual reproduction which creates the potential for clonal specialization or reduces the probability of local extinction.

Many inhabitants of freshwater habitats are particularly notable in having both high levels of clonal reproduction and a sexual phase in the life cycle. For example, cladocerans and rotifers undergo extensive clonal replication by parthenogenesis during favourable conditions in lakes and ponds until a change in environmental conditions provokes a switch to sexuality. Empirical studies provide evidence for several advantages that can be conferred by the possession of two reproductive modes in these planktonic animals (see Hebert, 1987 and Carvalho, 1993 for review). For instance, such studies have demonstrated that in cladocerans mixis generates divergent, locally adapted genotypes that, through parthenogenesis, can achieve spatial and temporal exploitation of their habitat.

Perennial aquatic plants also utilize mixed reproductive modes and achieve extensive asexual growth by vegetative spread (Abrahamson, 1980; Eckert and Barrett, 1993). Benefits accruing from mixed life histories in aquatic plants have been less well characterized but may include enhanced competitive abilities of vegetatively derived ramets vs. sexually derived seedlings (Lovett Doust, 1981; Aspinwall and Christian, 1992) and clonal adaptation to microsite or niche heterogeneity (Ellstrand and Roose, 1987 and references therein).

Nearly all investigations of the ecological genetics of freshwater organisms with mixed reproductive modes have focused on planktonic cladocerans and rotifers and aquatic macrophytes. However, sessile freshwater invertebrates such as bryozoans and sponges are often abundant components of benthic habitats (see Okamura and Hatton-Ellis, 1995 for review) and also possess life histories that combine periods of sexual and asexual reproduction. The general purpose of this paper is to review the consequences of mixed life histories for the ecology and population genetics of a freshwater bryozoan in southern England and to compare and contrast these consequences with those observed for planktonic cladocerans and rotifers and for attached aquatic macrophytes. However, as freshwater bryozoans are relatively unfamiliar, it is necessary to briefly review their general biology and life cycle before comparisons may be made with the more familiar aquatic groups.

Life histories of freshwater bryozoans and comparisons with other groups

Bryozoans that inhabit freshwaters include some 60 known species in the exclusively freshwater class, the Phylactolaemata, and a few species from the largely marine class, the Gymnolaemata. The following discussion summarizes the biology of phylactolaemates, although the biology of freshwater gymnolaemates is broadly similar in outline (for distinctions see Wood, 1991). During favourable conditions, freshwater bryozoans grow as colonies composed of genetically identical zooids produced by budding. Colonies can be found adhering to a variety of substrata, including aquatic vegetation, submerged branches, and rocks in all major faunal regions of the world (Bushnell, 1973). Like their marine counterparts, freshwater bryozoans extend a hollow, ciliated, tentacular crown (the lophophore) to feed on suspended phytoplankton, but unlike their marine counterparts, they produce resistant propagules called statoblasts that survive adverse conditions. Statoblasts are small (≤ 1 mm), multicellular, amictic resting stages that are produced in great numbers (Brown, 1933). They develop initially within the body cavity as undifferentiated vegetative tissue that is budded from a strand of tissue which joins the gut to the body wall (the funiculus). Mature statoblasts incorporate a central mass of undifferentiated, yolk-rich, germinative cells that are enclosed by resistant chitinous valves. Upon the return of favourable conditions, the valves separate and a small, recently differentiated colony, emerges and attaches to the substratum. In temperate regions this occurs in the late spring following overwintering by statoblasts. Colonies become simultaneously hermaphroditic in the early part of the growing season, and sexual reproduction results in the production of short-lived, free swimming larvae. Recent evidence confirms that some larvae are products of outcrossing although selfing cannot be ruled out for others (Jones et al., 1994). However, as discussed later, a variety of evidence indicates sexual reproduction is often foregone in local populations.

As the above description indicates, benthic freshwater bryozoans possess life histories that show both similarities and differences to the life histories of the more familiar aquatic macrophytes and planktonic cladocerans and rotifers (see Tab. 1). Thus, although freshwater bryozoans undergo extended periods of amictic reproduction, this results in colony growth through budding, the production of statoblasts, and, in some cases, colony proliferation through fission. Bryozoans also undergo a short sexual phase, but, unlike cladocerans and rotifers, this is not triggered by unfavourable conditions. As in aquatic macrophytes, amictic and mictic forms of reproduction occur in bryozoans. Thus colonies may simultaneously produce larvae and statoblasts and increase their size by budding new zooids. By contrast, parthenogenesis in cladocerans and rotifers is associated with separate bouts of sexual or asexual reproduction. These animals are therefore cyclically clonal (the exception being obligately parthenogenetic populations) whilst bryozoans and macrophytes are more or less continuously so. Aquatic macrophytes may overwinter as sexual (seeds) and/ or asexual (e.g., mature plants/rhizomes/apomictic seeds) stages. Finally, fertilization in bryozoans is achieved indirectly by sperm spawned into the

Table 1. Similarities and differences in various life history features of cladocerans and rotifers, aquatic macrophytes, and freshwater bryozoans

Life history features	Cladocerans and rotifers	Aquatic macrophytes	Freshwater Bryozoans
Mechanisms of amixis:	Parthenogenesis	Vegetative spread	Zooidal budding; statoblasts
Timing of sexuality:	End of season in general (although sometimes several bouts within a season in cladocerans)	Variable	Early-mid season
Sex invoked by deteriorating conditions:	Yes	No	No
Clonality:	Cyclic	Continuous	Continuous
Overwintering stage:	Sexual resting eggs	Sexual seeds; overwintering plants, rhizomes	Asexual statoblasts
Fertilization:	Direct (copulation)	Indirect (pollen release)	Indirect (free-spawned sperm)

water column. This is similar to the release of pollen in aquatic macrophytes and contrasts with direct fertilization by copulation in cladocerans and rotifers.

The above-described similarities and differences in the life histories of bryozoans and the other better studied freshwater groups beg the question of what the ecological and population genetic consequences of such a mixed life history may be in these sessile, colonial forms. This chapter reviews results from ongoing studies of populations of the freshwater bryozoan, *Cristatella mucedo*, with this question in view. Synthesis of these results indicates that the ecology of *C. mucedo* at least in southern England is characterized by: 1) great genetic similarity; 2) persistent clonality in the face of parasitism; 3) a metapopulation structure. This synthesis leads to the conclusion that a life history involving extensive clonality by vegetative growth, limited mobility and indirect fertilization entails consequences that are more similar to those of aquatic macrophytes than of planktonic cladocerans and rotifers and also raises the question of how commonly metapopulation structure allows the maintenance of high levels of clonality.

The natural history of *Cristatella mucedo*

Colonies of *C. mucedo* can be found adhering to substrata such as roots, submerged branches, aquatic vegetation, and rocks in shallow regions of

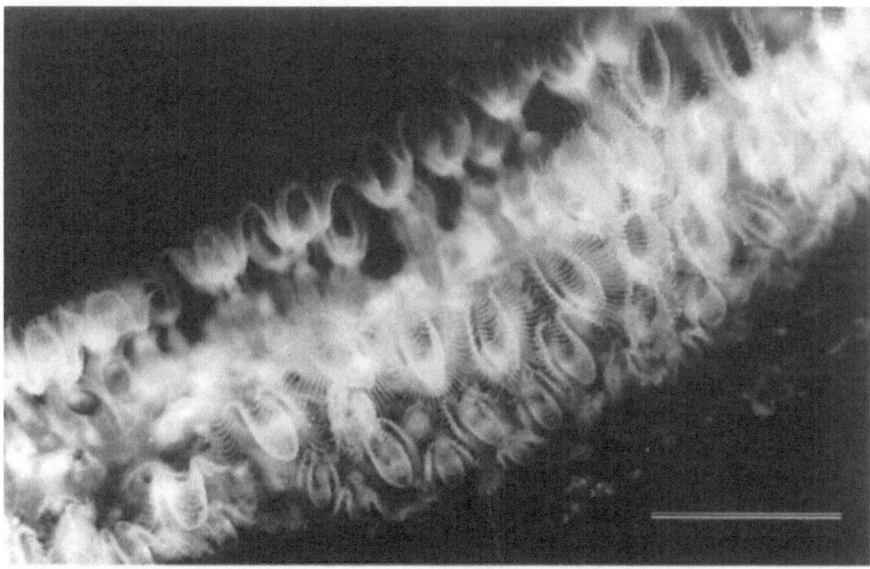

Figure 1. Part of a colony of *Cristatella mucedo*. Colonies grow as linear series of zooids that are budded on either side of a central region. U-shaped lophophores can be seen extended in suspension feeding. The gut of each bryozoan zooid extends downwards from the mouth (which lies in the center of the lophophore) and is deflected towards the anus (not discernible) situated outside the lophophore. Rounded fecal pellets can be seen forming near the distal end of the intestine prior to defecation. Scale bar = 2 mm.

lakes and ponds throughout the Holarctic (Bushnell, 1973). Colonies grow as a linear series of zooids budded on either side of a central region giving the colony an elongate, caterpillar-like appearance (Fig. 1). Each zooid in the colony possesses a U-shaped lophophore that is used in suspension feeding on phytoplankton. In southern England colonies may be found from early summer through the autumn, however, their occurrence and abundance are highly variable on both spatial and temporal scales. Thus they may be abundant in some sites but absent or rare in other apparently similar sites, they may be abundant within a site in some years but absent or rare in others, and the period of time in which they may be found varies amongst sites both within and between years (B. Okamura, unpublished data). Such variation is typical of freshwater bryozoan populations (e.g., Jónasson, 1963; Bushnell, 1973; Wood, 1989; Joo et al., 1992).

Colony growth is achieved by budding of new zooids while colony fission and subsequent movement apart of the two daughter colonies permit a given clonal genotype a degree of microhabitat exploitation (Buss, 1979; Rubin, 1987). Greater dispersal of clones is provided by statoblasts. *C. mucedo* produces statoblasts that possess gas-filled cells which confer flotation, thus statoblasts can be dispersed widely within a site. In addition,

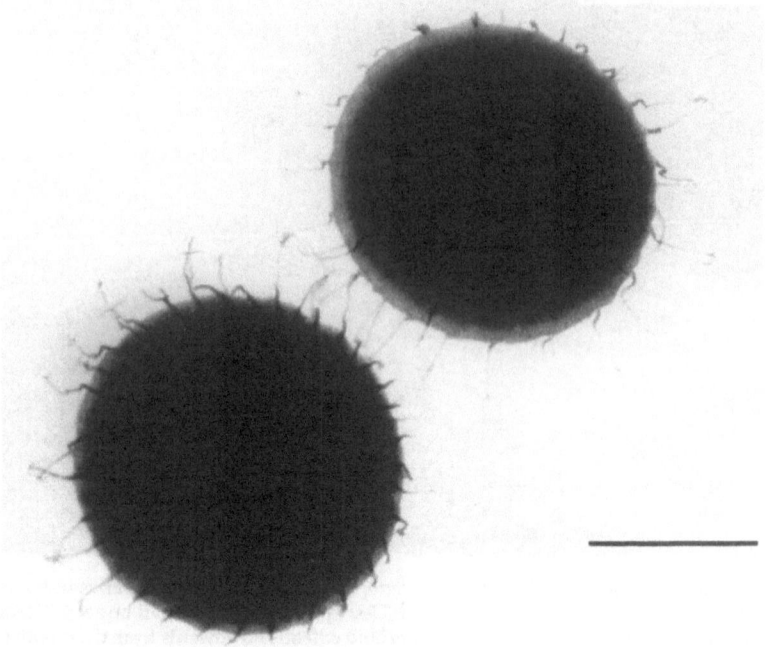

Figure 2. Two statoblasts of *C. mucedo*. Note spines protruding from statoblast margins and the two, recurved extensions at the tip of each spine that act as hooks for attachment. Scale bar = 0.5 mm.

hooks on spines that extend from statoblast margins (see Fig. 2). are inferred to result in entanglement in fur and feathers, thereby effecting dispersal amongst sites by animal vectors. However, such hooks also promote retention within habitats through entanglement with other statoblasts and with vegetation growing along the margins of water bodies, especially amongst fine roots. This can sometimes lead to large accumulations of statoblasts adhering as clumps to marginal vegetation (B. Okamura, personal observation).

Apart from an initial period when colonies recently hatched from overwintering statoblasts undergo rapid growth and become established, statoblasts are generally produced throughout the growing season. Their release occurs through specialized pores (Mukai, 1982) and also possibly when colonies undergo fission. At the end of the growing season colonies degenerate and essentially become bags of statoblasts (B. Okamura, personal observation) that are released following colony degradation. The production of statoblasts in great numbers (Brown, 1993) is presumably necessary to ensure successful colonization in the following year. However, statoblasts released during the summer by *C. mucedo* may result in multiple statoblast-derived generations per year as occurs in some other species of freshwater bryozoans (Mukai, 1982; Wöss, 1994).

Like all freshwater bryozoans, sexual reproduction in *C. mucedo* is restricted to the earlier phases of the growing season (late spring to mid-summer) when ciliated, actively swimming larvae are released. Phylactolaemates brood their embryos and release short-lived larvae that are present in the water column for only a few hours before they settle and a ciliated mantle is folded back to expose a small adult colony (Mukai, 1982). They presumably provide only a limited degree of dispersal.

Extensive clonality in time and space

As the foregoing discussion implies, the life cycle of *C. mucedo* imparts the potential for extended clonality. In this section I review how results obtained from ecological sampling and molecular studies provide evidence that clonality is indeed maintained over broad spatial and temporal scales in populations of *C. mucedo* in southern England.

Populations are composed of closely-related genotypes

To investigate levels of clonality in *C. mucedo* RAPD PCR (*R*andom *A*mplification of *P*olymorphic *D*NA via the *P*olymerase *C*hain *R*eaction) has been used as it provides a high level of genetic resolution and is thus appropriate for discriminating amongst highly inbred systems such as closely related clonal types. Briefly, the technique employs a series of arbitrary primers in the polymerase chain reaction to amplify fragments of DNA (Welsh and McClelland 1990; Williams et al., 1990). Fragments are then separated by gel electrophoresis, and clones can be identified by the pattern of fragments generated through amplification by multiple primers.

Results from a RAPD PCR study conducted on three populations in southern England using 13 primers indicated the presence of multiple clones in each of three sites during the summer of 1992 (Okamura et al., 1993). Dorchester Fishing Lake and Beale Bird Park Lake are approximately 17 km apart in the Thames Valley region (central, southern England) whilst Thompson Water is some 270 km to the northeast in West Norfolk (East Anglia). The former two sites are gravel pits both less than 50 years old (Okamura, 1994) while Thompson Water is one of the natural water bodies that overlies chalk in the Breckland District of East Anglia (Rackham, 1986). Eight primers provided polymorphisms (both population-specific and within-site markers) while five primers gave monomorphic results (see Fig. 3). Six distinct clones were characterized in Dorchester Fishing Lake, five in Beale Bird Park Lake, and eight in Thompson Water.

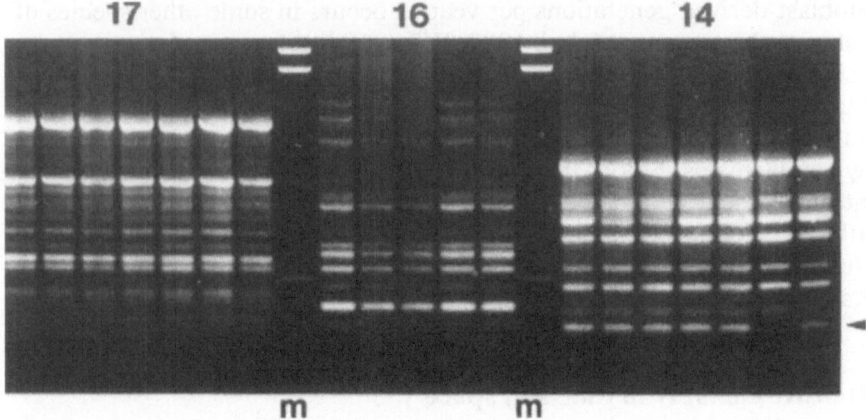

Figure 3. RAPD-PCR gel showing amplification by three different arbitrary primers of colonies of *C. mucedo* collected from Beale Bird Park Lake on 23 July 1992. Each lane represents a different colony amplified by each primer. Primers 17 and 16 gave monomorphic results while primer 14 shows one polymorphic fragment (see arrowhead). A molecular weight standard (m) separates colonies amplified by different primers.

Some clones were represented by only one colony, but in each case a particular clone was dominant despite the relatively small sample sizes (n = 14 colonies per site) and the care given to maximizing collection of genetically distinct colonies.

Although distinct clones could be recognized, the clones were genetically very similar and a high degree of relatedness obtained both within and amongst the three sites. This is reflected in the very high similarity coefficients which are based on the total proportion of shared fragments between clones. The overall mean pairwise similarity across the three sites was 0.982 (mean within-site pairwise similarity coefficients ranged from 0.9908–0.9936) (Okamura et al., 1993). Similarity coefficients derived by RAPD PCR analysis of other populations of animals are substantially lower. For instance, similarity coefficients obtained for the grasshoppers *Melanoplus sanguinipes* and *M. fermurrubrum* range from 0.417–0.473 and 0.213–0.687, respectively (Chapco et al., 1992). This contrast reflects differences in only one or two fragments generated by primers that reveal such polymorphisms for *C. mucedo* compared with the much greater number of fragment differences typically revealed in RAPD surveys of populations of other organisms including two species of marine bryozoans (T. Hatton-Ellis and B. Okamura, unpublished data; J. Porter, personal communication) and some populations of cyclically clonal *Daphnia* (Fig. 2 in Schierwater et al., 1994).

Despite the high genetic similarity amongst sites, cluster analysis of genetic distance values was able to separate the three populations of

C. mucedo with respect to geographic region. Thus two major branches resolved the Thames Valley populations from the Norfolk population and the two Thames Valley populations were further distinguished from each other (Okamura et al., 1993). Nested AMOVA (*A*nalysis of *M*olecular *V*ariance; Excoffier et al., 1992) of the data provides further evidence for differences amongst populations. All three variance components (among regions, among populations within regions, and within populations) in the nested analysis were significant at $p < 0.01$ (T. Hatton-Ellis and B. Okamura, unpublished data). Both the among regions and among populations within regions variance components were proportionately large (responsible for 38.7% and 40.7% of the variance, respectively) whilst the smaller (although still substantial) within populations variance component (20.6%) indicates less genetic variation within sites than among sites within regions or between regions. However, despite the ability of these analyses to distinguish populations and to detect significant levels of variation, the high within-site and overall genetic similarity values are nonetheless indicative of great relatedness both within and amongst all three sites.

The high degree of relatedness of clonal types over a spatial scale of some 270 km is accompanied by similar evidence for the maintenance of a high degree of relatedness over time. A temporal study of a localized population at Beale Bird Park Lake was conducted by collecting colonies haphazardly from one particular area (some 5 × 5 m) of the shoreline of a small cove on the lake during the period of 1992–1994 (Vernon et al., 1996). Larger sample sizes in this study (n = 29–33 colonies for each of three sampling dates during the summer of 1992 and for one sampling date in the summer of 1994) continued to confirm the generality of the presence of multiple, highly related clones at any one time with domination by a single clone. Although clonal representation was found to vary with time, high band-sharing coefficients amongst clones, clonal persistence, and lack of invasion by new clones all indicated that this population was characterized by a high degree of relatedness through time. Similar patterns of temporal stability in clonal genotypes may be found in some large-lake populations of *Daphnia* (Mort and Wolf, 1985).

It should be noted that the geographic study of *C. mucedo* populations revealed the presence of five clonal genotypes at Beale Bird Park Lake early in the summer of 1992 which were distinct from the four clonal genotypes detected by Vernon et al. (1996) later in the same season and in 1994. There are several possible explanations for these differences. Firstly, sampling of colonies in the geographic study was not restricted to one localized area of the lake. Secondly, certain clonal genotypes present earlier in the growing season may have become rare or extinct by later sampling dates. Thirdly, slight variation in protocols employed by different laboratory workers may have produced different profiles (see Grosberg et al., 1996 for review). However, the adoption of similar rigorous and conservative procedures in both studies provides confidence in the general validity of their results. Each study thus independently confirms the existence of high clonality and relatedness within the population.

Ongoing RAPD PCR work based on a larger number (n = 13) of primers providing polymorphisms and including bryozoan populations in some nine additional sites in southern England continues to support the view that *C. mucedo* occurs as a series of highly clonal populations in this region showing great degrees of relatedness over both spatial and temporal scales (T. Hatton-Ellis and B. Okamura, unpublished data). As colonies in these populations were collected from a range of substrata (e.g., branches, roots, rocks and aquatic macrophytes other than *Elodea canadensis*), the genetic similarity of colonies collected exclusively from *E. canadensis* in the above-described studies does not appear to reflect habitat specialization of clones.

Although RAPD PCR may not reveal all variation present and the adoption of other molecular approaches may provide further information, the clonal nature revealed by RAPD PCR is nonetheless consistent with both ecological sampling (see below) and knowledge of the life cycle of *C. mucedo*. Also, as RAPD PCR provides much greater levels of genetic variation for marine bryozoan populations (T. Hatton-Ellis and B. Okamura, unpublished data; J. Porter, personal communication) and illustrates substantial variation between populations of *C. mucedo* in the U.K. vs. North America (T. Hatton-Ellis and B. Okamura, unpublished data), the picture of high clonality obtained for British populations of *C. mucedo* provided by RAPD PCR would seem robust.

Relatively limited investment in sexual reproduction

Further support for the maintenance of extensive clonality is provided by evidence that *C. mucedo* populations in southern England invest little in sexual relative to clonal reproduction. Thus periodic sampling of the population at Beale Bird Park Lake indicated that larvae were only produced from early July to early August although colonies were collected through mid-October (B. Okamura, unpublished data). More sporadic monitoring suggests that sexual activity is foregone in many local populations at least in some years (see Tab. 2). Thus larvae were not observed in three of five populations in 1992 and 1993 and in one of two populations sampled in 1994, while in one population sexual reproduction was noted in one year but not another (South Cerney Pit # 3 in 1991 and 1992). In addition, when sexual reproduction was observed, generally only a small percentage of *C. mucedo* colonies were producing larvae in most sites (the mean maximum percentage of colonies producing larvae in sexual populations from 1991–1994 = 13.2%, SD = 14.5, n = 7, range = 4.2–41.9%; B. Okamura, unpublished data). Finally, even when colonies are producing larvae they are generally also simultaneously producing asexual statoblasts in similar numbers (B. Okamura, unpublished data).

Table 2. Sites in which sexual reproduction was observed and not observed during 1991–1994. Sexual reproduction noted by the presence of brooded larvae within colonies

Year	No sexual reproduction	Sexual reproduction
1991		Beale Bird Park Lake South Cerney Pit # 3
1992	Shillingford Carp Lake Hinksey Lake South Cerney Pit #3	Beale Bird Park Lake Dorchester Fishing Lake
1993	Whiteknights Lake Standlake Young Lake 1 Blenheim Lake	Dorchester Fishing Lake Anonymous Lake
1994	Blenheim Lake	Anonymous Lake

Lack of sexual activity has been noted in populations of other freshwater bryozoan species as well (see below). In some cases this may be the result of meiotic nondisjunction arising from extensive chromosomal heteromorphism which may preclude sexual reproduction (Backus and Mukai, 1987), but whether this is true for some populations of *C. mucedo* is not known.

Evidence that sex generates little genetic variation

While ecological sampling indicates that sexual reproduction occurs irregularly and often in a low proportion of resident colonies, the presence of highly related clonal types within a site suggests that when sex does occur, it may be ineffective in generating significant levels of genetic variation. This is supported by molecular characterization of larvae and their known parental colonies. A RAPD PCR study by Jones et al. (1994) took advantage of the propensity of sexually-active colonies to release larvae overnight which allowed collection of larvae from a known parent colony. Although the presence of extra-parental bands in some larvae confirmed outcrossing in five of 10 different families (nine families were from one site and the 10th from another), results indicated that only a limited amount of genetic variation was generated in these outcrossed larvae. Indeed, with one exception, larvae were characterized by profiles identical to their parents, or their profiles could be matched with those previously characterized in the population at large. Larvae that lacked extraparental fragments could have been produced by selfing or by outcrossing between genetically similar clonal stock. It should be noted that families analyzed came from the two sites that have shown the highest maximum levels of sexual reproduction.

In summary, the short period of sexual activity, the generally low proportion of sexually active colonies, the similar numbers of asexual and sexual propagules (statoblasts and larvae) produced simultaneously, the apparent lack of genetic variation resulting from outcrossing, the great emphasis on clonal modes of reproduction via colony growth, fission, and prolonged statoblast production, and the high levels of relatedness of clonal types within and amongst habitats all indicate that clonality is extensive and predominant in populations of *C. mucedo* in southern England.

Wesenburg-Lund (1907) suggested that freshwater bryozoan populations are entirely clonal in northerly regions and that sex occurs in more central regions of species' ranges. Genetic studies real similar geographic trends in clonal aquatic plant (Eckert and Barrett, 1993) and cladoceran populations (Hebert et al., 1988). Indeed, the large body of comparative evidence for such geographical parthenogenesis in both animals (Bell, 1982; Hebert, 1987) and plants (Bierzychudek, 1987) suggests that sex is not an advantage in physically harsh or unpredictable environments (Stearns, 1987). Alternatively, clonality may be associated with genetic isolation and stochastic effects on genetic variability in range extremes (Eckert and Barrett, 1993 and references therein). There is also evidence for plants that harsh conditions reduce the likelihood of establishment of sexual progeny. Thus the growing season may be too short for seedlings to become large enough to overwinter but it is long enough for clonal progeny to attain large size and extensive root systems (Eckert and Barrett, 1993).

None of the above explanations are likely to explain the high levels of clonality of *C. mucedo* populations in southern England. In Britain, *C. mucedo* occurs as far north as the Shetland Islands (Lacourt, 1968) whilst on the European mainland its distribution extends nearly to the Arctic Circle in the Kola Peninsula of Russia (Lacourt, 1968). The documentation of sexuality in Finnish populations (Uotila and Jokela, 1995) suggests that sexually produced larvae become established before winter in at least some more northerly populations. An alternate explanation is that the high degree of relatedness and extensive clonality of *C. mucedo* populations in southern England reflect a founder effect with colonization from perhaps a more genetically variable mainland population followed by inbreeding and clonal reproduction on the relatively isolated land mass of Great Britain. However, evidence for meiotic nondisjunction in a population of *Pectinatella gelatinosa* in Japan (Backus and Mukai, 1987), lack of larval production in a careful study of growth and development of colonies of *Fredericella sultana* and *Plumatella casmiana* in Colorado (Wood, 1973), and common observations of testes resorption and lack of gametogenesis, ovarian development, embryos, and larvae (Wesenberg-Lund, 1907; Brien and Mordant, 1956; Wood, 1973; Toriumi, 1974; Mukai et al., 1979) all suggest that extensive clonality may not be restricted to populations in isolated habitats or to those subject to extreme physical conditions in this group. The issue of how generally high levels of clonality and genetic relatedness obtain in *C. mucedo* can only be resolved by further study across broader spatial and temporal scales.

Clonality and parasitism

One of the main hypotheses accounting for the maintenance of sex is the Red Queen which predicts that when parasites decrease host fitness, coevolution between parasites and hosts results in a regime of frequency-dependent selection. Briefly, the hypothesis proposes that the generation of novel genotypes by sex provides a mechanism that allows organisms to persist in the face of such rapidly evolving biological enemies as pathogens and parasites (Jaenike, 1978; Hamilton, 1980). In this coevolutionary arms race novel offspring generated by sex in the host population escape attack by common genotypes in a parasite population, and a regime of negative frequency dependent selection results in which cycles of hosts evolve resistance (through sex) and parasites co-evolve high specificity attack. Recent review of the Red Queen is provided by Lively (1996) and Ebert and Hamilton (1996), whilst Johnson, Lively and Schrag cover the topic further in this volume.

A basic premise of the Red Queen hypothesis is the existence of genetic variation for resistance in host populations, and an expectation is the production of genetically variable offspring. However, the previous section indicated that sex may not generate genetic novelty in populations of *C. mucedo* in southern England due to high levels of inbreeding that occur amongst closely related clones. The genetic similarity of these populations therefore raises the question of whether such a highly clonal host is parasitized and, if so, what the patterns of parasitism may be.

The presence of myxozoan parasites

Recently, a number of populations of *C. mucedo* in the Thames Valley region have been found to harbour a parasitic myxozoan which represents the new genus and species, *Tetracapsula bryozoides* (Canning et al., 1996, Okamura, 1996). Although long considered to be protists, new molecular and ultrastructural evidence (Smothers et al., 1994; Siddall et al., 1995) confirms myxozoans are proper metazoans although their relationships to other metazoan groups are currently debated. 18S ribosomal gene sequence analyses have suggested affinities with both the nematodes (Smothers et al., 1994) and a parasitic clade of the cnidarians (Siddall et al., 1995).

T. bryozoides forms unattached sac-like structures that can be easily observed using a dissection microscope as they tumble about in the body cavity of *C. mucedo* (see Fig. 4) due to ciliary-driven circulation of body cavity fluids by the host. Large numbers of infective spores develop within these sacs. Parasitism by *T. bryozoides* adversely affects the fitness of *C. mucedo* colonies by significantly compromising statoblast production and causing malformation, slowing of response times, and general degeneration (Okamura, 1996). However, parasitism by myxozoans does not

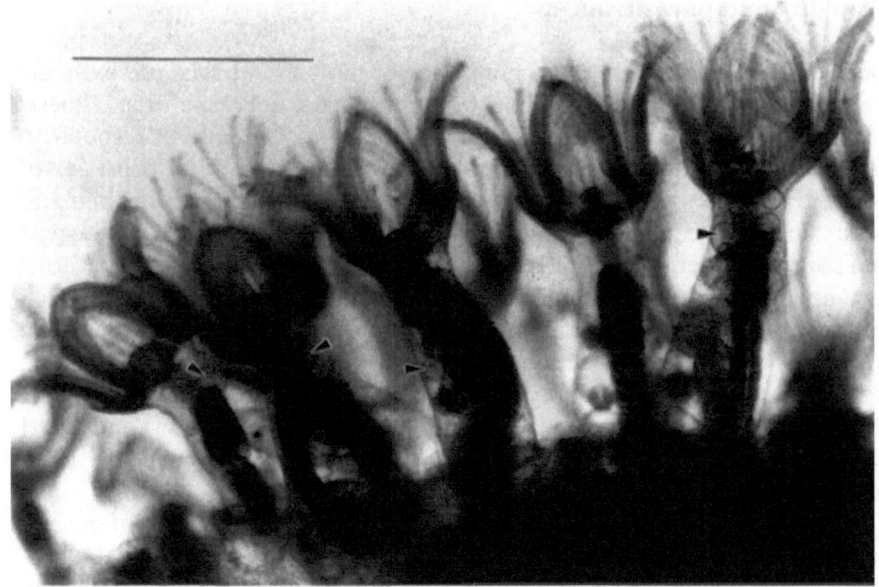

Figure 4. Colony of *C. mucedo* infected with the myxozoan parasite, *Tetracapsula bryozoides*. Numerous parasitic sacs (see arrowheads) can be seen on either side of the gut below the extended lophophores. For interpretation of bryozoan morphology see Figure 1. Scale bar = 1 mm.

compromise the production of larvae, and thus sexual function is unaffected (Okamura, 1996). Although some myxozoans have two hosts in their life cycle (Wolf and Markiw, 1984), it is not known whether two hosts are obligatory. The life cycle of *T. bryozoides* is under current investigation.

Temporal patterns of clonal representation and parasite prevalence

The presence of myxozoan parasites that exert a negative effect on the fitness of *C. mucedo* prompted Vernon et al. (1996) to ask the following questions: 1) How does myxozoan prevalence vary through time? 2) What is the temporal pattern of clonal representation in a parasitized host population? 3) Do myxozoans differentially infect common host genotypes within the population and therefore provide evidence for negative frequency dependent selection in host/parasite interactions as predicted by the Red Queen?

Myxozoan prevalence in the bryozoan population at Beale Bird Park Lake was variable over the summer of 1992. Infection levels of colonies collected consistently from a local area ranged from about a third of colonies to less than 10% (Vernon et al., 1996) while samples pooled from a number of localities within the site indicated that prevalence levels overall

Table 3. Prevalence of myxozoans (the percentage of colonies with myxozoan infections) in parasitized populations of *C. mucedo* in sites in the Thames Valley region from the period of 1992–1995 (n = total number of colonies sampled). Data reported are highest prevalence values if site was sampled more than once. Data from Okamura (1996)

Site	Year	Prevalence of myxozoans
Backwater of River Thames	1992	16.1% (n = 87)
	1995	18.5% (n = 27)
Beale Bird Park Lake	1992	49.2% (n = 59)
	1994	7.9% (n = 38)
	1995	6.9% (n = 87)
South Cerney Gravel Pit # 3	1992	30.0% (n = 10)
Standlake Young Lake 1	1993	36.4% (n = 22)
Blenheim Lake	1993	18.8% (n = 37)
	1994	15.6% (n = 27)
Whiteknights Lake	1993	26.5% (n = 34)
Anonymous Lake	1994	4.2% (n = 24)
Tufty's Corner	1995	2.6% (n = 39)

reached 49.2% of colonies in the lake in 1992 (Okamura, 1996). The mean prevalence over the entire season at the local site was 22.7% (SD = 9.3, n = 7). Prevalence data obtained for other parasitized populations of *C. mucedo* in the Thames Valley region from 1992 to 1994 indicate infection levels at Beale Bird Park Lake were relatively high in 1992 (see Tab. 3) (Okamura, 1996). These data provide a mean prevalence for parasitized populations at other sites of 18.8% (SD = 11.1, n = 9). At yet other sites in the Thames Valley region myxozoans have not been detected (n = 3) (Okamura, 1996). Prevalence of myxozoans in bryozoans at Beale Bird Park Lake in subsequent years also suggest the 1992 levels of infection were relatively high (see Tab. 3). Prevalence could not be determined in 1993 due to a crash in the host population in that year (Okamura, 1996).

The relatively high prevalence of myxozoans in bryozoans at Beale Bird Park Lake in 1992 may have had some influence in the crash in the host population in 1993 (Vernon et al., 1996). As discussed earlier, myxozoans adversely affect *C. mucedo*, and no other obvious causes of a population crash were evident. Although apparent extinctions of *C. mucedo* in other sites (Okamura, unpublished data) have not been specifically related to the presence of myxozoan parasites, these extinctions occurred before myxozoan infections were recognized and routinely monitored. However, it is also notable that *C. mucedo* populations have persisted in other sites in years following relatively high prevalence of myxozoan infections (T. Hatton-Ellis and B. Okamura, unpublished data). While populations of freshwater bryozoans, in general, are notorious for showing great variation in density over time (discussed in Okamura and Hatton-Ellis, 1995) the reasons for such variability are poorly understood. Population crashes and extinctions may therefore be explained by a variety of factors acting independently or in concert. Although in 1994 and 1995 colonies of *C. mucedo* were again abundant in Beale Bird Park Lake, they had not escaped infection by myxozoans (Okamura, 1996; Vernon et al., 1996).

RAPD PCR indicated that four highly-related clones were present in Beale Bird Park Lake throughout the period of 1992–1994 and that all were susceptible to myxozoan infection (Vernon et al., 1996). During this period the two common clones showed a notable reversal in their relative abundance with the significantly more common clone (clone A) in 1992 becoming the significantly less common clone in 1994. However, the reversal in relative abundance of the two clones did not appear to be explained by parasitism. On the two dates for which sample sizes of both clones were sufficient for statistical analysis (using a Fisher Exact Test which provides a powerful test of dichotomous, nominally scaled data when sample sizes are small) parasitism did not depend on clonal genotypes (i.e., the frequencies of parasitized and unparasitized clonal genotypes did not differ). Thus on the 23rd of July 1992, 47% (8/17 colonies) of clone A colonies were parasitized compared to 25% (2/8 colonies) of clone B (Fisher Exact Test; $p = 0.36$). At a similar period of the summer 2 years later there was no evidence that infection levels (44.4% of clone A [4/9 colonies] vs. 22.7% of clone B [5/22 colonies]) were influenced by clonal type ($p = 0.16$).

Since clonal genotypes are so closely related, it is not surprising that myxozoans apparently do not distinguish amongst clones and exert similar effects on them. This is further supported by trends in clonal abundance and infection levels. If the eventual switch in clonal abundance was mediated by differential virulence of the myxozoans to clones, it would be anticipated that the clone that decreased in relative abundance should show increased levels of parasitism during this period. However, infection levels in this clone were highly variable over the period of the study (Vernon et al., 1996).

Lack of evidence for disproportionate infection of common clones counters predictions of negative frequency dependent selection under the Red Queen. This contradiction may reflect the general problem associated with assessing the frequency dependence of host-parasite arms races due to time-lags between parasite adaptations and the parasite-mediated decline of hosts (Dybdahl and Lively, 1995; Ebert and Hamilton, 1996). However, the following observations suggest that time-lags are unlikely in this system. First, myxozoans were similarly invasive as evidenced by lack of disproportionate infection levels. Second, extensive persistence of clones of *C. mucedo* over time should greatly increase the probability of exploiting genetically homogeneous hosts even if an alternate host is required in the life cycle of the myxozoan. Thus it is unlikely that a complex life cycle in the parasite would produce a time-lag sufficiently long to preclude the development of negative frequency dependent selection in host and parasite populations, particularly since evidence for other myxozoans suggests that cycling through two hosts occurs in a period of less than 1 year (Yokoyama et al., 1993). Third, as discussed above, the reversal in relative abundance of the two common clones does not appear to be explained by differential virulence of the myxozoan to the two clones since the clone that decreased in abundance did not show concomitant increased levels of parasitism. Finally, the high genetic similarity of the host clonal genotypes suggests that the myxozoans likely do not distinguish amongst potential host clones and hence time lags associated with parasite adaptations and host decline would not be expected.

In summary, the most parsimonious interpretation of the data is that at least in southern England, myxozoan parasites do not distinguish between genetically similar bryozoan host clones and therefore that traditional Red Queen expectations do not apply. Below I argue that such high clonality is sustainable, at least in southern England, because *C. mucedo* occurs as a metapopulation in a series of isolated water bodies.

Evidence and significance of metapopulation structure

The metapopulation concept developed from an appreciation that many organisms are distributed in a series of local populations that are linked by immigration and emigration (Gotelli, 1995). An outcome of such movement is that, although populations in local sites may go extinct, subsequent immigration may result in recolonization. A metapopulation therefore is composed of a series of local populations which can undergo local extinction and re-colonization. In such a system, stability applies at regional scales that encompass a series of connected sites whilst local populations themselves are potentially unstable. Thus only some sites amongst a range of potential sites may be occupied at any one point in time (for review see Hanski, 1996; Harrison and Hastings, 1996). The isolation of lakes and ponds combined with dispersal via statoblasts provides great potential for metapopulation structure to develop amongst local populations of *C. mucedo*. This is supported by genetic evidence, data on site occupancy and suitability, and transplant experiments as discussed below.

High genetic relatedness amongst populations

The high degree of genetic relatedness amongst populations of *C. mucedo* in southern England provides evidence that populations are linked by dispersal. However, both cluster analysis and AMOVA indicate higher relatedness of populations from Thames Valley sites relative to the population in Norfolk (see earlier discussion) suggesting that dispersal is more frequent within the catchment area of the River Thames. This is not surprising since statoblasts are likely dispersed by waterfowl, and waterfowl can be expected to visit water bodies within a catchment area most consistently. Nonetheless, the high levels of genetic relatedness overall indicate that longer distance dispersal within southern England occurs as well. This implies that probabilities of dispersal and thus of local extinction and colonization are variable amongst sites and that a spatial structure applies in this metapopulation system (Gotelli, 1995; Hanski 1996).

It should be noted that alternative explanations for the maintenance of high relatedness amongst sites in southern England do not depend on widespread dispersal amongst these sites. These are that rates of divergence following post-glacial colonization some 8000–9000 years ago are low, perhaps due to similar selection pressures amongst sites and considerable inbreeding, and/or that populations in southern England derive from a common source population, perhaps on the European mainland which must itself be highly inbred and clonal. The latter case would suggest that *C. mucedo* populations in Britain and Europe may conform more to a source-sink model resulting from heterogeneity in habitat quality (see Dias, 1996 and Pulliam, 1996 for review) or to a mainland-island metapopulation (Hanski, 1996). The evidence indicates, however, that dispersal amongst sites within Britain must occur with greater frequency than dispersal between mainland Europe and Britain. Populations of *C. mucedo* in many recently created sites (Okamura, 1994) and colonization of previously uninhabited sites (see below) indicate high levels of dispersal within England, while proximity of sites within southern England should be associated with more frequent dispersal amongst these sites than between the European mainland and Britain. While the genetic composition of *C. mucedo* populations in southern England is explained by both present-day and historical events, the widespread genetic similarity amongst sites in southern England is best explained by regular and frequent dispersal amongst these British populations.

Patterns of site occupancy and suitability

Monitoring the presence or absence of *C. mucedo* in a series of local sites in the Thames Valley region provides evidence that local populations are unstable and undergo extinction and colonization (B. Okamura, unpublished data). In particular, apparent extinctions were noted in four out of 24 inhabited sites while colonization events were observed in two out of 18 uninhabited sites over the consecutive year period of 1989–1990. Further extinctions in a smaller number of sites over subsequent consecutive year periods have also been noted. Although specific reasons for extinctions are not known, it may be that high levels of genetic similarity preclude local persistence of *C. mucedo* populations when temporarily adverse biotic and abiotic conditions are encountered.

As noted above, at any one point in time, suitable sites in a metapopulation may be uninhabited. The colonization events reported above indicate this to be the case for *C. mucedo*. Data on water chemistry (dissolved oxygen content, conductivity, pH, temperature) and site size were collected from a number of sites in the Thames Valley region. Canonical discriminant function analysis of these five variables did not discriminate significantly between sites from which *C. mucedo* was present (n = 26) and absent (n = 18) respectively, suggesting that *C. mucedo* does not inhabit all potentially suitable water bodies (B. Okamura, unpublished data).

Transplant experiments and site suitability

Transplant experiments conducted in a pilot study provide additional evidence that unoccupied but suitable sites are present in the Thames Valley region (B. Okamura, unpublished data). Small colonies newly hatched from statoblasts from a source population some 3 km away were transplanted into a site in which *C. mucedo* was absent. Despite substantial initial mortality, a small number of transplanted colonies underwent prolific growth and fission and produced numerous statoblasts. The production of statoblasts provides reasonable evidence for establishment potential since overwintering statoblasts hatch into new colonies in the following season.

The importance of metapopulation structure

The variety of both genetic and ecological data presented here consistently indicate that *C. mucedo* is distributed as a metapopulation in southern England that is composed of closely related, highly clonal sub-populations in which sex is limited and apparently ineffective in regularly generating novel genotypes. Such genetic homogeneity may preclude local persistence. This is supported both by extinction events and by notable variation in local abundance from year to year and site to site. The evidence suggests that persistence is instead maintained at the regional scale through high rates of asexual reproduction that allows dispersal of statoblasts to other temporarily favourable sites within the metapopulation and/or local persistence as cryptic low density populations within sites. Determining whether this is a general strategy employed by *C. mucedo* would require conducting comparative studies of populations across broader spatial scales.

Synthesis and general discussion

Maintaining genetic similarity and escaping the Red Queen

Evidence indicates that populations of *C. mucedo* in southern England emphasize asexual methods of reproduction and show high levels of genetic similarity even in the face of parasites that exert a negative effect on fitness. It is proposed that rather than achieving persistence through the production of novel genotypes, *C. mucedo* escapes the Red Queen race through dispersal within the metapopulation. Indeed, such dispersal in time and space has been suggested to provide an alternative to sex and to allow organisms escape from parasites (Ghiselin, 1974; Tooby, 1982). Recent modeling supports this view showing that under negative frequency dependent selection due to parasites dispersal in space amongst subdivided populations can allow asexuals to escape parasites and to outcompete sexuals (Ladle et al., 1993; Judson, 1995).

If *C. mucedo* is able to escape the Red Queen through dispersal within a metapopulation, then a variety of expectations under the Red Queen may not be met. There is limited evidence that this is the case. Thus, there is no evidence that myxozoan parasites attack common host clones, but rather they are apparently indiscriminate. In other words, there is no evidence for the basic assumption of the Red Queen that there is genetic variation in host resistance. The Red Queen also predicts that natural host populations suffering intense parasite pressure should maintain higher levels of genetic variability than unparasitized host populations (Ebert and Hamilton, 1996). However, the population at Beale Bird Park Lake with a high prevalence of myxozoan parasites in 1992 had a similar number of clones (n = 5) to the unparasitized population at Dorchester Fishing Lake (n = 6) in the same year (Okamura et al., 1993). Finally, the prediction that parasitism should promote sexual reproduction is also unsupported: there were no differences in the frequencies of sexually-active colonies that were infected by myxozoans and sexually-active colonies that were uninfected (Okamura, 1996).

Lively (1992) and Howard and Lively (1994) have proposed that high and fluctuating selection pressures exerted by parasites may select for clonal diversity in asexuals for the same reasons that genetic diversity is expected to be maintained by parasitism in sexuals. Judson (1995) provides theoretical evidence that, if a metapopulation structure applies, high levels of clonal diversity can also be maintained given certain levels of host migration amongst sites and differential parasite pressures within subpopulations. Such explanations may underlie the high levels of clonal diversity observed in many populations of asexual organisms (see references in Judson, 1995). *C. mucedo* populations in southern England, however, present a rather different picture in two respects: 1) they maintain limited sex that results in inbreeding and are therefore not entirely asexual; 2) they show little clonal diversity.

Low levels of genetic variation in populations of *C. mucedo* may result from limited sex and/or lack of differential parasite pressures. Alternatively, extremes of both total and low levels of host migration can result in low genetic diversity in host sub-populations (Judson, 1995) while the Howard and Lively model (1994) would predict low clonal diversity may result if mutation rates in clonal lineages are low and/or if parasites do not exert strong effects on host fitness. The latter converges on May and Anderson's (1983) prediction that antagonistic coevolution favours sex only when parasites have severe effects. Elucidation of which explanation(s) apply to the low genetic diversity in *C. mucedo* sub-populations requires further investigation.

Hurst and Peck (1996) point out that persistent asexuality may be maintained and the Red Queen thus avoided if rates of host reproduction are so rapid that the ability of parasites to "catch up" is limited. A related issue is the relative extension of a parasite's life cycle by a phase infecting alternate hosts. *C. mucedo* can show prolific rates of colony growth, fission, and statoblast production (B. Okamura, unpublished data). Although the comparative growth rate and complete life cycle of *T. bryozoides* are unknown, the extensive genetic similarity of *C. mucedo* populations in both time and space should promote parasite adaptations. Hurst and Peck (1996) note that rapid rates of increase combined with cell division may be of significance in initiating parasite-free lineages in asexual unicellular organisms. The role of fission and statoblast production as a means of generating uninfected daughter colonies in *C. mucedo* merits further investigation in this respect.

In many ways, the question reduces to why sex is retained at all in *C. mucedo*. The only other potentially important function of larvae is that of site selection. It is, however, improbable that this is a crucial function in the short term. Larve are produced sporadically, in low numbers, and adult colonies are mobile. Alternatively, perhaps sex is important in other parts of the range of *C. mucedo*, thus the dynamics of populations in southern England represent those of peripheral isolates in which sex has not been entirely lost. Finally, perhaps only a small amount of occasional sex with the rare genetically-distinct migrant introduces enough variation in the long term. Recent modeling indicates that a very little sex can go a very long way (see discussion in Hurst and Peck, 1996), although the extensive genetic similarity of *C. mucedo* provides little evidence that sex has introduced significant variation at least in the very recent past. Obvious areas for future research are to compare genetic variability and levels of sexuality of *C. mucedo* populations in mainland Europe with British populations and to conduct long-term studies of populations in southern England.

Comparisons and contrasts with cladocerans and rotifers

The evidence presented here indicates a contrast between the consequences of extensive clonality to the population structure of freshwater zooplankton and bryozoans. In cyclically parthenogenetic populations of *Daphnia* large gene frequency differences generally occur among local populations (see Hebert, 1987 and Carvalho, 1993 for review) and thousands of genotypes can coexist within populations in permanent habitats (Carvalho and Crisp, 1987). Obligately parthenogenetic populations of *Daphnia* also show high levels of clonal diversity (Weider, 1989) whilst the coexistence of many clones within populations of rotifers has similarly been documented (King and Zao, 1987). Such genetic differentiation may, in part, reflect local adaptation. Thus differences in thermal tolerances and food requirements may promote coexistence of clonal types within habitats (Carvalho, 1987; Weider, 1993) while certain genotypes may be specialized for particular sites (DeMeester, 1991). Hybridization amongst *Daphnia* spp. results in similar high levels of genetic diversity of clonal types (see Schwenk and Spaak, 1995 for review). By contrast, *C. mucedo* populations in southern England are remarkably genetically uniform, showing very little divergence between sites over a spatial scale of some 270 km and the presence of a few highly related clones within local populations.

Carvalho and Crisp (1987) propose that the seasonal succession of *Daphnia* clones in sites of high clonal diversity may be explained by two factors. Parthenogenetic females may survive unfavourable conditions in the winter and so carry on the lineage in the following season. In addition, mating between ecologically similar clones may prevent divergence of clonal types, especially if there is a temporal separation of sexual reproduction amongst clones. High clonal diversity in *Daphnia* in both cases reflects the genetic variation generated by sexual reproduction that initiates new parthenogenetic lineages. In *C. mucedo* fertilization occurs between closely related clones and so results in high levels of inbreeding within populations (Jones et al., 1994). Since the sexual season occurs relatively early in the season and is brief in duration, there is little potential for a temporal separation of sexual reproduction amongst clones. Thus the generation of differentiated, specialized clonal types by temporal segregation of fertilization amongst seasonally-adapted clones as suggested for *Daphnia* is unlikely in *C. mucedo*.

The relative timing of sexual reproduction may be determined by differences in the mechanisms of fertilization associated with these planktonic and sessile forms. The temporal restriction of the sexual phase in *C. mucedo* may be a consequence of its sessile nature. Bryozoans must spawn sperm into the water column to effect fertilization. For *C. mucedo* the chances of successful cross fertilization should be greatest at the early part of the growing season for two reasons. First, genotypes will have been mixed during the winter by within- and between-site dispersal of floating statoblasts. Second, as the season progresses, colonies nearby will tend to be genetically identical due to colony growth and fission. Since sperm are short-lived and their flagellar locomotion will allow limited dispersal, especially as currents are trivial in lakes and ponds, fertilization rates will decrease with distance. Thus as the season progresses the probability of selfing should greatly increase.

By contrast, the timing of sexual reproduction in cladocerans may arise as a consequence of fertilization that is achieved through copulation in these mobile animals. Since the chances of copulation will be increased by a phase of parthenogenesis, the general pattern of a delayed sexual phase in cladocerans may reflect the greater mating opportunities afforded by high population densities. Temporal flexibility in the timing of sex amongst clones could then be superimposed over this general delay to effect mating amongst seasonally adapted clonal types as proposed by Carvalho and Crisp (1987) for *Daphnia*. Similar evidence exists for seasonally isolated clones of rotifers (King, 1977).

Comparisons and contrasts with aquatic macrophytes

Although extensive clonality occurs in the life cycles of many plants, vegetative reproduction has been proposed to be particularly dominant in submersed freshwater plant populations (Cook, 1987; Titus and Hoover, 1991). Genetic studies have revealed wide variation in genotypic diversity in aquatic plant species. Thus, considerable genetic variability can occur both within and between sites for some species (Harris et al., 1992; Lokker et al., 1994). However, populations of other species may be monoclonal in some sites and possess high levels of clonal diversity in other sites (Eckert and Barrett, 1993; Piquot et al., 1996). In some species certain clones are widespread and common while others are represented locally by only a few individual plants (Lokker et al., 1994). In some cases, a few widespread clones are associated with marginal sites while clonal diversity is restricted to central sites within the species range (Stewart and Excoffier, 1996 and references therein), however in other species widespread genotypes are apparently absent (Piquot et al., 1996). Hollingsworth et al. (1995) found evidence that genetic variation is decreased in sites of recent origin in British populations of *Potamogeton coloratus*. *C. mucedo* populations show a mixture of these traits. High genetic relatedness both within and between sites indicates the presence of widespread, closely-related clonal stock. However, no populations studied to date are characterized by a diversity of divergent clonal types, and there is little evidence that established, natural sites are centres for high clonal diversity.

Establishment of sexual propagules in aquatic plants can be rare (Eckert and Barrett, 1993; Lokker et al., 1994) as is presumably the case with *C. mucedo*. Furthermore, the inefficiency of the pollination mechanism that may contribute to low seedling recruitment in submersed macrophytes (Les, 1988) is similar to the problem of achieving successful fertilization by free-spawned sperm in *C. mucedo*. Evidence for low fertilization success in free spawning benthic marine invertebrates provides further support for the generality of this problem for aquatic organisms with such indirect fertilization mechanisms (e.g., Pennington, 1985; Denny and Shibata, 1989; Levitan et al., 1992).

Ellstrand and Roose (1987) identify a distinct category of plants in which seed are rarely genetically different from the parent. In such populations, although outcrossing among individuals may occur at a low rate, crosses among individuals of the same genotype will yield offspring identical to their parents. Thus recombination should be a relatively rare source of clonal diversity. Evidence suggests that the production of novel genotypes by outcrossing is similarly rare in *C. mucedo* populations (Jones et al., 1994).

An apparent association between clonal reproduction and complete sexual sterility in aquatic plants has been noted by many workers (Sculthorpe, 1967; Hutchinson, 1975; Les, 1988). Although the basis of such sterility is not known, founder events, genetic drift in marginal populations, and/or inbreeding could lead to the chance fixation of certain genes which are deleterious to sexual function but do not affect vegetative growth. Sexual function could thus be lost gradually through the accumulation of such somatic mutations in clonal lineages (Klekowski, 1988, cited in Eckert and Barrett, 1993). As *C. mucedo* employs analogous vegetative mechanisms to achieve clonality as those employed by aquatic clonal plants there is no theoretical reason that sexual function could not gradually be lost by the same mechanism. The rarity of novel sexual progeny and the great emphasis on clonal reproduction in *C. mucedo* and some clonal aquatic plant populations suggests the possibility that sexuality is gradually being lost in these systems. A question to resolve therefore, is whether the rare production of novel genotypes is of any significance.

Conclusions

Studies of populations of *C. mucedo* in southern England suggest that the ecological and population genetic consequences of mixed life histories in sessile, colonial freshwater invertebrates are more similar to those of attached aquatic plants than of planktonic cladocerans and rotifers. These similarities may result because sexuality is not invoked by deteriorating conditions and because indirect fertilization and vegetative mechanisms of clonal reproduction are employed, although other unidentified explanations may also be involved. However, for all these groups the distribution of populations in isolated habitats may be an important precondition for both the maintenance of extensive clonality and the occasional loss of sex. Further empirical study may reveal just how commonly metapopulation structure is associated with maintaining high levels of clonality, and in particular whether this is of special significance for systems of low genetic diversity such as that illustrated by *C. mucedo* in southern England.

Acknowledgements
The research described here has been supported by grants from the Natural Environment Research Council to B. Okamura and L.R. Noble (GR3/8961) and to E.U. Canning and B. Okamura (GR3/09956), and by the E.P. Abraham fund to L.R. Noble and B. Okamura. I thank G. Carvalho, J. Jokela and T. Städler for their very helpful comments on the manuscript; C.S. Jones, J.G. Vernon, and T. Hatton-Ellis for their molecular skills and contributions; R. Manuel for the photograph of *C. mucedo* colony in Figure 1, and T. Hatton-Ellis for AMOVA results.

References

Abrahamson, W.G. (1980) Demography and vegetative reproduction. *In:* O.T. Solbrig (ed.): *Demography and Evolution in Plant Populations.* Blackwell Scientific Publications, Oxford, pp 89–106.

Aspinwall, N. and Christian, T. (1992) Clonal structure, genotypic diversity, and seed production in populations of *Filipendula rubra* (Rosaceae) from the northcentral United States. *Am. J. Bot.* 79:294–299.

Backus, B.T. and Mukai, H. (1987) Chromosomal heteromorphism in a Japanese population of *Pectinatella gelatinosa* and karyotypic comparison with some other phylactolaemate bryozoans. *Genetica* 73:189–196.

Bell, G. (1982) *The Masterpiece of Nature – The Evolution and Genetics of Sexuality.* University of California Press, Berkeley, California.

Bierzychudek, P. (1987) Patterns in plant parthenogenesis. *In:* S.C. Stearns (ed.): *The Evolution of Sex and Its Consequences.* Birkhäuser Verlag, Basel, pp 197–217.

Brien, P. and Mordant, C. (1956) Relations entre les reproduction sexuée et asexuée a propose de phylactolaemates. *Annls Soc. R. Zool. Belg.* 86:169–189.

Brown, C.J.D. (1933) A limnological study of certain fresh-water Polyzoa with special reference to their statoblasts. *Trans. Am. Microsc. Soc.* 52:271–313.

Bushnell, J.H. (1973) The freshwater Ectoprocta: A zoogeographical discussion. *In:* G.P. Larwood (ed.): *Living and Fossil Bryozoa.* Academic Press, London, pp 503–521.

Buss, L.W. (1979) Habitat selection, directional growth and spatial refuges: Why colonial animals have more hiding places. *In:* G. Larwood and B.R. Rosen (eds): *Biology and Systematics of Colonial Organisms.* Academic Press, London, pp 459–497.

Canning, E.U., Okamura, B. and Curry, A. (1996) Development of a myxozoan parasite *Tetracapsula bryozoides* n.g., n.sp. in *Cristatella mucedo* (Bryozoa, Phylactolaemata). *Folia Parasitol.* 43:249–261.

Carvalho, G.R. (1987) The clonal ecology of *Daphnia magna* (Crustacea: Cladocera) II. Thermal differentiation among seasonal clones. *J. Anim. Ecol.* 56:469–478.

Carvalho, G.R. (1994) Evolutionary genetics of aquatic clonal invertebrates: Concepts, problems and prospects. *In:* A. Beaumont (ed.): *Genetics and Evolution of Aquatic Organisms.* Chapman and Hall, London, pp 291–323.

Carvalho, G.R. and Crisp, D.J. (1987) The clonal ecology of *Daphnia magna* (Crustacea: Cladocera). I. Temporal changes in the clonal structure of a natural population. *J. Anim. Ecol.* 56:453–468.

Chapco, W., Ashton, N.W., Martel, R.K.B. and Antonishyn, N. (1992) A feasibility study of the use of randomly amplified polymorphic DNA in the population genetics and systematics of grasshoppers. *Genome* 35:569–574.

Cook, R.E. (1987) Vegetative growth and genetic mobility in some aquatic weeds. *In:* K.M. Urbanska (ed.): *Differentiation Patterns in Higher Plants.* Academic Press, London, pp 217–225.

DeMeester, L. (1991) An analysis of the phototactic behaviour of *Daphnia magna* clones and their sexual descendants. *Hydrobiologia* 225:217–227.

Denny, M.W. and Shibata, M.F. (1989) Consequences of surf-zone turbulence for settlement and external fertilization. *Amer. Nat.* 117:838–840.

Dias, P.C. (1996) Sources and sinks in population biology. *Trends Ecol. Evol.* 11:326–330.

Dybdahl, M.F. and Lively, C.M. (1995) Host-parasite interactions: Infection of common clones in natural populations of a freshwater snail (*Potamopyrgus antipodarum*). *Proc. R. Soc. Lond. B* 260:99–103.

Ebert, D. and Hamilton, W.D. (1996) Sex against virulence: The coevolution of parasitic diseases. *Trends. Ecol. Evol.* 11:79–82.

Eckert, C.G. and Barrett, S.C.H. (1993) Clonal reproduction and patterns of genotypic diversity in *Decodon verticillatus* (Lythraceae). *Am. J. Bot.* 80:1175–1182.

Ellstrand, N.C. and Roose, M.L. (1987) Patterns of genotypic diversity in clonal plant species. *Am. J. Bot.* 74:123–131.

Excoffier, L., Smouse, P.E. and Quattro, J.M. (1992) Analysis of molecular variance inferred from metric distances among DNA haplotypes: Application to human mitochondrial DNA restriction data. *Genetics* 131:479–491.

Ghiselin, M.T. (1974) *The Economy of Nature and the Evolution of Sex.* University of California Press, Berkeley, California.

Gotelli, N.J. (1995) *A Primer of Ecology.* Sinauer Associates, Inc., Sunderland, Massachusetts.

Grosberg, R.K., Levitan, D.R. and Cameron, B.B. (1996) Characterization of genetic structure and genealogies using RAPD-PCR markers: A random primer for the novice and nervous. *In:* J.D. Ferraris and S.R. Palumbi (eds): *Molecular Zoology: Advances, Strategies and Protocols.* Wiley-Liss, New York, pp 67–100.

Hamilton, W.D. (1980) Sex versus non-sex versus parasite. *Oikos* 35:282–290.

Hanski, I. (1996) Metapopulation ecology. *In:* O.E. Rhodes, Jr., R.K. Chesser and M.H. Smith (eds): *Population Dynamics in Ecological Space and Time.* University of Chicago Press, Chicago, pp 13–43.

Harris, S.A., Maberly, S.C. and Abbott, R.J. (1992) Genetic variation within and between populations of *Myriophyllum alterniflorum* DC. *Aquat. Bot.* 44:1–21.

Harrison, S. and Hastings, A. (1996) Genetic and evolutionary consequences of metapopulation structure. *Trends Ecol. Evol.* 11:180–183.

Hebert, P.D.N. (1987) Genotypic characteristics of cyclic parthenogens and their obligately asexual derivatives. *In:* S.C. Stearns (ed.): *The Evolution of Sex and Its Consequences.* Birkhäuser Verlag, Basel, pp 175–195.

Hebert, P.D.N., Ward, R.D. and Weider, L.J. (1988) Clonal-diversity patterns and breeding system variation in *Daphnia pulex*, an asexual-sexual complex. *Evolution* 42:147–159.

Hollingsworth, P.M., Gornall, R.J. and Preston, C.D. (1995) Genetic variability in British populations of *Potamogeton coloratus* (Potamogetonaceae). *Plant Syst. Evol.* 197:71–85.

Howard, R.S. and Lively, C.M. (1994) Parasitism, mutation accumulation and the maintenance of sex. *Nature* 367:554–557.

Hughes, R.N. and Cancino, J.M. (1985) An ecological overview of cloning in Metazoa. *In:* J.B.C. Jackson, L.W. Buss and R.E. Cook (eds): *Population Biology and Evolution of Clonal Organisms.* Yale University Press, New Haven, Connecticut, pp 153–186.

Hurst, L.D. and Peck, J.R. (1996) Recent advances in understanding of the evolution and maintenance of sex. *Trends Ecol. Evol.* 11:46–52.

Hutchinson, J.E. (1975) *A Treatise on Limnology, Vol. 3 Limnological Botany.* John Wiley and Sons, New York.

Jackson, J.B.C. (1985) Distribution and ecology of clonal and aclonal benthic invertebrates. *In:* J.B.C. Jackson, L.W. Buss and R.E. Cook (eds): *Population Bioloy and Evolution of Clonal Organisms.* Yale University Press, New Haven, Connecticut, pp 297–355.

Jaenike, J. (1978) An hypothesis to account for the maintenance of sex within populations. *Evol. Theory* 3:191–194.

Jónasson, P.M. (1963) The growth of *Plumatella repens* and *P. fungosa* (Bryozoa Ectoprocta) in relation to external factors in Danish eutrophic lakes. *Oikos* 14:121–137.

Jones, C.S., Okamura, B. and Noble, L.R. (1994) Parent and larval RAPD fingerprints reveal outcrossing in freshwater bryozoans. *Mol. Ecol.* 3:193–199.

Joo, G.-J., Ward, A.K. and Ward, G.M. (1992) Ecology of *Pectinatella magnifica* (Bryozoa) in an Alabama oxbow lake: Colony growth and association with algae. *J. N. Am. Benthol. Soc.* 11:324–333.

Judson, O.P. (1995) Preserving genes: A model of the maintenance of genetic variation in a metapopulation under frequency-dependent selection. *Genet. Res.* 65:175–191.

King, C.E. (1977) Genetics of reproduction, variation and adaptation in rotifers. *Arch. Hydrobiol. Beih.* 8:187–201.

King, C.E. and Zao, Y. (1987) Coexistence of rotifer (*Brachionus plicatilis*) clones in Soda Lake, Nevada. *Hydrobiologia* 147:57–64.

Lacourt, A.W. (1968) A monograph of the freshwater Bryozoa – Phylactolaemata. *Zoologische Verhandelingen,* No. 93. Rijksmuseum von Nat. Hist., Leiden.

Ladle, R.J., Johnstone, R.A. and Judson, O.P. (1993) Coevolutionary dynamics of sex in a meta-population: Escaping the Red Queen. *Proc. R. Soc. Lond. B* 253: 155–160.

Les, D.H. (1988) Breeding systems, population structure, and evolution in hydrophilous angiosperms. *Annls Miss. Bot. Gard.* 75:819–835.

Levitan, D.R., Sewall, M.A. and Chia, F.-S. (1992) How distribution and abundance influence fertilization success in the sea urchin *Strongylocentrotus franciscanus. Ecology* 73:248–254.

Lively, C.M. (1992) Parthenogenesis in a freshwater snail: Reproductive assurance versus parasitic release. *Evolution* 46:907–913.

Lively, C.M. (1996) Through the Looking-Glass: Host-parasite coevolution and sex. *BioScience* 46:107–114.

Lokker, C., Susko, D., Lovett Doust, L. and Lovett Doust, J. (1994) Population genetic structure of *Vallisneria americana,* a dioecious clonal macrophyte. *Am. J. Bot.* 81:1004–1012.

Lovett Doust, L. (1981) Population dynamics and local specialization in a clonal perennial (*Ranunculus repens*). I. The dynamics of ramets in contrasting habitats. *J. Ecol.* 69:743–755.

May, R.M. and Anderson, R.M. (1983) Epidemiology and genetics in the coevolution of parasites and hosts. *Proc. R. Soc. Lond. B* 219:281–313.

Mort, M.A. and Wolf, H.G. (1985) Enzyme variability in large-lake *Daphnia* populations. *Heredity* 55:27–36.

Mukai, H. (1982) Development of freshwater bryozoans. *In:* F.W. Harrison and R.R. Cowden (eds): *Developmental Biology of Freshwater Invertebrates.* Alan R. Liss, New York, pp 535–576.

Mukai, H., Karasawa, T. and Matsumoto, Y. (1979) Field and laboratory studies on the growth of *Pectinatella gelatinosa* Oka, a freshwater bryozoan. *Sci. Rep. Fac. Educ. Gunma Univ.* 28:27–57.

Okamura, B. (1994) Variation in local populations of the freshwater bryozoan *Cristatella mucedo. In:* P.J. Hayward, J.S. Ryland, and P.D. Taylor (eds): *Biology and Palaeobiology of Bryozoans.* Olsen and Olsen, Fredensborg, Denmark, pp 145–149.

Okamura, B. (1996) Occurrence, prevalence and effects of the myxozoan *Tetracapsula bryozoides* Canning, Okamura et Curry parasitic in the freshwater bryozoan *Cristatella mucedo* Cuvier (Bryozoa, Phylactolaemata). *Folia Parasitol.* 43:262–266.

Okamura, B. and Hatton-Ellis, T. (1995) Population biology of bryozoans: Correlates of sessile, colonial life histories in freshwater habitats. *Experientia* 51:510–525.

Okamura, B., Jones, C.S. and Noble, L.R. (1993) Randomly amplified polymorphic DNA analysis of clonal population structure and geographic variation in a freshwater bryozoan. *Proc. R. Soc. Lond. B* 253:147–154.

Pennington, J.T. (1985) The ecology of fertilization of echinoid eggs: The consequences of sperm dilution, adult aggregation, and synchronous spawning. *Biol. Bull. Woods Hole* 183:417–430.

Piquot, Y., Saumitou-Laprade, P., Petit, D., Vernet, P. and Epplen, J.T. (1996) Genotypic diversity revealed by allozymes and oligonucleotide DNA fingerprinting in French populations of the aquatic macrophyte *Sparganium erectum. Mol. Ecol.* 5:251–258.

Pulliam, H.R. (1996) Sources and sinks: Empirical evidence and population consequences. *In:* O.E. Rhodes, Jr., R.K. Chesser and M.H. Smith (eds): *Population Dynamics in Ecological Space and Time.* University of Chicago Press, Chicago, pp 45–69.

Rackham, O. (1986) *The History of the Countryside.* J.M. Dent and Sons, Ltd., London.

Rubin, J.A. (1987) Growth and refuge location in continuous, modular organisms: Experimental and computer simulation studies. *Oecologia (Berlin)* 72:46–51.

Schierwater, B., Ender, A., Schwenk, K., Spaak, P. and Streit, B. (1994) The evolutionary ecology of *Daphnia. In:* B. Schierwater, B. Streit, G.P. Wagner and R. DeSalle (eds): *Molecular Ecology and Evolution: Approaches and Applications.* Birkhäuser Verlag, Basel, pp 495–508.

Schwenk, K. and Spaak, P. (1995) Evolutionary and ecological consequences of interspecific hybridization in cladocerans. *Experientia* 51:465–481.

Sculthorpe, C.D. (1967) *The Biology of Aquatic Vascular Plants.* Edward Arnold, London.

Siddall, M.E., Martin, D.S., Bridge, D., Desser, S.S. and Cone, D.K. (1995) The demise of a phylum of protists: Phylogeny of myxozoan and other parasitic Cnidaria. *J. Parasitol.* 81:961–987.

Silander, J.A., Jr. (1985) Microevolution in clonal plants. *In:* J.B.C. Jackson, L.W. Buss and R.E. Cook (eds): *Population Biology and Evolution of Clonal Organisms*. Yale University Press, New Haven, Connecticut, pp 107–152.

Smothers, J.F., Von Dohlen, C.D., Smith, L.H., Jr. and Spall, R.D. (1994) Molecular evidence that the myxozoan protists are metazoans. *Science* 265:1719–1721.

Stearns, S.C. (1987) Why sex evolved and the differences it makes. *In.:* S.C. Stearns (ed.): *The Evolution of Sex and its Consequences*. Birkhäuser Verlag, Basel, pp 15–31.

Stewart, C.N. Jr. and Excoffier, L. (1996) Assessing population genetic structure and variability with RAPD data: Application to *Vaccinium macrocarpon* (American Cranberry). *J. Evol. Biol.* 9:153–171.

Titus, J.E. and Hoover, D.T. (1991) Toward predicting reproductive success in submersed freshwater angiosperms. *Aquat. Bot.* 41:111–136.

Tooby, J. (1982) Pathogens, polymorphism, and the evolution of sex. *J. Theor. Biol.* 97: 557–576.

Toriumi, M. (1974) Analysis of interspecific variation in *Lophopodella carteri* (Hyatt) from the taxonomical viewpoint. XXII. General considerations. *Bull. Mar. Biol. Stn Asamushi* 15:1–12.

Uotila, L. and Jokela, J. (1995) Variation in reproductive characteristics of colonies of the freshwater bryozoan *Cristatella mucedo*. *Freshw. Biol.* 34:513–522.

Vernon, J.G., Okamura, B., Jones, C.S. and Noble, L.R. (1996) Temporal patterns of clonality and parasitism in a population of freshwater bryozoans. *Proc. R. Soc. Lond. B* 263: 1313–1318.

Weider, L.J. (1989) Spatial heterogeneity and clonal structure in Arctic populations of apomictic *Daphnia*. *Ecology* 70:1405–1413.

Weider, L.J. (1993) Niche breadth and life history variation in a hybrid *Daphnia* complex. *Ecology* 74:935–943.

Welsh, J. and McClelland, M. (1990) Fingerprinting genomes using PCR with arbitrary primers. *Nucl. Acids Res.* 18:7213–7218.

Wesenberg-Lund, C. (1907) On the occurrence of *Fredericella sultana* Blumenbach and *Paludicella ehrenbergi* von Bened. in Greenland. *Med. Grønl.* 34:63–75.

Williams, J.G.K., Kubelik, A.R., Livak, K.J., Rafolski, J.A. and Tingey, S.V. (1990) DNA polymorphisms amplified by arbitrary primers are useful as genetic markers. *Nucl. Acids Res.* 18:6531–6535.

Wöss, E.R. (1994) Seasonal fluctuations of bryozoan populations in five water bodies with special emphasis on the life cycle of *Plumatella fungosa* (Pallas). *In:* P.J. Hayward, J.S. Ryland and P.D. Taylor (eds): *Biology and Palaeobiology of Bryozoans*. Olsen and Olsen, Fredensborg, Denmark, pp 211–214.

Wolf, K. and Markiw, M.E. (1984) Biology contravenes taxonomy in the Myxozoa: New discoveries show alternation of invertebrate and vertebrate hosts. *Science* 225:1449–1452.

Wood, T.S. (1973) Colony development in species of *Plumatella* and *Fredericella* (Ectoprocta: Phylactolaemata). *In:* R.S. Boardman, A.H. Cheetham and W.A. Oliver, Jr. (eds): *Development and Function of Animal Colonies through Time*. Dowden, Hutchinson and Ross, Stroudsberg, Pennsylvania, pp 395–432.

Wood, T.S. (1989) Ectoproct bryozoans of Ohio. *Bulletin of the Ohio Biological Survey New Series*, Vol. 8, No. 2. The Ohio State University, Columbus, Ohio.

Wood, T.S. (1991) Bryozoans. *In:* J.H. Thorp and A.P. Covich (eds): *Ecology and Classification of North American Freshwater Invertebrates*: Academic Press, San Diego, pp 481–499.

Yokoyama, H., Ogawa, K. and Wakabayashi, H. (1993) Involvement of *Branchiura sowerbyi* (Oligochaeta: Annelida) in the transmission of *Hoferellus carassii* (Myxosporea: Myxozoa), the causative agent of kidney enlargement disease (KED) of goldfish *Carassius auratus*. *Gyobyo Kenkyu* 28:135–139.

Evolutionary process following colonizations

Evolutionary Ecology of Freshwater Animals
ed. by B. Streit, T. Städler and C. M. Lively
© 1997 Birkhäuser Verlag Basel/Switzerland

Evolutionary consequences of postglacial colonization of fresh water by primitively anadromous fishes

M. A. Bell and C. A. Andrews

Department of Ecology and Evolution, State University of New York, Stony Brook, New York, 11794-5245, USA

Summary. Anadromous fish breed in fresh water and spend most of their life cycle in the ocean. Thus, they have ample opportunity to colonize fresh water, which is favored when the cost of migration exceeds the value of marine food resources. Deglaciation of the boreal Holarctic has created countless opportunities to colonize fresh water, where strongly contrasting environmental conditions have favored rapid endemic radiation. We focus on radiation of the threespine stickleback (*Gasterosteus aculeatus*), but review radiation of lampreys (Petromyzontiformes), several salmonids (char, trout, salmon, whitefish), rainbow smelt (*Osmerus mordax*), and other groups.

Repeated colonization of freshwater and rapid phenotypic evolution are general features of these radiations. Trophic diversification is typically the most important component of radiation, but life history, body form (in relation to locomotion), and predator avoidance phenotypes also diversify. Lake fish often specialize for use of either plankton or benthic prey, and it sometimes results in formation of species pairs or trophic polymorphism. Loss of parasitism is a special feature of lamprey radiation, and diversification of bony armor phenotypes is peculiar to sticklebacks. Low diversity of sympatric species from postglacial freshwater radiations probably does not just reflect the youth of boreal habitats, but may result from the small number of climatic cycles since deglaciation. Climatic cycles cause fusion and fragmentation of lakes and create opportunities for allopatric speciation and secondary contact. Complexes of anadromous fishes and their freshwater isolates probably form phylogenetic racemes in which anadromous populations are a phenotypically stable, phylogenetically continuous ancestor from which predictable sets of divergent, freshwater phenotypes have evolved repeatedly.

Introduction

Evolutionary or adaptive radiation is the increase in numbers of species and breadth of resource utilization, with concomitant phenotypic diversification, within a monophyletic group. It often results from colonization of an isolated, depauperate habitat in which the colonizing species has access to resources dominated by other species elsewhere. Directional selection on descendants of the colonizing species may cause reproductive isolation and specialization for use of different subsets of these underutilized resources. Analysis of endemic radiations has been instrumental in the development of evolutionary thought. Radiations of Australian marsupials (Keast, 1972) or South American mammals (Simpson, 1980), and numerous, smaller, insular radiations (Stern, 1971; Carlquist, 1974; Grant, 1986) attest to the power of natural selection to fashion adaptive fits of species to environments. Insular radiations have also provided crucial insights into speciation (Mayr, 1942; Schluter, 1996a). The theoretical importance of endemic radiations reflects the close correspondence between phenotypic divergence and habitat variation in simple insular ecosystems.

Endemic radiations are common in freshwater fishes but have had less influence on evolutionary theory than terrestrial radiations (but see Echelle and Kornfield, 1984; Schluter and McPhail, 1993; Robinson and Wilson, 1994). Two types of ecological transitions stimulate endemic radiations of freshwater fishes: colonization of lakes by fluvial species (e.g., Fryer and Iles, 1972; McCune, 1996; most examples in Echelle and Kornfield, 1984) and colonization of recently deglaciated regions that are relatively inaccessible through fresh water but easily reached by anadromous fishes. Anadromous fishes spend most of the life cycle in marine water, but breed and experience larval growth in fresh water, which they readily colonize (McDowall, 1988). As in more well-known terrestrial cases, radiations of freshwater fishes require a suitable balance between isolation and accessibility and the presence of novel ecological opportunities.

In this review we show that the postglacial colonization of freshwater environments by primitively anadromous fishes has produced endemic radiations with recurrent trends that can yield theoretical insights. We emphasize radiation of freshwater *Gasterosteus aculeatus* (threespine stickleback) because it is extensive and well studied (reviewed in Wootton, 1976, 1984; Paepke, 1983; van den Assem and Sevenster, 1985; Ziuganov, 1991; Bell and Foster, 1994a; Bakker and Sevenster, 1995). Postglacial radiation of other primitively anadromous fishes may exhibit less diversity and be less well studied, but recognition of the importance of this phenomenon for conservation of biodiversity has become widespread (Nielsen, 1995). We highlight recurrent trends and important phenomena in several other primitively anadromous groups, but speciation receives limited attention because radiation of anadromous fishes in fresh water is largely allopatric (see also Schluter, 1996b for a recent review of speciation).

The threespine stickleback phenotype

Many of the most characteristic features of *G. aculeatus* deviate from the norm in some populations (Fig. 1). It is generally fusiform and less than 6 cm standard length (i.e., tip of snout to end of last vertebra). It has a pair of large, locking, free dorsal spines and a third small spine immediately anterior to the dorsal fin. The large pectoral fins are important for labriform locomotion and egg brooding. The pelvis is robust and complex and supports a stout locking spine on either side (Fig. 2; Bell, 1987, 1988). Up to about 35 bony lateral plates form a row (one per myomere) from head to tail along the flanks (Bell, 1981, 1984).

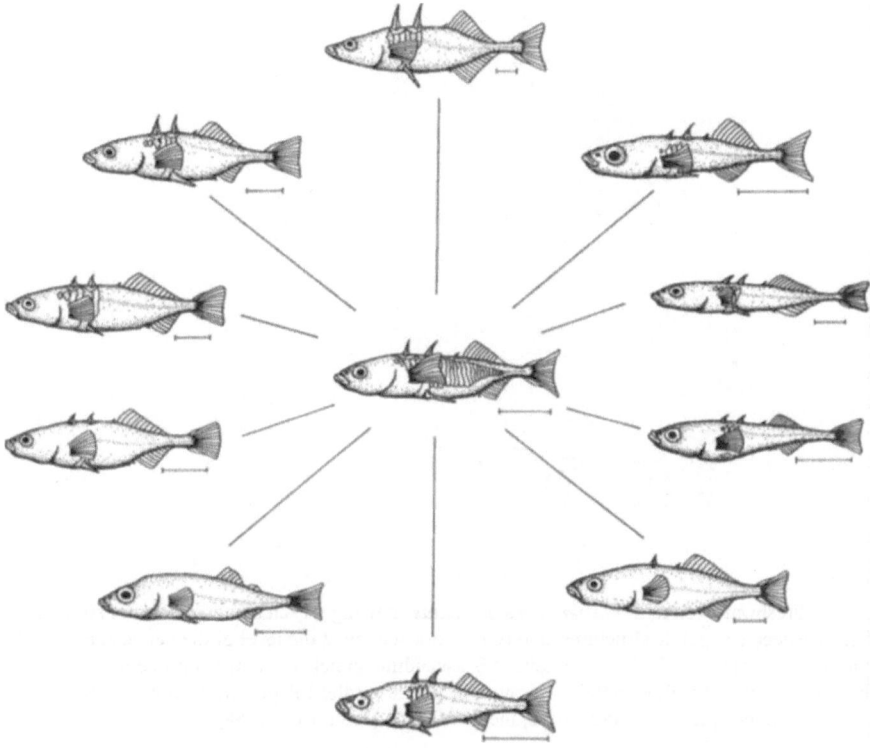

Figure 1. Phenotypic variation of western North American populations of *Gasterosteus aculeatus*. The central figure represents the ancestral phenotype of marine and anadromous populations, and peripheral figures represent postglacial phenotypic diversity. The scale bars are 1 cm. (From Bell and Foster, 1994b).

G. aculeatus is well known for territoriality, courtship, and reproductive behavior (reviewed by Rowland, 1994). Males defend territories in shallow water and court females, which may follow them to the nest and spawn. Males brood the eggs and fry in the nest. Threespine stickleback can spawn in hypoxic habitats that are unsuitable for most boreal fishes, and their relatively large egg size may be needed for fry to grow large enough the first short summer to survive the winter (Mina, 1991; Conover, 1992).

G. aculeatus is widespread throughout the Holarctic and has two principal lifestyles. Marine and anadromous populations are pelagic during most of the life cycle (Jones and John, 1978; Cowen et al., 1991) but spawn in near-shore subtidal and freshwater habitats, respectively. Anadromous populations spawn in streams and occasionally in lakes (Narver, 1969; Ziuganov and Bugayev, 1988). Resident freshwater populations occur in diverse habitats, including lakes, ponds, margins of large rivers, small streams, and swamps.

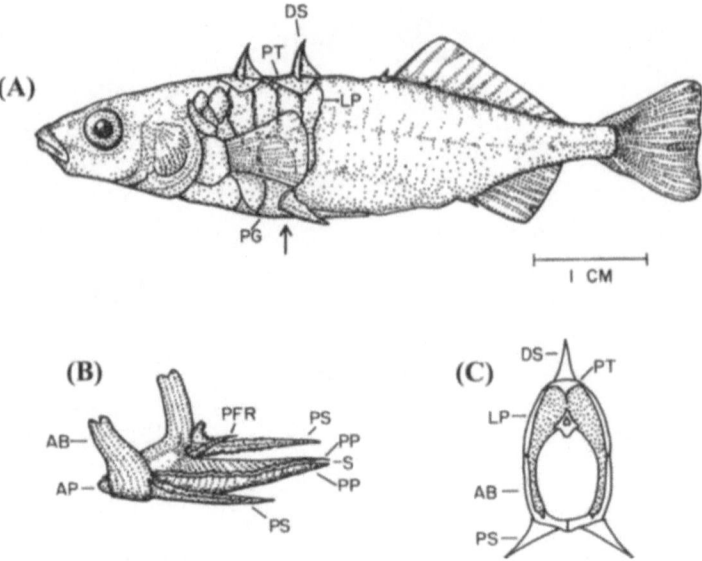

Figure 2. Freshwater (stream) *Gasterosteus aculeatus* showing (A) characteristic armor structures in lateral aspect, (B) pelvic structures, and (C) a cross-section at the level of the pelvic and second dorsal spines (arrow in [A]). Abbreviations: AB, ascending branch; AP, anterior process; DS, dorsal spine; LP, lateral plate; PFR, pelvic fin ray; PG; pelvic girdle; PP, posterior process; PS, pelvic spine; PT predorsal pterygiophore, and S, median suture. (From Bell, 1988).

Repeated invasion of freshwater by marine and anadromous populations of *Gasterosteus aculeatus*

Bell and Foster (1994b) summarized evidence that freshwater *G. aculeatus* has evolved repeatedly from marine and anadromous populations since deglaciation of the boreal Holarctic (see also e.g., McPhail and Lindsey, 1970; Bell, 1976, 1984; McPhail, 1994). Most stickleback genera and related taxa are strictly marine, indicating an ancestral marine habitat. Furthermore, marine and anadromous *G. aculeatus* can disperse long distances through the ocean and readily colonize fresh water. It frequently occurs in freshwater habitats containing few or no species of strictly freshwater fishes, but many species with marine affinities. It is widespread on recently deglaciated islands and in isolated continental habitats that are inaccessible through fresh water. Genetic variation of anadromous and adjacent freshwater populations is also consistent with repeated local colonization of fresh water. The systematics, zoogeography, natural history, and population genetics of *G. aculeatus* all indicate that it repeatedly colonized fresh water via marine dispersal routes.

Phylogeny of *Gasterosteus aculeatus* populations

Molecular data indicate that *G. aculeatus* includes two major clades that diverged in the north Pacific (Haglund et al., 1992; Buth and Haglund, 1994; Ortí et al, 1994). The "Japanese clade" occurs alone in the seas of Japan and Okhotsk and sympatrically with the "Euro-American clade" elsewhere in Japan, where they form separate biological species (Higuchi and Goto, 1996). However, mitochondrial markers for these clades often occur together within samples from northwestern North America, which probably represent populations of hybrid ancestry (Deagle et al., 1996). The Euro-American clade occurs alone in western North America, south of

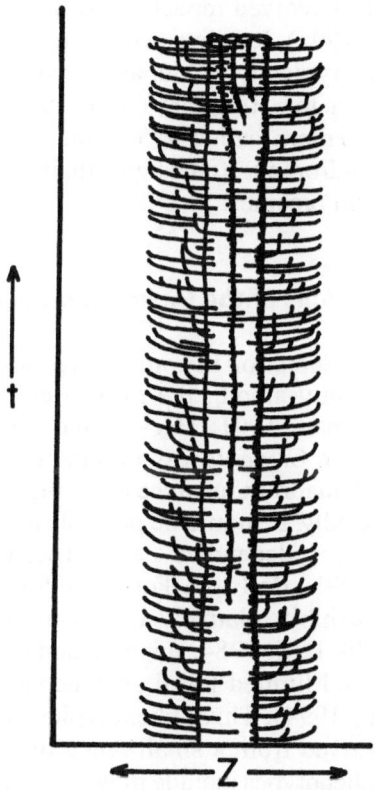

Figure 3. Phylogenetic raceme for the *Gasterosteus aculeatus* species complex. The broad central stem of the raceme represents ancestral states of marine and anadromous populations. The axis bifurcates repeatedly, forming separate Japanese and Euro-American clades, subsequent division of the latter into western North American and Atlantic clades, and finally, division of the Atlantic clade into eastern North American and European clades. The stems for these major clades actually overlap in phenotypic space. Further cladogenesis is not shown, except for representative freshwater isolates, which diverge rapidly from the stem. The vertical axis is time (t), and the horizontal axis (Z) represents multivariate phenotypic variation.

British Columbia (Haglund et al., 1992; E.B. Taylor, personal communication 1996). It dispersed from the Pacific to the Atlantic through the Arctic Ocean when the Bering Strait opened, about 3.5 million years ago, and formed separate European and eastern North American clades in the Atlantic (Ortí et al., 1994).

Marine and anadromous populations within each clade form a set of relatively uniform, persistent, ancestral lineages from which innumerable isolated, divergent, freshwater populations have been derived repeatedly (Fig. 3). Freshwater isolates diverge rapidly from the phenotypic states of their marine and anadromous ancestors toward those of freshwater populations, and isolating mechanisms may evolve rapidly (Francis et al., 1985; Klepaker, 1993; McPhail, 1993; Ziuganov, 1995). Many phenotypic states that marine populations lack are widespread in freshwater populations of both clades and must have evolved repeatedly (e.g., Bell et al., 1993; Bell and Foster, 1994b; Bell, 1995). Iterative evolution of divergent, freshwater populations from a phenotypically stable, ancestral lineage has characterized stickleback evolution for 10^7 years (Bell, 1987, 1988; Bell, 1994), forming a phylogenetic pattern called a "raceme" (Williams, 1992). The phylogenies of many anadromous fishes and their postglacial freshwater derivatives must also form racemes.

Epistemological implications of the *Gasterosteus aculeatus* raceme

Phenotypes used to infer adaptation in comparative studies must have been free to evolve independently among the groups compared (reviewed by e.g., Brooks and McLennan, 1991; Harvey and Pagel, 1991; Ricklefs, 1996). Because freshwater *G. aculeatus* populations have been derived locally from marine and anadromous ancestors, their derived phenotypes must have evolved repeatedly. It should be possible to treat such freshwater phenotypes as statistically independent cases (Harvey and Pagel, 1991; Martins and Garland, 1991; Bell and Foster, 1994b). Marine and anadromous *G. aculeatus* exhibit relatively little variation throughout their range (Gross, 1977, 1978; Bell, 1984, unpublished data; Reimchen et al., 1985; Bańbura, 1994), and limited fossil evidence suggests that this uniformity is ancient (Bell, 1994). Thus, phenotypic diversity of freshwater populations has been derived from a small range of ancestral phenotypes. Possible clade-specific phenotypes include minor armor differences between the Japanese and Euro-American clades in Japan (Haglund et al., 1992; Higuchi and Goto, 1996), and restriction of "leiura-with-keel" to Europe (Ziuganov, 1983; Bańbura and Bakker, 1995) and modal plate counts of seven plates per side to western North America (Hagen and Gilbertson, 1972; Gross, 1977; Klepaker, 1995). Nevertheless, the major differences of freshwater *G. aculeatus* from marine and anadromous stickleback and variation among freshwater populations show little effect of common ancestry and strongly reflect local adaptation (Reimchen, 1994; McPhail, 1994; Bell and Foster, 1994b). However, the minimum geographical separation needed to justify the assumption that phenotypic variation among freshwater populations is phylogenetically independent is unknown.

The ecology of colonization of fresh water by threespine stickleback

Freshwater isolates should evolve from anadromous ancestors if migration becomes impeded or the ratio of freshwater productivity to the fitness cost of migration is high (McDowall, 1988; Wood, 1995). *G. aculeatus* is ubiquitous in lowland boreal fresh waters, indicating that this ratio is generally exceeded, but freshwater colonization still must represent a drastic ecological change. Anadromous and marine stickleback spend a brief portion of their life cycle in shallow marine of freshwater habitats and migrate as fry to offshore areas (Cowen et al., 1991) for 1 or 2 years (Baker, 1994). They are presumably exposed at sea to diverse gape-limited predators, which, however, have numerous, less well-armored, alternative prey. Similarly, marine plankton is much more diverse than freshwater plankton.

In contrast, freshwater stickleback encounter great ecological variation that generally contrasts with the ancestral marine environment. They do not migrate between breeding and foraging habitats. The diversity of both predators and prey will be lower in fresh water, but there may be few or no alternative prey for piscivores. Benthic invertebrates are a major added class of freshwater prey. Life in lotic habitats may impose additional hydrodynamic demands not faced by marine and anadromous populations. These sharp ecological differences affect selection on diverse phenotypic characters.

Major dimensions of radiation in freshwater *Gasterosteus aculeatus*

In contrast to marine and anadromous *G. aculeatus*, freshwater populations exhibit extraordinary phenotypic variation. States of many freshwater phenotypes covary with habitat type, are heritable and have limited phenotypic plasticity, and may cause differences in rates of growth, survival, and reproduction. Threespine stickleback radiation is manifestly adaptive (Hagen and McPhail, 1970; Bell, 1984; Bell and Foster, 1994b; McPhail, 1994; Reimchen, 1994). We will focus on more thoroughly studied or instructive phenotypes of *G. aculeatus* and must omit some traits for which there is significant information, including vertebral number (Swain and Lindsey, 1984; Swain, 1992a, 1992b; Reimchen and Nelson, 1987), life history (reviewed by Baker, 1994), and male nuptial coloration (e.g., Reimchen, 1989; Blouw and Hagen, 1990; McDonald et al., 1995).

Armor and predator avoidance behavior

The spines, lateral plates, and pelvic girdle form a flexible, spine-studded armor ring that shields the abdomen from compression and laceration by predatory fishes and birds (Hoogland et al., 1957; Reimchen, 1983, 1992, 1994). Little is known about predation in the ocean, but marine and anadromous populations have robust armor, suggesting that they experience significant gape-limited predation (Gross, 1977, 1978). Diverse predators eat freshwater *G. aculeatus* (Reimchen, 1994), and armor evolution has received more attention than any other aspect of phenotypic variation (reviewed by Bell, 1976, 1984; Reimchen, 1994, 1995). More is known about the effects of fishes as selection agents than about birds or other predators, and little is known about the effects of insect predation (but see Reimchen, 1980, 1983, 1994; Reist, 1980; Ziuganov and Zotan, 1995).

Armor varies greatly among populations, it is generally heritable (e.g., Lindsey, 1962; Hagen, 1973; Hagen and Gilbertson, 1973a; Blouw and Boyd, 1992; Bańbura and Bakker, 1995; Baumgartner, 1995), and similar phenotype-environment associations occur in Europe (Gross, 1977, 1978; Campbell, 1984) and western North America (e.g., Hagen and Gilbertson, 1972; Moodie and Reimchen, 1976). Three or four major lateral plate morphs are common in western Europe and western North America (Hagen and Moodie, 1982; Bańbura and Bakker, 1995), and the low morph has less than 10 plates per side restricted to the abdomen (reviewed by Bell, 1976, 1984; Wootton, 1984; Campbell, 1984; Reimchen, 1983, 1994). Low morph populations subjected to intense fish predation have high frequencies of six- and seven-plated (per side) phenotypes and high mean plate counts (e.g., Hagen and Gilbertson, 1972; Moodie and Reimchen, 1976; Reimchen, 1994), long spines (e.g., Hagen and Gilbertson, 1972; Moodie, 1972; Gross, 1978), and robust pelvic structure (Gross, 1978; Reimchen, 1983; Bell, 1987). Plate number is lower (< 5/side) in populations with low fish predation (Reimchen, 1994; 1995), and low ionic concentration may also favor low plate counts (Giles, 1983; Bourgeois et al., 1994). Absence of predatory fishes is also associated with short spines (e.g., Hagen and Gilbertson, 1972; Reimchen, 1983) and weakness of the pelvic structure, which may be vestigial or lost in sites with low ionic concentration (Giles, 1983; Bell, 1987, 1988; Bell et al., 1993).

The role of vertebrate predation in the evolution of stickleback armor has been confirmed by field and laboratory studies. Hagen and Gilbertson (1973b), Moodie (1972), and Gilbertson (1980) showed that stickleback with seven plates per side are underrepresented in salmonid stomachs. Laboratory studies to confirm the advantage of seven-plated fish with respect to fish predation were inconclusive (Moodie et al., 1973), but Bell and Haglund (1977) demonstrated that five-plated fish were nearly invulnerable to snake predation in the laboratory. Similarly, complete morphs in pike stomachs had slightly fewer plates (X = 31.46) than the population at large (X = 32.03; Bańbura et al., 1989).

Hoogland et al. (1957) demonstrated the importance of stickleback spines in deterring fish predation, and the ratio between the gape of predatory fish and the distance between the tips of the dorsal and pelvic spines strongly influences the probability of escape after capture (Reimchen, 1991a, 1994). Stickleback in pike stomachs have relatively small size-adjusted pelvic spines and girdles (Bańbura et al., 1989). Reimchen (1980) proposed that insect predation selects for pelvic girdle reduction, and experiments using other stickleback species support this hypothesis (Reist, 1980; Ziuganov and Zotan, 1995).

Intrapopulation spatial and temporal variation also indicate the importance of vertebrate predation in armor evolution. Stickleback from a lake without predatory fishes had more (dorsal plus pelvic) spines in mid-lake, where birds feed, than near shore, where cover precludes feeding (Reimchen, 1980). There was a seasonal increase (toward 5/side) in mean plate number of juveniles of another population when fish predation was more important and a decrease (toward 4/side) when bird predation predominated (Reimchen, 1995).

There is limited evidence concerning differential reproductive success and sexual selection of low-morph plate-number. Kynard (1972, 1978, 1979) estimated several correlates of male and female reproductive success in a lacustrine population. These estimates cannot be combined, but higher plate counts were usually favored because of differences in breeding season and phenotypic correlations between morphology and behavior. Limited results for another lake population (Moodie, 1972) and a stream population with lower average plate counts (Haglund, 1981) are similar.

Population differentiation for predator avoidance behavior also occurs (Huntingford et al., 1994). A multivariate measure of "boldness" toward a stalking pike in the laboratory was generally high in stickleback from six sites with a high risk of fish predation and low in those from seven low-risk habitats (Huntingford, 1982). Andraso and Barron (1995) observed a reciprocal relationship between armor and the startle response in brook stickleback (*Culaea inconstans*), but the relationship between predator avoidance behavior and morphology is unknown in *G. aculeatus*, and boldness has been estimated in only a few populations.

Foraging behavior and trophic morphology

Contrasting stickleback from large lakes and small streams, Hagen and Gilbertson (1972) developed initial evidence for adaptive differentiation of trophic morphology in *G. aculeatus*. Populations from large lakes had high gill raker counts (19.0–21.7), and those from small streams had low counts (14.6–17.9); adjacent populations may differ considerably (see also

Moodie, 1972; Reimchen et al., 1985; McPhail, 1994). Moodie and Reimchen (1976) and Lavin and McPhail (1985) observed that gill-raker number in lacustrine G. aculeatus increases with lake surface area, presumably reflecting reliance on plankton as food, and Gross and Anderson (1984) showed that gill-raker number, length, and spacing are related to the prey size. Gill-raker number is heritable (Hagen, 1973), and phenotypic plasticity is inconsequential (Day et al., 1994), so habitat-specific differences must represent evolutionary divergence in relation to prey size.

Lavin and McPhail (1985, 1986, 1987) extended analysis of trophic diversification to other characters in solitary lacustrine populations, and Walker (1995, 1996, 1997) detected additional body form differences related to food type. Tooth arrangement (Caldecutt et al., unpublished data) and several cranial bones that function in feeding also vary among populations (W. J. C. Caldecutt and D. C. A. Adams, unpublished data). However, McPhail's (1984, 1992, 1993, 1994) demonstration of six "species pairs" on islands in the Strait of Georgia, British Columbia, and their use by Schluter (1993, 1994, 1995, 1996; Schluter and McPhail, 1992; Day et al., 1994) to investigate the ecology of speciation have provided the clearest insights into the causes of trophic divergence in threespine stickleback.

The species pairs resemble long-known pairs in lacustrine char and whitefish (see below), in which a "limnetic" morphotype specialized for planktivory forages in open water, and a bottom-foraging "benthic" morphotype takes larger prey, such as aquatic insect larvae and ostracods. Divergence between members of each species pair within Enos and Paxton Lake for allozyme and nucleotide markers indicates that they are separate species (McPhail, 1984, 1992; Taylor et al., 1997). Ridgway and McPhail (1984) observed strong positive assortative mating in the Enos Lake pair, which may be based on color and morphological differences, nest location, and the tempo of courtship. Schluter (1993, 1994) observed in cage and pond experiments that morphologically intermediate hybrids feed less efficiently and grow slower than members of either species pair. McPhail (1993, 1994) argued that lacustrine species pairs are restricted to the Strait of Georgia because this region has experienced two episodes of postglacial marine transgression, permitting successive invasions of anadromous populations about 2000 years apart. He argued that benthics, which more closely resemble solitary species in similar lakes, evolved from the first invasion, and that planktivores evolved from the second. Molecular divergence (Taylor et al., 1997) and differences in salinity tolerance of eggs (Kassen et al., 1995) are consistent with this argument.

Limnetic and benthic stickleback pairs have characteristic differences related to feeding. In addition to having more gill rakers, limnetic stickleback are characterized by longer gill rakers, narrower mouths, longer and narrower snouts, shallower bodies, and more dorsal and anal fin rays, differences that are retained in laboratory crosses and are intermediate in hybrids (McPhail, 1984; 1992, 1994). Phenotypic plasticity accentuates genetic differences between species in a pair under appropriate conditions (Day et al., 1994). Feeding observations using species of the Enos (Bentzen and McPhail, 1984) and Paxton Lake (Schluter, 1993) pairs showed that each form uses its own food type more efficiently than it does the other's. Each species of the Paxton Lake pair grew faster than the other in cages located within its own microhabitat (Schluter, 1995). Similarly, specimens with more benthic-like phenotypes in a solitary lacustrine population grew faster and survived better in competition with the Paxton Lake limnetic than did those with more limnetic-like phenotypes (Schluter, 1994). McPhail (1993) introduced hybrids between the Paxton Lake species pair to a fishless lake, and within 10 generations they diverged significantly toward benthic morphology for each of nine traits. Differences between McPhail's benthic-limnetic species pairs are adaptive, and, despite claims to the contrary (Skúlason and Smith, 1995), they are separate biological species (see Bell, 1996).

Locomotion and body form

The locomotory demands on fishes for migration (or lack thereof), feeding, predator avoidance, and reproduction determine selection for burst and sustained swimming and, in turn, on body form (e.g., Weihs, 1989; Webb, 1994). Freshwater stickleback do not typically undertake sustained migrations, and their reproductive behavior appears to be sufficiently conservative that it should not be a major factor in body form diversification. Diversification for feeding and predator avoidance are more likely causes for body form differentiation (Walker, 1995). The lateral surface area of the caudal portion of the body, including both the body and median fins, are most important for thrust generation during burst swimming. If burst swimming is important to escape from predatory fishes, the caudal portion of the body of prey fish species should be deep and median fins should be larger and shifted posterially. Additionally, greater relative body depth in the abdomen contributes to separation of the tips of dorsal spines from pelvic spines, making ingestion more difficult (Reimchen, 1983, 1991a, 1994). Benthic foraging in structured habitats should place a premium on turning ability, which is facilitated by deep body form, and cruising in search of plankton should favor a shallow body to minimize drag (Walker, 1995).

Again, McPhail's (e.g., 1994) studies of species pairs provided important insights into divergence for body form; body depth of limnetic planktivores averages only about 80% that of sympatric benthics, and body depth experiences character displacement between sympatric species pairs (Schluter and McPhail, 1992). Multivariate analysis of stickleback body form in relation to habitat differences indicates that stickleback from lakes with predatory fish have longer dorsal and anal fins but not greater body depth than those from lakes without them (Walker, 1995, 1997). These effects are greater in lakes with relatively small littoral zones, in which benthic shelter from predatory fish should be less accessible. This morphology appears to represent a compromise between the need for rapid fast-starts to escape predation and retention of streamlining for efficient cruising in search of plankton in lakes with small littoral zones. Walker (1995, 1997) also observed that stickleback from lakes with native predatory fishes have large pelvic skeletons and the dorsal spines are shifted forward. Robust pelvic structure contributes to armoring (see above), and anterior placement of dorsal spines may make the spines more effective (Reimchen, 1991b). A robust pelvis, however, may interfere with fast start performance (Andraso and Barron, 1995). The interaction between predation and food resource availability provides the potential for extensive body form differentiation in *G. aculeatus*. This potential is reflected in the diversity of body form within limited areas of the Queen Charlotte Islands, British Columbia (Reimchen et al., 1985) and Cook Inlet, Alaska (Walker, 1995). These interactions also provide the potential for serious errors in ecomorphological studies that fail to incorporate numerous anatomical landmarks and to consider multiple environmental variables.

Social and reproductive behavior

Stickleback reproductive behavior has been studied for decades by ethologists (van den Assem and Sevenster, 1985), but research on interpopulation behavioral differentiation is more recent (Foster, 1994). This research is hampered by the intrinsic phenotypic plasticity of behavior; accordingly, it is desirable to use laboratory-reared individuals with known learning experience, and each individual should be used only once (Bakker, 1994). Bakker's (1994) study of divergence of juvenile territoriality and aggression between a freshwater and anadromous population of *G. aculeatus* comes closest to meeting this high standard. Huntingford et al. (1994) observed that paternal behavior of males affects behavior of their fry. These results confirm the danger of using field-caught stickleback to assess behavioral population differentiation.

Relaxing this standard, Foster (1994, 1995) has observed differentiation of male reproductive behavior that is clearly adaptive, if, as seems likely (reviewed by Foster, 1995), it represents genetic differentiation. Egg cannibalism is important for early mortality of *G. aculeatus* (Whoriskey and FitzGerald, 1985; Foster, 1988, 1995; Ridgway and McPhail, 1988; Hyatt and Ringler, 1989), and Foster (1988, 1994, 1995) observed that it is common in benthic populations but almost non-existent in limnetics. In cannibalistic populations courtship is less conspicuous (e.g., less zigzagging, more dorsal pricking), and brooding males perform diversionary displays when conspecific schools approach their territories (Foster, 1994, 1995). Conspicuous courtship may attract cannibals, and diversionary displays often succeed in diverting them from the nest (Foster, 1994). Anti-cannibal reproductive behavior apparently is primitively present, tends to be reduced in freshwater stickleback, and has been lost repeatedly in limnetics (Foster, 1994, 1995).

The "white stickleback," a marine sibling species of *G. aculeatus*, is endemic to Nova Scotia, Canada (Blow and Hagen, 1990) and represents an extreme for reproductive behavior. Instead of brooding the eggs and fry, males disperse the eggs throughout their territories, where they may even be exposed to air at low tide (MacDonald et al., 1995).

Phenotypic integration

The general practice of atomizing phenotypes into functional units that represent the domains of separate subdisciplines may conceal broad functional interactions among traits. Natural selection clearly produces adaptive fits between individual traits of *G. aculeatus* and environmental conditions, and some sets of functionally related traits are associated with environmental factors. However, other seemingly unrelated traits of *G. aculeatus* may also covary among populations due to their functional interactions or common environmental associations. Failure to make comparisons among such sets of traits may conceal "phenotypic integration," which goes beyond mere correlation among elements of a single functional unit (e.g., spine lengths, plate counts, and pelvic structure; see Baumgartner, 1995) to produce correlation among less-obviously related traits (Fig. 4; Foster et al., 1992; Bell and Foster, 1994b; Foster and Bell, 1994).

There is already limited evidence of phenotypic integration in lacustrine *G. aculeatus*. Absence of predatory fishes favors armor reduction and planktivory in lacustrine stickleback. Planktivory favors high gill raker counts, long and narrow snouts, and streamlining. It also is associated with retention of elaborate courtship and loss of diversionary displays because planktivores do not engage in egg cannibalism. Low clutch volume, which should maintain streamlining of gravid females, is also associated with planktivory. Although no study has sought an association among all these phenotypic factors, some seemingly unrelated functional units appear to exhibit phenotypic integration (Foster et al., 1992; Walker, 1995, 1997). Further research will be needed to determine the importance of phenotypic integration in *G. aculeatus*, but initial results are promising.

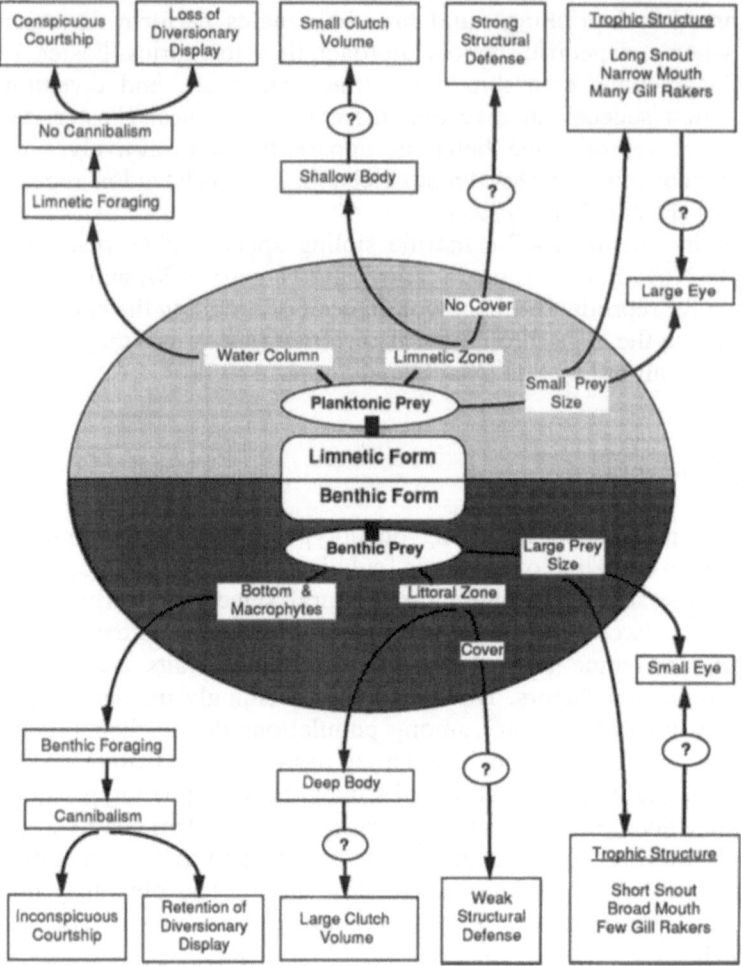

Figure 4. Phenotypic integration in lacustrine *Gasterosteus aculeatus* in relation to divergence for plantivory and foraging on benthic prey (limnetic and benthic forms, respectively, *sensu* McPhail, 1984). Relationships marked by arrows with question marks are hypothetical or little studied. (From Bell and Foster, 1994b).

Other sticklebacks

Diversification of other sticklebacks will not be discussed here because information is limited. The ninespine stickleback, *Pungitius pungitius*, includes anadromous populations (McDowall, 1988) which probably have colonized fresh water repeatedly throughout its north Holarctic distribution (Wootton, 1984). Pelvic reduction has evolved repeatedly in *Pungitius* (Nelson, 1971), but its evolution in fresh water has received little attention (see e.g., McPhail, 1963). Blouw and Hagen (1981, 1984a, 1984b, 1984c, 1984d; Hagen and Blouw, 1983) studied differentiation in marine fourspine stickleback, *Apeltes quadracus*. It is endemic to northeastern North America and occasionally colonizes freshwater. McDowall (1988) did not list this species as anadromous, and there is little information on its evolutionary diversification in fresh water.

Other fishes

McDowall (1988) summarized the distribution of anadromy among fishes, including three families and nine species each of lampreys (Petromyzontiformes) and sturgeons (Acipenseriformes), and 14 families and 90 species of teleosts. With 27 species each, the majority of anadromous teleost species are in two groups, the Salmonidae (trout, salmon, char, whitefish) and Clupeidae (herring). We will focus on the lampreys and salmoniforms, about which considerable information is available.

Lampreys

The Petromyzontiformes includes about 40 species of anadromous and freshwater lampreys. About half are ectoparasites on teleost fishes, and the remainder are small, non-parasitic forms that do not feed as adults (Hardisty, 1986). Parasitic lampreys are either anadromous or freshwater, but non-parasitic forms are always restricted to fresh water (Vladykov and Kott, 1979). All species are semelparous (Vladykov, 1985).

Non-parasitic or "satellite" species (Vladykov and Kott, 1979) of lampreys were first described by Zanandrea (1959) and have evolved repeatedly from parasitic ancestors. All but three of 10 lamprey genera include non-parasitic species (Hardisty and Potter, 1971; Vladykov and Kott, 1979; Vladykov, 1985; Hardisty, 1986; see Bailey, 1980 for a discussion of lamprey systematics). They are most frequent in the Holarctic Petromyzontidae, in which six of eight genera contain 12 parasitic and 20 satellite species (Vladykov and Kott, 1979). The geographic ranges of the satellite species tend to be included within those of the parasitic ancestor. However, non-parasitic forms may be allopatric to congeneric parasites, suggesting local or global extinction of the ancestor (Hardisty, 1986). Non-parasitic lampreys are consistently divergent from the ancestral parasitic phenotype in relative duration of larval (ammocete) life, gonadal development, body size, fecundity, body proportions, gut morphology, and dentition (Hardisty and Potter, 1971; Vladykov and Kott, 1979; Hardisty, 1986).

The average life spans of related non-parasitic and parasitic species are similar, but the ammocete stage is longer in the former. In the European species-pair, *Lampetra fluviatilis-Lampetra planeri,* for example, total life span averages seven years, but non-parasitic *L. planeri* has a larval phase of about $6^{1}/_{4}$ years compared to $4^{1}/_{2}$ years for *L. fluviatilis* (Hardisty and Potter, 1971; Vladykov, 1985; Hardisty, 1986). The extended larval period in non-parasitic lampreys translates into a larger average size at metamorphosis and precocious gonadal development (Hardisty and Potter, 1971; Potter, 1980; Vladykov, 1985). Nevertheless, non-parasitic species are smaller at spawning than related parasites because the latter experience rapid post-metamorphic growth as parasites (Hardisty and Potter, 1971; Vladykov and Kott, 1979; Hardisty, 1986; McDowall, 1988). Hardisty and Potter (1971) suggested that selection of non-parasitic phenotypes results from low larval mortality and avoidance of high mortality during migration.

In addition to reduced body size, non-parasitic lampreys exhibit characteristic morphological divergence from their parasitic ancestors. Body-form differences are concentrated anterior to the branchial region and include eye size and oral disk diameter (Hardisty and Potter, 1971). Non-parasitic forms do not feed on blood after metamorphosis, and their gut atrophies (Hardisty and Potter, 1971; Vladykov and Kott, 1979; Vladykov, 1985; Hardisty, 1986), which also would preclude osmoregulation in a marine environment if they migrated (Hardisty, 1986). The horny tubercles of the oral disk (hereafter referred to as "teeth") of parasitic lampreys are generally larger and sharper than those of their non-parasitic satellites (Hardisty, 1986). Compared to their parasitic ancestors, tooth number may be reduced (e.g., *Lampetra aepyptera*, a derivative of parasitic *L. ayresii*; Hardisty, 1986), increased (e.g., *Lampetra planeri*, a derivative of *L. fluviatilis*, *Lethenteron reissneri*, a derivative of *L. japonicum;* Hardisty, 1986), or show greater variability (e.g., *L. planeri* vs. *L. fluviatilis*, *Mordacia praecox* vs. *M. mordax*; Hardisty and Potter, 1971), suggesting reduced stabilizing selection in non-parasitic species.

Glaciation may have promoted evolution of non-parasitic lampreys by blocking migratory routes and preventing anadromy (Hardisty and Potter, 1971; Hardisty, 1986). For example, migration of *L. fluviatilis* of the eastern Baltic may have been terminated by the last glaciation, giving rise to resident freshwater *L. planeri* (Hardisty and Potter, 1971; Hardisty, 1986). However, non-parasitic species, such as *Ichthyomyzon gagei*, *I. hubbsi*, and *Lampetra aepyptera*, occur south of the Wisconsin ice margins in North America, limiting the role of glaciation in their divergence (Hardisty and Potter, 1971). Pluvial conditions (Smith, 1978) may have facilitated dispersal of lampreys into regions where interpluvial desiccation caused subsequent isolation and the evolution of satellites.

Lampreys are semelparous and have a restricted breeding season. Combined with the necessity of metamorphosis for breeding, changes in lifecycle length must occur in whole years and produce discontinuous phenotypic changes. The most obvious example of this is the substantial and consistent size reduction between parasitic ancestors and their satellites (Hardisty and Potter, 1971). Because successful mating depends on exact positioning of the male's tail with respect to the female's cloaca, spawning efficiency is highest when the mates have similar lengths. Thus, evolution of satellites should be facilitated by length-based assortative mating (Hardisty and Potter, 1971; see also Schluter and Nagel, 1995). Iterative evolution of non-parasitic lampreys therefore appears to be the result of similar modifications occurring to similar life-histories in the various genera in which satellite species have arisen.

There is general agreement that some satellite species have arisen repeatedly from freshwater parasites (e.g., in the North American genus, *Ichthyomyzon*; Vladykov and Kott, 1979), but there is disagreement concerning the immediate ancestry of non-parasitic lampreys whose closest known relatives are anadromous. They may have evolved directly from them (e.g., Zanandrea, 1959; Hardisty and Potter, 1971) or from a resident freshwater intermediate that is dimorphic for parasitic and non-parasitic life histories (Beamish, 1987). Alternatively, non-parasitic lampreys may evolve in two-steps from small, anadromous, parasitic "praecox" forms, which spend 1 year less feeding at sea and are less fecund than normal anadromous forms (Hardisty and Potter, 1971).

Hardisty (1986) noted that satellites rarely evolve from the largest anadromous, parasitic lampreys (but see Vladykov and Kott, 1979), but from species that are intermediate in size between the largest lampreys and typical satellites. Parasitic species that already show a tendency toward restricted migration are most likely to produce satellites. Hardisty (1986) suggested that such species are in a delicate balance between the fitness advantages of being large, fecund, and anadromous and those of being small and less fecund, but avoiding the fitness costs of migration. Environmental change may shift this balance toward non-parasitic, resident-freshwater habits by increasing the costs of migration or reducing the benefits of parasitism (Hardisty, 1986). However, a recently landlocked anadromous population of *Lampetra tridentata* failed to adopt a freshwater life cycle (Beamish and Northcote (1989), showing that this evolutionary response may be unattainable for some populations.

Satellite species have evolved independently from different parasitic species and repeatedly from the same species in different river systems, but their phenotypes are very similar even among genera. The genetic basis for the observed suite of divergent phenotypes may be widespread throughout the Petromyzontiformes (Hardisty and Potter, 1971). Alternatively, although many characters have similar states in different non-parasitic species (Vladykov and Kott, 1979), they may represent elements of three functional complexes, feeding, growth, and reproduction, and there may be only three complex characters involved in evolution of non-parasitic habits.

Family Salmonidae

Genus Salvelinus

The Holarctic genus *Salvelinus* exhibits remarkable morphological and ecological variation throughout its range, and up to four morphotypes may occur within a single watershed. The greatest variability occurs in Arctic char, S. *alpinus*, which includes anadromous and resident freshwater populations (Nordeng, 1983; reviewed by Crane et al., 1994). Anadromous Arctic char have repeatedly invaded postglacial freshwater systems, where low interspecific competition for food may have favored evolution of sympatric trophic morphs (i.e., dwarf planktivores, "normal-sized" benthic foragers, and occasionally piscivores) in numerous lakes (Hindar and Jonsson, 1982; Walker et al., 1988; Skúlason et al., 1989; Svedäng, 1990; Parker and Johnson, 1991; Hindar and Jonsson, 1993; Skúlason and Smith, 1995). Four morphs occur in the Icelandic lake, Thingvallavatn (Skúlason et al., 1989; Snorrason et al., 1994), but there are rarely more than two per lake (Griffiths, 1994). In addition to overall body size and trophic habits, sympatric char morphotypes differ consistently for numerous attributes, including age at sexual maturity, sex ratio, spawning coloration, flesh color, spawning time and location, morphometric and meristic characters, and migration habits, (Ferguson, 1981; Hindar and Jonsson, 1982; Nordeng, 1983; Hindar et al., 1986; Partington and Mills, 1988; Walker et al., 1988).

Griffiths (1994) described assemblages of dwarf and normal forms as bimodal populations, rather than species pairs, and suggested that seasonal food availability is the most important factor favoring bimodality. Phenotypic variation in S. *alpinus* reflects both genetic and environmental factors, and some morphometric differences are due to allometry and size difference (Griffiths, 1994). The importance of these factors appears to differ among populations, depending on the magnitude of morph specialization (Skúlason and Smith, 1995), and bimodality is favored in large, deep lakes with few species (Griffiths, 1994). Hindar and Jonsson (1982) proposed that the number of char morphs in a population depends on the number of "available niches" in a habitat. Riget et al. (1986) accepted this proposal and pointed out that there appear to be distinct size frequency peaks in a Greenland population of S. *alpinus*, such that intraspecific feedback mechanisms might regulate the growth of small char into a group of large "established" char. The age and length composition of char in this population indicated that such transformations occur (Riget et al., 1986). Dwarf and normal char have similar growth patterns until age 2+, when the dwarf begins to mature sexually and grow more slowly (Svedäng, 1990). Dwarfing may be facultative, such that slow-growing individuals become dwarfs and fast-growing individuals delay maturation until they are larger (Jonsson and Hindar, 1982; Hindar and Jonsson, 1993). Cannibalism could result from a food shortage for much of the year.

It should favor increased growth of larger cannibals and restrict smaller individuals to refuge habitats, where sparse food and greater competition further reduce growth.

Genetic crosses using populations throughout the north Atlantic, "common garden" rearing experiments, and transplants suggest that coexistence of *S. alpinus* morphotypes usually represents intraspecific polymorphism rather than speciation and divergence (Nordeng, 1983; Sparholt, 1985; Svedäng, 1990; Hindar and Jonsson, 1993). Indeed, a single individual may manifest three morphs (i.e., dwarf, normal, anadromous) in succession during its lifetime. Habitat choice based on growth rate and age at sexual maturity, annual food variation, and genetic variation all seem to contribute to trophic polymorphism in *S. alpinus*.

Although allozyme studies also indicate that morphotypes of *S. alpinus* are usually conspecific (Ferguson, 1981; Partington and Mills, 1988), they suggest that some sympatric morphotypes are reproductively isolated. Differentiation of allozyme and mtDNA haplotype frequencies of two char morphs in Loch Rannoch, Scotland indicated reproductive isolation, possibly maintained by morph-specific spawning seasons and sites (Gardner et al., 1988; Walker et al., 1988; Hartley et al., 1992a). The Loch Rannoch char morphotypes are distinguishable at any size or age, suggesting greater genetic differentiation between these morphs than between those within populations in which the morphs diverge later in life (Walker et al., 1988). Family studies also indicate that differences among the Thingvallavatn morphs are heritable (Skúlason et al., 1989).

The mode of origin of sympatric morphs of arctic char is unclear. Walker et al. (1988) reviewed three possibilities for char in Loch Rannoch: 1) multiple invasions by different stocks that still persist today in allopatry, sympatry, and with various levels of introgression, 2) phenotypic differentiation at an early ontogenetic stage, and 3) long-term differentiation from a common ancestor either by sympatric divergence or by allopatric divergence in different parts of the drainage system. The existence of separate races and multiple invasion have been proposed to explain variation of *S. alpinus* in England (Partington and Mills, 1988), North America (reviewed by Fraser and Power, 1989), Scandinavia (Andersson et al., 1983), and more widely throughout Europe (Hartley et al., 1992b). Sympatric differentiation may have occurred in some lakes, however. Divergence of mtDNA RFLP's of less than 0.2% among the four morphs in Thingvallavatn, Iceland, indicates that they diverged within the last 10000 to 60000 years, since the lake formed (Danzmann et al., 1991). Dwarf and normal morphs within lakes are generally more similar to each other than to corresponding morphs from different lakes (Hindar et al., 1986). Sympatric morphotypes of *S. alpinus* may represent intraspecific polymorphism, phenotypic plasticity, or some combination of these mechanisms, as well as intralacustrine speciation or successive colonization by populations that diverged in allopatry.

Genus Salmo

Salmo trutta, the brown trout, and *S. salar*, the Atlantic salmon, are both classified as anadromous, but the migratory cycle of *S. trutta* appears to be much more facultative than that of *S. salar* (reviewed by McDowall, 1988). Both include resident freshwater populations that appear to be polyphyletically derived from an anadromous ancestor (reviewed by McDowall, 1988; Ståhl, 1987; Ferguson, 1989; Foote et al., 1989). However, there is no evidence of significant morphological differentiation among allopatric non-anadromous populations of *S. salar* (Claytor and MacCrimmon, 1988; Claytor and Verspoor, 1991).

S. trutta, a native of Europe (McDowall, 1990a), is more variable than *S. salar* and can be subdivided into several glacial races based on an eye-specific lactate dehydrogenase locus (*Ldh-5;* Ryman et al., 1979; Ferguson and Mason, 1981; Hamilton et al., 1989; Ferguson and Taggart, 1991). Resident freshwater and anadromous *S. trutta* do not form separate monophyletic groups, indicating recurrent origin of freshwater populations in different drainages (reviewed by Ferguson, 1989), and anadromous populations of *S. trutta* can be derived from freshwater populations (reviewed by McDowall, 1988).

Three morphotypes of *S. trutta*, gillaroo, sonaghen, and ferox, coexist in Lough Melvin, Ireland. They differ in allozyme frequencies, breeding site, head morphology, tooth number, gill-raker length, and coloration, suggesting that they are separate biological species (Ferguson and Mason, 1981; Ferguson and Taggart, 1991). Ferguson and Taggart (1991) suggested that these populations have coexisted for about 13 000 years, since deglaciation. The ferox form appears to represent early post-glacial colonization, based on its high frequency of the hypothesized ancestral allele of *Ldh-5*, while the gillaroo and sonaghen forms seem to be more recent invaders. Ferguson and Taggart (1991) could not distinguish between allopatric divergence followed by invasion versus sympatric speciation to explain the origin of the gillaroo and sonaghen forms, but favored the latter. Lough Melvin is one of the few localities in which the ancestral and modern races are known to coexist (Ferguson and Taggart, 1991), possibly because separate breeding-season and trophic niches are present (Hamilton et al., 1989). All three Lough Melvin morphs eat insect larvae, but only ferox eat fish, gillaroo eat molluscs, and sonaghen eat plankton (Ferguson and Mason, 1981).

Allozyme divergence (Ryman et al., 1979) suggests that two reproductively isolated demes of *S. trutta* diverged in allopatry and postglacially entered Lake Bunnersjoarna, Sweden. Although they differ significantly in growth rate, they are not morphologically divergent (including trophic traits; Ryman et al., 1979), but more research on ecological divergence is needed. The brown trout readily colonizes fresh water and ecologically divergent sets of biological species that parallel cases in other salmonids occur.

Genus Oncorhynchus

Oncorhynchus nerka
Oncorhynchus nerka is a Pacific salmon with anadromous (sockeye) and resident freshwater (kokanee) populations. It occurs between 41 and 61° north latitude in Japan, the Russian Far East, and western North America. Kokanee have evolved repeatedly from anadromous sockeye postglacially throughout the range (Ricker, 1940; Nelson, 1968; Foote et al., 1989; Wood, 1995; Taylor et al., 1996). A major phenotypic difference between the two forms is reduced adult body length of kokanee due to lower freshwater productivity (Foote and Larkin, 1988; Foote et al., 1989). The two forms differ in gill raker number, a heritable trait (Leary et al., 1985) related to food size (Lindsey, 1981), and in morphological characters related to swimming performance, with kokanee having deeper bodies, shorter caudal regions, and lower vertebral numbers than sockeye of the same size (Wood and Foote, 1990; Taylor and Foote, 1991).

Because of the great commercial importance of sockeye salmon, there is extensive molecular genetic data on its population structure and the origin of kokanee (e.g., Foote, et al., 1989; Taylor et al., 1996; reviewed by Wood, 1995). There is strong genetic structure, including differentiation among drainages and even lakes within drainages, and a strong influence of postglacial colonization. Allopatric kokanee occur in lakes from which migration has become impeded, but they may occur sympatrically with sockeye in lakes with high productivity (Nelson, 1968; Foote et al., 1989; Wood, 1995). In sympatry, sockeye and kokanee usually have different spawning times and locations, though they sometimes interbreed (reviewed by McDowall, 1988; Foote et al., 1989). Sockeye males mate nearly exclusively with sockeye females, but kokanee males may try, with limited success, to fertilize eggs of sockeye females (Foote et al., 1989), which discriminate against small males (Foote and Larkin, 1988). However, kokanee smolts may be able to make the transition to marine water, and first-generation kokanee adults can return from the sea (reviewed by McDowall, 1988; Foote et al., 1989).

Kokanee clearly have been derived repeatedly from sockeye salmon, a process strongly influenced by the balance between the fitness cost of migration and the gain of high freshwater productivity. Kokanee exhibit characteristic morphological adaptations to feeding on lacustrine plankton and loss of migratory habits, and populations sympatric with anadromous sockeye are generally reproductively isolated, even if some hybridization occurs (reviewed in Wood, 1995). Repeated evolution of kokanee from sockeye contrasts with the less polyphyletic origin of resident freshwater subspecies in congeneric Pacific trouts (see Laudenslager and Gall, 1980; Leary et al., 1989), and probably results from breeding of anadromous sockeye in lakes, where they more readily become isolated, rather than in rivers or streams (Wood et al., 1994; Wood, 1995).

Kokanee are generally allopatric or sympatric with sockeye, but two reproductively isolated kokanee (*O. nerka kennerlyi*) occur in Kronotskiy Lake, Kamchatka, Russia (Kurenkov, 1978). They differ in spawning time and location, gill-raker number and length, length of the base of the anal fin, and trophic habits. As in several other groups, a planktivore morph has longer, more numerous gill rakers, while fewer, shorter ones occur in the benthic feeder. Kurenkov (1978) suggested that the two forms evolved from anadromous spring- and summer-spawning sockeye that used the lake before it was cut off from the sea 10 000 to 15 000 years ago.

Subfamily Coregoninae

Genus Coregonus

Coregonus has a continuous north circumpolar distribution and has traditionally been divided into two species complexes, the European whitefish (*C. lavaretus*) and lake whitefish (*C. clupeaformis*), though they may be a single variable species (reviewed by Bernatchez and Dodson, 1994). Anadromous and freshwater *Coregonus* occur (Bernatchez and Dodson, 1991), and, as in *Salvelinus*, freshwater populations exhibit great phenotypic diversity. The phylogeographic structure of this genus suggests a history of episodic fragmentation and dispersal, much of it through fresh water (Bernatchez and Dodson, 1994).

MtDNA and allozyme variation have recently been used to study phylogenetic relationships among whitefish, revealing that the traditional taxonomy of the genus includes several polyphyletic assemblages (Bernatchez and Dodson, 1991, 1994; Bodaly et al., 1992; Bernatchez et al., 1996). There appear to be three major glacial races, the Beringian, Atlantic, and Mississippian races, plus the minor Arcadian race, in North America (Bernatchez and Dodson, 1991; Bodaly et al., 1992), and Foote et al. (1992) suggested the existence of an additional one, the Nahanni. Estimated divergence times for *Coregonus* races correlate well with the chronology of Pleistocene glaciation (Bernatchez and Dodson, 1991). MtDNA variation indicates that whitefish from the Atlantic refugium recolonized northern North America through coastal rather than continental routes, possibly because the Wisconsin deglaciation produced low-salinity conditions along the Atlantic coast (Bernatchez and Dodson, 1991). Hypothesized phylogenetic relationships between Eurasian and North American whitefish indicate that lakes in Alpine and central Europe were postglacially colonized by two genetically distinct Beringian groups that probably evolved allopatrically and subsequently integrated at a low level (Bernatchez and Dodson, 1994).

Svärdson (e.g., 1961, 1979) first drew attention to trophic differentiation of lacustrine *Coregonus* in Europe, and many similar observations have been made elsewhere. Lake whitefish, *C. clupeaformis*, and other coregonids exhibit what Kirkpatrick and Selander (1979) called "a dichotomous 'canalization' of life history," in which dwarf morphotypes are frequently generated during adaptive radiation. As in *Salvelinus*, the normal morphology appears to be selectively favored where food is more abundant, while the dwarf phenotype, with slower growth and a shorter maturation time and life span, is favored in marginal environments (Bodaly, 1979; Kirkpatrick and Selander, 1979; Vuorinen et al., 1993). Dwarfs have relatively long, numerous gill rakers, large terminal mouths, large eyes, compressed bodies, and silvery coloration (Lindsey, 1981). Kirkpatrick and Selander (1979) also noted a relationship among whitefish morphology, lake size, and species composition of the lake, as Griffiths (1994) did for *Salvelinus alpinus*.

It is more widely accepted that sympatric morphotypes of *C. clupeaformis* are separate species than it is for *Salvelinus* (Bodaly et al., 1992; Bernatchez et al., 1996). Eight sympatric nominal species of whitefish in the Great Lakes of North America are largely isolated by differences in breeding season and, to a lesser extent, location, which probably reflect divergence in maturation time, migration route, social behavior, and seasonal availability of prey for the young (summarized by Smith and Todd, 1984). Allozyme differences suggest reproductive isolation between members of several species pairs. It is generally accepted that the dwarf phenotype is polyphyletic (Kirkpatrick and Selander, 1979; Lindsey, 1981; Bodaly et al., 1992), but there is disagreement over whether members of species pairs arose sympatrically (or microallopatrically) within the same glacial refugia (Kirkpatrick and Selander, 1979; Bodaly et al., 1992; Foote et al., 1992) or in different refugia (e.g., Bernatchez and Dodson, 1990, 1991; Vuorinen et al., 1993; Bernatchez et al., 1996). Both mtDNA RFLP and allozyme (nuclear gene) data for *C. clupeaformis* from three southern Yukon lakes suggested different modes of divergence in different drainages (Bernatchez et al., 1996). Similarly, Smith and Todd (1984) proposed that two monophyletic groups of *Coregonus* colonized the Great Lakes of North America postglacially, but that some of the eight nominal species arose within lakes through microallopatry and allochrony of breeding sites while others arose allopatrically in different lakes.

Interspecific competition appears to be important in the evolution of the planktivorous dwarf form of *C. clupeaformis* (Lindsey, 1981; Bernatchez and Dodson, 1990). Gill-raker number in *Coregonus* is known to have a substantial genetic component, but some phenotypic plasticity has been demonstrated (reviewed by Lindsey, 1981), and differences in the length and number of gill rakers are associated with prey differences (Bergstrand, 1982). Sympatric benthic (few, short gill rakers) and planktivorous (many, long gill rakers) forms of whitefish have been reported in European populations (e.g., Svärdson, 1961; Bergstrand, 1982), and similar differentiation occurs throughout North America, but only where planktivorous ciscoes (e.g., *Coregonus artedii*) are absent. If ciscoes are present in a lake with *C. clupeaformis*, the latter exhibits the benthic phenotype (Lindsey, 1981).

Genus Prosopium

Related round whitefish, genus *Prosopium*, do not include anadromous populations (Nelson, 1994), but their divergence in the Bear Lake basin, western USA, is notable because it fits the pattern of trophic differentiation seen in other groups so well (Smith and Todd, 1984). Three species occur within Bear Lake: insectivorous *P. spilonotum* is large with a blunt mouth and few gill rakers; *P. abyssicola* is about half as large with similar head shape and gill raker counts, but it eats mostly ostracods; and smaller still, *P. gemmiferum* is a planktivore with a long, shallow snout, and about twice as many gill rakers. The fourth species in the basin, *P. williamsoni*, occurs in a stream tributary to Bear Lake. Its size, head shape, and meristic characters are similar to those of *P. spilonotum*, the largest lacustrine benthic feeder, but its snout is noticeably down-turned, presumably for feeding in a current. Thus, the four species are ecologically discrete, and the lacustrine species differ either in spawning time or depth. Topographic complexity of the Bear Lake basin and fossil evidence indicate that members of this endemic radiation arose in allopatry (Smith and Todd, 1984).

Family Osmeridae

The rainbow smelt, *Osmerus mordax*, is native to tributaries of the western Atlantic from New Jersey, USA, north to Labrador, Canada (reviewed by Taylor and Bentzen, 1993a). Postglacial rebound appears to have isolated landlocked populations through repeated invasion by anadromous ancestors, probably from a single Atlantic coastal refugium (Black et al., 1986; Taylor and Bentzen, 1993a). Analysis of mtDNA from four landlocked populations suggested that freshwater and anadromous populations form separate clades (Baby et al., 1991), but a similar study using 16 populations showed that two major groups of *O. mordax* include both freshwater and anadromous populations (Taylor and Bentzen, 1993a).

Resident freshwater and anadromous smelt may be sympatric, and freshwater forms have diversified into sympatric normal (benthic) and dwarf (planktivorous) forms that differ genetically and may be reproductively isolated (Taylor and Bentzen, 1993a, 1993b). The morphological differences between these freshwater forms conform to the usual differences between benthic-limnetic pairs, and the dwarf form is believed to be polyphyletic in origin (Baby et al., 1991; Taylor and Bentzen, 1993).

Other anadromous colonists of fresh water

Postglacial colonization of fresh water has occurred in several other primitively anadromous groups. As in the cases discussed in detail above, colonization of fresh water generally appears to be polyphyletic, planktivore and benthic-foraging forms often evolve allopatrically or sympatrically in lakes, and the isolates sometimes represent biological species. Pacific trouts (family Salmonidae), *Oncorhynchus mykiss* (steelhead and rainbow) and *O. clarki* (cutthroat), often colonize streams, and the latter may have produced a pair of freshwater radiations (Loudenslager and Gall, 1980). Another salmonid, *Brachymystax lenok* (lenok) is restricted to fresh waters tributary to the Arctic and Pacific basins of Asian Russia and China. It includes allopatric and reproductively isolated sympatric blunt and sharp-snouted forms that are divergent for ecological properties and numerous trophic and locomotory traits (Mina, 1991). Mina (1991) favored the view that the two morphs are separate monophyletic groups, but more information is needed. The relationship between morphology and resource use by lenok also requires further documentation. *Myoxocephalus thompsoni* (family Cottidae), the deepwater sculpin, is generally considered to be a polyphyletic glacial relict derived from the widespread brackish and freshwater *M. quadricornis*, the fourhorn sculpin (Johnson, 1964; Parker, 1988; Houston, 1990). Differences between the *M. quadricornis* and its freshwater derivative involve mostly loss of spines and tubercles in the latter (McPhail and Lindsey, 1970). *M. thompsoni* and morphologically similar populations are widely distributed in fresh waters adjacent to the coastal range of the fourhorn sculpin (Crossman and McAllister, 1986), from which some populations have been derived very recently (Gyllensten and Ryman, 1988). The Galaxiidae and Retropinnidae are related families of cool-temperate, Southern Hemisphere fishes. Several species of the diadromous (migratory between marine and fresh water) genus *Galaxias*, or whitebait, have produced freshwater isolates (McDowall, 1988), some of which have probably been derived repeatedly (McDowall, 1970, 1990a). Freshwater derivatives may differ little from their diadromous ancestors (McDowall, 1970; Ovenden and White, 1990), but they often differ for traits related to the demands of locomotion in rapid streams (McDowall, 1970) and, to a lesser extent, for feeding (see McDowall, 1990a). McDowall (1970, 1988) attributed many of these differences to phenotypic plasticity, but genetic analyses are lacking, and molecular genetic data suggest that some monophyletic complexes include several biological species (Allibone et al., 1996). *Retropinna retropinna,* the common southern smelt, is primitively anadromous and has produced many freshwater isolates that exhibit changes in life history and trophic and locomotory morphology that parallel those of kokanee salmon (McDowall 1990b). Although the genetic basis for these changes and the status of the isolates as species are unclear (McDowall, 1988,

1990a), Northcote and Ward (1985) have demonstrated reproductive isolation in one case.

Discussion

The transition from anadromy to freshwater residence has occurred within each of the groups discussed, often a huge number of times. Colonization of fresh water by anadromous fishes represents a radical ecological transition with multiple fitness effects. Long-distance migration is usually eliminated, the environment tends to be more ecologically heterogeneous and physically structured but less productive, and predator and prey assemblages change radically. Fresh waters are insular, imposing genetic population structure and isolation.

The evolutionary response to selection based on this ecological transition is rapid and predictable in some respects. Aside from evolutionary changes related to year-round residence in fresh water (e.g., reduced salinity tolerance, loss of migratory behavior), freshwater colonization is commonly accompanied by decreased body size and reproductive isolation from the migratory ancestor. Further diversification within fresh water may involve formation of distinctive fluvial and lacustrine forms and, within lakes, of pairs of benthic foragers and planktivores (Fig. 5). Divergence of lacustrine isolates into discrete benthic and limnetic forms is conspicuous in stickleback, salmonids and smelt. It must reflect consistent patterns of natural selection in depauperate lakes, in which specialization for use of benthic and limnetic prey may reduce competition with a minimal loss of resource availability (Lindsey, 1981; Smith and Todd, 1984; Schluter and McPhail, 1992, 1993; Robinson and Wilson, 1994). Formation of such pairs may be suppressed by presence of predators or competitors, and additional prey specialists may be piscivores or molluscivores. Intralacustrine diversification may involve multi-character trophic polymorphism (especially in char), phenotypic plasticity, or speciation. If speciation is involved, it usually seems to require a period of allopatry, often in separate basins (see also below), though seasonal and ecological isolation may suffice. Although intralacustrine species pairs may be endemic to one lake, sympatric speciation hypotheses rarely withstand critical testing. Insularity and ecological diversity of freshwater habitats plus the striking ecological changes that accompany freshwater colonization provide excellent opportunities for adaptive radiation.

Freshwater threespine stickleback populations appear to exhibit exceptional geographic variation in recently deglaciated regions. Three factors favor population differentiation: 1) good dispersal ability, 2) a propensity for isolation, and 3) divergent selection among isolates. The threespine stickleback resembles most primitively anadromous boreal fishes in dispersal ability (McDowall, 1988), but differs by its ability to breed in relatively

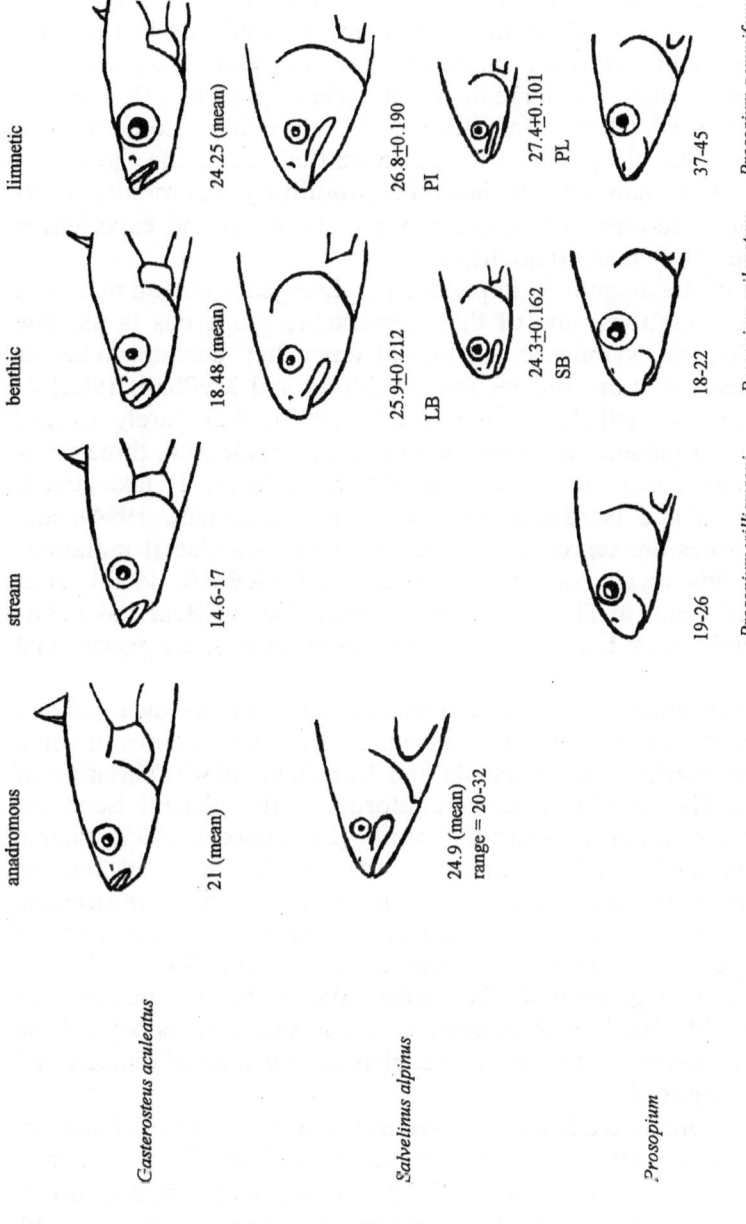

Figure 5. Heads of anadromous threespine stickleback (*Gasterosteus*) and Arctic char (*Salvelinus*), and their freshwater derivatives with different feeding habits, and comparable forms of round whitefish (*Prosopium*). Numbers indicate typical gill-raker counts. Arctic char are from Thingvallavatn, Iceland, where large (LB) and small benthic (SB) and large (PI, piscivorous) and small limnetic (PL, plankti-vorous) species occur. (Redrawn fish heads and gill-rakers counts from various sources.)

warm hypoxic waters of lakes, where it is likely to become isolated. Echelle and Echelle (1984) argued for the importance of residence in lakes for radiation of endemic Mexican atherinids. The ability of *G. aculeatus* to coexist with salmonids, which are major stickleback predators (Reimchen, 1994) and to flourish where salmonids are absent provides another important ecological axis along which it has diversified. Thus, the threespine stickleback may be more diverse than other primitively anadromous, postglacial colonists because it is more prone to isolation and experiences greater ecological diversity than they do.

Regardless of the magnitude of phenotypic divergence among members of postglacial radiations, one of their remarkable properties is the low number of discrete, sympatric, ecological units they contain. Whether these ecological units are species (e.g., Schluter and McPhail, 1992) or conspecific morphs (Skúlason and Smith, 1995), they rarely exceed two sympatric or parapatric forms. In threespine stickleback, three types of species pairs occur: 1) anadromous-stream resident, 2) lake-stream resident, and 3) and lacustrine limnetic-benthic (McPhail, 1994), and these dichotomies are represented in many of the postglacial radiations discussed in this review (see also Schluter and McPhail, 1993). Four forms of Arctic char in Thingvallavatn, Iceland, four of Bear Lake *Prosopium* and eight Great Lakes *Coregonus* in North America are exceptional cases.

Even these exceptional cases of postglacial radiation, however, pale by comparison to the huge numbers of endemic fishes in radiations of some temperate and tropical lakes (Echelle and Kornfield, 1984). Variation of intralacustrine fish species number conforms to the general trend for increasing species number toward the tropics (Rosenzweig, 1995). Pianka (1966) summarized possible causes for this relationship, and Terborg (1973) argued that it is due to presence of greater habitat area in the tropics. This explanation, however, cannot account for presence of hundreds of cichlid fish species in Lake Victoria, with an area of 68 100 km², and only eight species of *Coregonus* in all five Great Lakes of North America, with a total area of 245,300 km². A general review of this issue is beyond the scope of this paper, but some potential explanations can be eliminated, and a new one is proposed.

"Key innovations" have been proposed as the stimulus for evolutionary radiation (see Liem, 1974). It seems unlikely that all anadromous boreal fishes would lack key innovations needed for evolution of sympatric diversity, and one group, the cottids, has radiated elsewhere (Smith and Todd, 1984). It also seems unlikely that all of the boreal fish taxa that have postglacially colonized fresh water lack the complex mating systems believed to predispose fishes to speciation (Dominey, 1984). Positive assortative mating typically contributes to isolation between pairs of sympatric species of lampreys, sticklebacks, and salmonids. It is very unlikely that either ecological or reproductive properties limit the potential for boreal anadromous fishes to give rise to large numbers of ecologically divergent, reproductively isolated, sympatric freshwater species.

Repeated lake fission and fusion, however, appears to be important for evolution of sympatric species in many fish groups (examples in Echelle and Kornfield, 1984; Verheyen et al., 1996). Cycles of 21000, 42000 and 100000 years are an ancient feature of global climate (Olsen, 1986) and can cause periodic lake fission and fusion at low and mid-latitudes (Smith, 1978). Thousand-year climatic "flickering" has also occurred in the Pleistocene (Roy et al., 1996). However, for at least 10 million years, climatic cycles have contributed to glaciation of boreal habitats (Plafker and Addicott, 1976). Thus, climatic cycles that cause fission and fusion of lakes and fish speciation at low latitudes may cause fish extinction in high-latitude lakes.

Current controversy over rates and patterns of evolution in nature and their underlying causes are as old as evolutionary theory itself (e.g., Simpson, 1944; Eldredge and Gould, 1972; Erwin and Anstey, 1995). The fossil record is usually too coarse and observation of extant populations too recent to address these issues. Although postglacial boreal fish radiations produce few sympatric species, their allopatric products may be highly divergent and must be very young. Thus, they can provide unsurpassed insights into evolutionary rates and patterns. Significant divergence can occur within a few decades in *G. aculeatus* (Francis et al., 1985; Klepaker, 1993; McPhail, 1993), and extreme phenotypic diversification must have evolved in several regions in less than 20000 years (e.g., Scotland, Campbell 1984; Alaska, Bell et al., 1993; Queen Charlotte Islands, Reimchen, 1994; southern British Columbia, McPhail, 1994). Evolution of even the most highly divergent freshwater, threespine stickleback phenotypes can be explained by neo-Darwinian mechanisms of drift and selection on polygenic traits, and the same argument can be made for other anadromous groups (see above).

The fossil record and biology of *G. aculeatus,* however, allow examination of evolutionary patterns in particular detail. Extraordinarily rapid phenotypic diversification has been occurring within this species complex for at least 10000000 years (Bell, 1994) but has not produced highly divergent stickleback taxa (Bell and Foster, 1994b). This history contrasts with the African rift lake cichlids, in which the older radiation in Lake Tanganyika exhibits greater morphological breadth than the younger Lake Victoria radiation, and morphological gaps, presumably caused by extinction, separate the older species of Lake Tanganyika (Greenwood, 1984). The reason for this contrast is that the most highly divergent sticklebacks are restricted to boreal lakes, where, siltation and glaciation limit lake age and cause extinction of lake specialists (Bell, 1987, 1994). Thus, conservative anadromous lineages colonize fresh waters, giving rise to isolates that evolve similar phenotypes during each glacial retreat (Fig. 3). Divergence is not progressive as in the Lake Tanganyika cichlids, because isolates do not diverge from other divergent isolates but always from the same primitive, anadromous ancestor. Results from simulated clades in which the probability of extinction increases with phenotypic divergence are consistent with this explanation, and are called phylogenetic racemes (Williams, 1992). Clades for other boreal anadromous fishes and their freshwater derivatives (Barlow, 1995) and other groups (Williams, 1992) are probably also racemic.

Phylogenetic racemes in boreal anadromous fishes have two other interesting implications. The marine environment is persistent, and the ranges of anadromous and marine stickleback can shift in response to environmental change (Bell, 1987; Bell and Foster, 1994b; see Pease et al., 1989 for a general treatment). In contrast, freshwater isolates are trapped within drainage basins and extinction is a probable outcome of climate change. The resulting process produces selection against divergent populations and species within the threespine stickleback complex (Stanley, 1975; Williams, 1992). This process also conforms to the "taxon cycle" proposed by Wilson (1961): (1) a generalist species colonizes insular habitats, for which it (2) evolves specializations, and (3) becomes extinct when the habitat patches for which it is specialized disappear. In boreal fresh waters, the taxon cycle is repeated during each glacial cycle, limiting formation of large endemic fish faunas, eliminating the most divergent specialists through species selection, and producing phylogenetic racemes.

The value of holding the effects of common ancestry constant in comparative studies has become widely recognized (e.g., Harvey and Pagel, 1991; Brooks and McLennan, 1991; Ricklefs, 1996). The ideal experiment to infer the evolutionary causes for a phenotype or correlations between traits of a species would use genetically identical replicate demes sampled simultaneously from a stock population. Although this ideal is only crudely approximated by the phylogenetic racemes of boreal anadromous fishes and their freshwater isolates, they probably come as close as natural populations can to meeting this standard. In the context of environmental data, comparisons of geographically separated postglacial derivatives of anadromous fishes offer an unusual opportunity to minimize the effects of common ancestry to investigate evolutionary causation (e.g., see Vladykov and Kott, 1979 for lampreys; Bell et al., 1993 and Reimchen, 1994 for threespine stickleback; and Wood, 1995 for sockeye salmon).

Phenotypic integration may help explain recurrence of multicharacter phenotypes among freshwater derivatives of many boreal anadromous fishes. It has received explicit attention only in *G. aculeatus* (Foster et al., 1992; Bell and Foster, 1994b; see above) but is apparent in other groups (e.g., lampreys [Vladykov and Kott, 1979]; sockeye/kokanee salmon [Wood, 1995]). Although seemingly unrelated traits may appear to covary, they can often be grouped into four functional sets, (1) feeding, (2) growth, (3) locomotion, and (4) reproduction, with a common underlying environmental cause. Adaptation to one environmental factor may create a cascade of adaptive responses that affect seemingly unrelated traits. The resulting iterative evolution of large sets of morphological characters may be mistaken for autapomorphies of species, leading to description of polyphyletic species.

Acknowledgements
We thank D.C. Adams, W.J. Caldecutt, M. Higouchi, T.E. Reimchen, D. Schluter, and E.B. Taylor for allowing us to read their unpublished manuscripts. T.C.M. Bakker and T. Klepaker contributed constructive criticism that improved the paper. This is contribution 982 from the Graduate Program in Ecology and Evolution at the State University of New York at Stony Brook.

References

Allibone, R.M., Crowl, T.A., Holmes, J.M., King, T.M., McDowall, R.M., Townsend, C.R. and Wallis, G.P. (1996) Isozyme analysis of *Galaxias* species (Teleostei: Galaxiidae) from the Taieri River, South Island, New Zealand: a species complex revealed. *Biol. J. Linn. Soc.* 57:107–127.

Anderson, L., Ryman, N. and Ståhl, G. (1983) Protein loci in the Arctic charr, *Salvenlinus alpinus* L.: electrophoretic expression and genetic variability patterns. *J. Fish Biol.* 23: 75–94.

Andraso, H.M. and J.N. Barron (1995) Evidence for a trade-off between defensive morphology and startle-response performance in the brook stickleback (*Culaea inconstans*). *Can. J. Zool.* 73:1147–1153.

Baby, M.-C., Bernatchez, L. and Dodson, J.J. (1991) Genetic structure and relationships among anadromous and landlocked populations of rainbow smelt, *Osmerus mordax* Mitchell, as revealed by mtDNA restriction analysis. *J. Fish. Biol.* 39 (Suppl. A):61–68.

Bailey, R.M. (1980) Comments on the classification and nomenclature of lampreys – an alternative view. *Can. J. Fish. Aquat. Sci.* 37:1626–1629.

Baker, J.A. (1994) Life history variation in female threespine stickleback. *In:* M.A. Bell and S.A. Foster (eds):*The Evolutionary Biology of the Threespine Stickleback.* Oxford University Press, Oxford, pp 144–187.

Bakker, T.C.M. (1994) Evolution of aggressive behaviour in the threespine stickleback. *In:* M.A. Bell and S.A. Foster (eds): *The Evolutionary Biology of the Threespine Stickleback.* Oxford University Press, Oxford, pp 345–380.

Bakker, T.C.M. and Sevenster, P. (eds) (1995) Sticklebacks as models for animal behaviour and evolution. *Behaviour*, vol. 132, parts 13–16.

Bańbura, J. (1994) Lateral plate morph differentiation of freshwater and marine populations of the three-spined stickleback, *Gasterosteus aculeatus*, in Poland. *Zool. Scr.* 18:303–309.

Bańbura, J. and Bakker, T.C.M. (1995) Lateral plate morph genetics revisited: evidence for a fourth morph in three-spined sticklebacks. *Behaviour* 132:1153–1171.

Bańbura, J., Prezyblyski, M. and Franiewicz, P. (1989) Selective predation of the pike *Esox lucius*: comparison of lateral plates and some metric features of the three-spined stickleback *Gasterosteus aculeatus. Zool. Scr.* 18:303–309.

Barlow, G.W. (1995) The relevance of behavior and natural history to evolutionary significant units. *In:* J.L. Nielsen (ed.): *Evolution and the Aquatic Ecosystem: Defining Unique Units in Population Conservation.* American Fisheries Society Symposium 17, Bethesda, pp 169–175.

Baumgartner, J.V. (1995) Phenotypic, genetic, and environmental integration of morphology in a stream population of the threespine stickleback, *Gasterosteus aculeatus. Can. J. Fish. Aquat. Sci.* 52:1307–1317.

Baumgartner, J.V., Bell, M.A. and Weinberg, P.H. (1988) Body form differences between the Enos Lake species pair of threespine sticklebacks (*Gasterosteus aculeatus* complex). *Can. J. Zool.* 66:467–474.

Beamish, R.J. (1987) Evidence that parasitic and non-parasitic life history types are produced by one population of lamprey. *Can. J. Fish. Aquat. Sci.* 44:1779–1782.

Beamish, R.J. and Northcote, T.G. (1989) Extinction of a population of anadromous parasitic lamprey, *Lampetra tridentata,* upstream of an impassable dam. *Can. J. Fish. Aquat. Sci.* 46:420–425.

Bell, M.A. (1976) Evolution of phenotypic diversity in *Gasterosteus aculeatus* superspecies on the Pacific coast of North America. *Syst. Zool.* 25:211–227.

Bell, M.A. (1981) Lateral plate polymorphism and ontogeny of the complete plate morph of threespine sticklebacks (*Gasterosteus aculeatus*). *Evolution* 35:67–74.

Bell, M.A. (1984) Evolutionary phenetics and genetics: the threespine stickleback, *Gasterosteus aculeatus,* and related species. *In:* B.J. Turner (ed.): *Evolutionary Genetics of Fishes.* Plenum, New York, pp 431–528.

Bell, M.A. (1987) Interacting evolutionary constraints in pelvic reduction of threespine sticklebacks, *Gasterosteus aculeatus* (Pisces, Gasterosteidae). *Biol. J. Linn. Soc.* 31:347–382.

Bell, M.A. (1988) Stickleback fishes: bridging the gap between population biology and paleobiology. *Trends Ecol. Evol.* 3:320–325.

Bell, M.A. (1994) Paleobiology and evolution of threespine stickleback. *In:* M.A. Bell and S.A. Foster (eds): *The Evolutionary Biology of the Threespine Stickleback.* Oxford University Press, Oxford, pp 438–471.

Bell, M.A. (1995) Intraspecific systematics of *Gasterosteus aculeatus* populations: implications for behavioral ecology. *Behaviour* 132:1131–1152.

Bell, M.A. (1996) The ecology of resource polymorphism in vertebrates. *Trends Ecol. Evol.* 11:25–26.

Bell, M.A. and Foster, S.A. (eds) (1994a) *The Evolutionary Biology of the Threespine Stickleback.* Oxford University Press, Oxford.

Bell, M.A. and Foster, S.A. (1994b) Introduction to the evolutionary biology of the threespine stickleback. *In:* M.A. Bell and S.A. Foster (eds): *The Evolutionary Biology of the Threespine Stickleback.* Oxford University Press, Oxford, pp 1–27.

Bell, M.A. and Haglund, T.R. (1977) Selective predation of threespine sticklebacks (*Gasterosteus aculeatus*) by garter snakes. *Evolution* 32:304–319.

Bell, M.A., Ortí, G., Walker, J.A. and Koenings, J.P. (1993) Evolution of pelvic reduction in threespine stickleback fish: a test of competing hypotheses. *Evolution* 47:906–914.

Bentzen, P. and McPhail, J.D. (1984) Ecology and evolution of sympatric sticklebacks (*Gasterosteus*): specialization for alternative trophic niches in the Enos Lake species pair. *Can. J. Zool.* 62:2280–2286.

Bergstrand, E. (1982) The diet of four sympatric whitefish species in Lake Parkijaure. *Rept. Inst. Freshwater Res. Drottningholm* 60:5–14.

Bernatchez, L. and Dodson, J.J. (1990) Allopatric origin of sympatric populations of lake whitefish (*Coregonus clupeaformis*) as revealed by mitochondrial DNA restriction analysis. *Evolution* 44:1263–1271.

Bernatchez, L. and Dodson, J.J. (1991) Phylogeographic structure in mitochondrial DNA of the lake whitefish (*Coregonus clupeaformis*) and its relation to Pleistocene glaciations. *Evolution* 45:1016–1035.

Bernatchez, L. and Dodson, J.J. (1994) Phylogenetic relationships among Palearctic and Nearctic whitefish (*Coregonus* sp.) populations as revealed by mitochondrial DNA variation. *Can. J. Fish. Aquat. Sci.* 51:240–251.

Bernatchez, L., Vuorinen, J.A., Bodaly, R.A. and Dodson, J.J. (1996) Genetic evidence for reproductive isolation and multiple origins of sympatric trophic ecotypes of whitefish (*Coregonus*). *Evolution* 50:624–635.

Black, G.A., Dempson, J.B. and Bruce, W.J. (1986) Distribution and postglacial dispersal of freshwater fishes of Labrador. *Can. J. Zool.* 64:21–31.

Blouw, D.M. and Boyd, G.J. (1992) Inheritance of reduction, loss, and asymmetry of the pelvis of *Pungitius pungitius* (ninespine stickleback). *Heredity* 68:33–42.

Blouw, D.M. and Hagen, D.W. (1981) Ecology of the fourspine stickleback, *Apeltes quadracus,* with respect to a polymorphism for dorsal spine number. *Can. J. Zool.* 59:1677–1692.

Blouw, D.M. and Hagen, D.W. (1984a) The adaptive significance of dorsal spine variation in the fourspine stickleback, *Apeltes quadracus.* I. Geographic variation in spine number. *Can. J. Zool.* 62:1329–1339.

Blouw, D.M. and Hagen, D.W. (1984 b) The adaptive significance of dorsal spine variation in the fourspine stickleback, *Apeltes quadracus*. II. Phenotype-environment correlations. *Can. J. Zool.* 62 : 1340–1350.

Blouw, D.M. and Hagen, D.W. (1984 c) The adaptive significance of dorsal spine variation in the fourspine stickleback, *Apeltes quadracus*. III. Correlated traits and experimental evidence on predation. *Heredity* 53 : 371–382.

Blouw, D.M. and Hagen, D.W. (1984 d) The adaptive significance of dorsal spine variation in the fourspine stickleback, *Apeltes quadracus*. IV. Phenotypic covariation with closely related species. *Heredity* 53 : 383–396.

Blouw, D.M. and Hagen, D.W. (1990) Breeding ecology and evidence of reproductive isolation of a widespread stickleback fish (Gasterosteidae) in Nova Scotia, Canada. *Biol. J. Linn. Soc.* 39 : 195–217.

Bodaly, R.A. (1979) Morphological and ecological divergence within the lake whitefish (*Coregonus clupeaformis*) species complex in Yukon Territory. *J. Fish. Res. Board Can.* 36 : 1214–1222.

Bodaly, R.A., Clayton, J.W., Lindsey, C.C. and Vuorinen, J. (1992) Evolution of lake whitefish (*Coregonus clupeaformis*) in North America during the Pleistocene genetic differentiation between sympatric populations. *Can. J. Fish. Aquat. Sci.* 49 : 769–779.

Bourgeois, J.F., Blouw, D.M., Koenings, J.P. and Bell, M.A. (1994) Multivariate analysis of geographic covariance between phenotypes and environments in the threespine stickleback, *Gasterosteus aculeatus*, in the Cook Inlet area, Alaska. *Can. J. Zool.* 72 : 1497–1509.

Brooks, D.R. and McLennan, D.A. (1991) *Phylogeny, Ecology, and Behavior.* University of Chicago Press, Chicago.

Buth, D.G. and Haglund, T.R. (1994) Allozyme variation in the *Gasterosteus aculeatus* complex. *In:* M.A. Bell and S.A. Foster (eds): *The Evolutionary Biology of the Threespine Stickleback.* Oxford University Press, Oxford, pp 261–284.

Campbell, R.N. (1984) Morphological variation in the three-spined stickleback (*Gasterosteus aculeatus*) in Scotland. *Behaviour* 93 : 161–168.

Carlquist, S. (1974) *Island Biology.* Columbia University Press, New York.

Claytor, R.R. and MacCrimmon, H.R. (1988) Morphometric and meristic variability among North American Atlantic salmon (*Salmo salar*). *Can. J. Zool.* 66 : 310–317.

Claytor, R.R. and Verspoor, E. (1991) Discordant phenotypic variation in sympatric resident and anadromous Atlantic salmon (*Salmo salar*) populations. *Can. J. Zool.* 69 : 2846–2852

Conover, D.O. (1992) Seasonality and the scheduling of life history differences. *J. Fish. Biol.* 41 (Suppl. B) : 161–178.

Cowen, R.K., Chiarella, L.A., Gomez, C.J. and Bell, M.A. (1991) Offshore distribution, size, age and lateral plate variation of late larval/early juvenile sticklebacks (*Gasterosteus*) of the Atlantic coast of New Jersey and New York. *Can. J. Fish. Aquat Sci.* 48 : 1679–1684.

Crane, P.A., Seeb, L.W. and Seeb, J.E. (1994) Genetic relationships among *Salvelinus* species inferred from allozyme data. *Can. J. Fish. Aquat. Sci.* 51 (Suppl 1) : 182–197.

Crossman, E.J. and D.E. McAllister (1986) Zoogeography of freshwater fishes of the Hudson Bay Drainage, Ungava Bay and the Arctic Archipelago. *In:* C. Hocutt and E. O. Wiley (eds): *Zoogeography of North American Freshwater Fishes.* Wiley, New York, pp 53–104.

Danzmann, R.G., Ferguson, M.M., Skúlason, S., Snorrason, S.S. and Noakes, D.L.G. (1991) Mitochondrial DNA diversity among four sympatric morphs of Arctic charr, *Salvelinus alpinus* L., from Thingvallavatn, Iceland. *J. Fish Biol.* 39 : 649–659.

Day, T., Pritchard, J. and Schluter, D. (1994) Ecology and genetics of phenotypic plasticity: a comparison of two sticklebacks. *Evolution* 48 : 1723–1734.

Deagle, B.E., Reimchen, T.E. and Levin, D.B. (1996) Origins of endemic stickleback from the Queen Charlotte Islands: mitochondrial and morphological evidence. *Can. J. Zool.* 74 : 1045–1056.

Dominey, W.J. (1984) Effects of sexual selection and life history on speciation: species flocks in African cichlids and Hawaiian *Drosophila. In:* A.A. Echelle and I. Kornfield (eds): *Evolution of Fish Species Flocks.* University of Maine, Orono, pp 231–249.

Echelle, A.A. and Echelle, A.F. (1984) Evolutionary genetics of a "species flock:" atherinid fishes of the Mesa Central of Mexico. *In:* A.A. Echelle and I. Kornfield (eds): *Evolution of Fish Species Flocks.* University of Maine, Orono, pp 93–110.

Echelle, A.A. and Kornfield, I. (eds) (1984) *Evolution of Fish Species Flocks.* University of Maine, Orono.

Eldredge, N. and Gould, S.J. (1972) Punctuated equilibria: an alternative to phyletic gradualism. *In:* T.J.M. Schopf (ed.): *Models in Paleobiology:* Freeman Cooper, San Francisco, pp 82–115.

Erwin, D.H. and Anstey, R.L. (1995) *New Approaches to Speciation in the Fossil Record.* Columbia University Press, New York.

Ferguson, A. (1981) Systematics of Irish charr as indicated by electrophoretic analysis of tissue proteins. *Biochem. Syst. and Ecol.* 9:225–232.

Ferguson, A. (1989) Genetic differences among brown trout, *Salmo trutta,* stocks and their importance for the conservation and management of the species. *Freshwater Biol.* 21: 35–46.

Ferguson, A. and Mason, F.M. (1981) Allozyme evidence for reproductively isolated sympatric populations of brown trout *Salmo trutta* L., in Lough Melvin, Ireland, *J. Fish. Biol.* 18: 629–642.

Ferguson, A. and Taggert, J.B. (1991) Genetic differentiation among the sympatric brown trout (*Salmo trutta*) populations of Lough Melvin, Ireland. *Biol. J. Linn. Soc.* 43:221–237.

Foote, C.J. and Larkin, P.A. (1988) The role of male choice in the assortative mating of anadromous and non-anadromous sockeye salmon *(Oncorhynchus nerka) Behaviour* 106:43–62.

Foote, C.J., Wood, C.C. and Withler, R.E. (1989) Biochemical genetic comparison of sockeye salmon and kokanee, the anadromous and non-anadromous forms of *Oncorhynchus nerka. Can. J. Fish. Aquat. Sci.* 46:149–158.

Foote, C.J., Clayton, J.W., Lindsey, C.C. and Bodaly, R.A. (1992) Evolution of lake whitefish (*Coregonus clupeaformis)* in North America during the Pleistocene: evidence for a Nahanni glacial refuge race in the northern Cordillera Region. *Can. J. Fish. Aquat. Sci.* 49:760–768.

Foster, S.A. (1988) Diversionary displays of paternal sticklebacks: defenses against cannibalistic groups. *Behav. Ecol. Sociobiol.* 22:335–340.

Foster, S.A. (1994) Evolution of the reproductive behaviour of threespine stickleback. *In:* M.A. Bell and S.A. Foster (eds): *The Evolutionary Biology of the Threespine Stickleback.* Oxford University Press, Oxford, pp 381–398.

Foster, S.A. (1995) Understanding the evolution of behavior in threespine stickleback: the value of geographic variation. *Behaviour* 132:1107–1129.

Foster, S.A. and Bell, M.A. (1994) Evolutionary inference: the value of viewing evolution through stickleback-tinted glasses. *In:* M.A. Bell and S.A. Foster (eds): *The Evolutionary Biology of the Threespine Stickleback.* Oxford University Press, Oxford, pp 472–486.

Foster, S.A., Baker, J.A. and Bell, M.A. (1992) Phenotypic integration of life history and morphology: an example from the three-spined stickleback, *Gasterosteus aculeatus* L. *J. Fish. Biol.* 41 (Suppl.):21–35.

Francis, R.C., Havens, A.C. and Bell, M.A. (1985) Unusual lateral plate variation of threespine sticklebacks (*Gasterosteus aculeatus*) from Knik Lake, Alaska. *Copeia* 1985:619–624.

Fraser, N.C. and Power, G. (1989) Influences of lake trout on lake-resident arctic char in northern Quebec, Canada. *Trans. Amer. Fish. Soc.* 118:36–45.

Fryer, G. and Iles, T.D. (1972) *The Cichlid Fishes of the Great Lakes of Africa.* Oliver and Boyd, Edinburgh.

Gardner, A.S., Walker, A.F. and Greer, R.B. (1988) Morphometric analysis of two ecologically distinct forms of Arctic charr, *Salvelinus alpinus* (L), in Loch Rannoch, Scotland. *J. Fish. Biol.* 32:901–910.

Gilbertson, L.G. (1980) *Variation and natural selection in an Alaskan population of the threespine stickleback* (Gasterosteus aculeatus L.). Ph.D. Dissertation, University of Washington, Seattle.

Giles, N. (1983) The possible role of environmental calcium levels during the evolution of phenotypic diversity in Outer Hebridean populations of the three-spined stickleback, *Gasterosteus aculeatus. J. Zool.* 199:535–544.

Grant, P.R. (1986) *Ecology and Evolution of Darwin's Finches.* Princeton University Press, Princeton.

Greenwood, P.H. (1984) African cichlids and evolutionary theories. *In:* A.A. Echelle and I. Kornfield (eds): *Evolution of Fish Species Flocks.* University of Maine, Orono, pp 141–154.

Griffiths, D. (1994) The size structure of lacustrine Arctic charr (Pisces: Salmonidae) populations. *Biol. J. Linn. Soc.* 51:337–357.

Gross, H.P. (1977) Adaptive trends of environmentally sensitive traits in the three-spined stickleback, *Gasterosteus aculeatus* L. *Zeit. zool. Syst. Evolutionsforsch.* 15:151–178.

Gross, H.P. (1978) Natural selection by predators on the defensive apparatus of the three-spined stickleback, *Gasterosteus aculeatus* L. *Can. J. Zool.* 56:398–413.

Gross, H.P. and Anderson, J.M. (1984) Geographic variation in the gill rakers and diet of European threespine sticklebacks. *Copeia* 1984:87–97.

Gyllensten, U. and Ryman, N. (1988) Biochemical genetic variation and population structure of fourhorn sculpin (*Myoxocephalus quadricornis*; Cottidae) in Scandinavia. *Hereditas* 108:179–185.

Hagen, D.W. (1973) Inheritance of numbers of lateral plates and gill rakers in stickleback. *Heredity* 30:275–281.

Hagen, D.W. and Blow, D.M. (1983) Heritability of dorsal spines in the fourspine stickleback (*Apeltes quadracus*). *Heredity* 50:275–281.

Hagen, D.W. and Gilbertson, L.G. (1972) Geographic variation and environmental selection in *Gasterosteus aculeatus* L. in the Pacific northwest, America. *Evolution* 26:32–51.

Hagen, D.W. and Gilbertson, L.G. (1973a) Selective predation and the intensity of selection acting upon the lateral plates of threespine sticklebacks. *Heredity* 30:273–287.

Hagen, D.W. and Gilbertson, L.G. (1973b) The genetics of plate morphs in freshwater threespine sticklebacks. *Heredity* 31:75–84.

Hagen, D.W. and McPhail, J.D. (1970) The species problem within *Gasterosteus aculeatus* on the Pacific Coast of North America. *J. Fish. Res. Bd. Canada* 27:147–155.

Hagen, D.W. and Moodie, G.E.E. (1979) Polymorphism for breeding colors in *Gasterosteus aculeatus*. I. Their genetics and geographic distribution. *Evolution* 33:641–648.

Haglund, T.R. (1981) *Differential reproduction among the lateral plate phenotypes of Gasterosteus aculeatus, the threespine stickleback.* Ph.D. Dissertation, University of California, Los Angeles.

Haglund, T.R., Buth, D.G. and Lawson, R. (1992) Allozyme variation and phylogenetic relationships of Asian, North American, and European populations of the threespine stickleback. *Copeia* 1992:432–443.

Hamilton, K.E., Ferguson, A., Taggart, J.B., Tómasson, T., Walker, A. and Fahy, E. (1989) Postglacial colonization of brown trout, *Salmo trutta* L.: *Ldh-5* as a phylogeographic marker locus. *J. Fish Biol.* 35:651–664.

Hardisty, M.W. (1986) General introduction to lampreys. *In*: J. Holcík (ed.): *The Freshwater Fishes of Europe. Volume 1, Part I. Petromyzontiformes.* AULA-Verlag, GmbH, Wiesbaden, pp 19–83.

Hardisty, M.W. and Potter, I.C. (1971) Paired species. *In:* M.W. Hardisty and I.C. Potter (eds): *The Biology of Lampreys*, vol. 1. Academic Press, London, pp 249–277.

Hartley, S.E., McGowan, C., Greer, R.B. and Walker, A.F. (1992a) The genetics of sympatric Arctic charr [*Salvelinus alpinus* (L.)] populations from Loch Rannoch, Scotland. *J. Fish Biol.* 41:1021–1031.

Hartley, S.E., Bartlett, S.E. and Davidson, W.S. (1992b) Mitochondrial DNA analysis of Scottish populations of Arctic charr, *Salvelinus alpinus* (L.) *J. Fish Biol.* 40:219–224.

Harvey, P.H. and Pagel, M.D. (1991) *The Comparative Method in Evolutionary Biology*. Oxford University Press, Oxford.

Higuchi, M. and Goto, A. (1996) Genetic evidence supporting the existence of two distinct species in the genus *Gasterosteus* around Japan. *Environm. Biol. Fish*; in press.

Hindar, K. and Jonnson, B. (1982) Habitat and food segregation of dwarf and normal Arctic charr (*Salvelinus alpinus*) from Vangsvatnet Lake, western Norway. *Can. J. Fish. Aquat. Sci.* 39:1030–1045.

Hindar, K. and Jonnson, B. (1993) Ecological polymorphism in Arctic charr. *Bio. J. Linn. Soc.* 48:63–74.

Hindar, K., Ryman, N. and Ståhl, G. (1986) Genetic differentiation among local populations and morphotypes of Arctic charr, *Salvelinus alpinus*. *Biol. J. Linn. Soc.* 27:267–285.

Hoogland, R., Morris, D. and Tinbergen, N. (1957) The spines of sticklebacks (*Gasterosteus* and *Pygosteus*) as a means of defense against predators (*Perca* and *Esox*). *Behaviour* 10:205–236.

Houston, J. (1990) Status of the fourhorn sculpin, *Myoxocephalus quadricornis* in Canada. *Can. Field-Natur.* 104:7–13.

Huntingford, F.A. (1982) Do inter- and intraspecific aggression vary in relation to predation pressure in sticklebacks? *Anim. Behav.* 30:909–916.

Huntingford, F.A., Wright, P.J. and Tierney, J.F. (1994) Adaptive variation in antipredator behaviour in threespine stickleback. *In:* M.A. Bell and S.A. Foster (eds): *The Evolutionary Biology of the Threespine Stickleback.* Oxford University Press, Oxford, pp 277–296.

Hyatt, K.D. and Ringler, N.H. (1989) Egg cannibalism and the reproductive strategies of threespine sticklebacks (*Gasterosteus aculeatus*) in a coastal British Columbia lake. *Can. J. Zool.* 67:2036–2046.

Johnson, L. (1964) Marine-glacial relicts of the Canadian Arctic islands. *Syst. Zool.* 13:76–91.

Jones, D.H. and John, A.W.G. (1978) The three-spined stickleback, *Gasterosteus aculeatus* L. from the north Atlantic. *J. Fish. Biol.* 13:231–236.

Kassen, R., Schluter, D. and McPhail, J.D. (1995) Evolutionary history of threespine sticklebacks (*Gasterosteus* spp.) in British Columbia; insights from a physiological clock. *Can. J. Zool.* 73:2154–2158.

Keast, A. (1972) Australian mammals: Zoogeography and evolution. *In:* A. Keast, F.C. Erk and B. Glass (eds): *Evolution, Mammals and Southern Continents.* State University of New York Press, Albany, pp 195–246.

Kirkpatrick, M. and Selander, R.K. (1979) Genetics of speciation in lake whitefishes in the Allegash Basin. *Evolution* 33:478–485.

Klepaker, T. (1993) Morphological changes in a marine population of threespine stickleback, *Gasterosteus aculeatus*, recently isolated in fresh water. *Can. J. Zool.* 71:1251–1258.

Klepaker, T. (1995) Postglacial evolution in lateral plate morphs in Norwegian freshwater populations of the threespine stickleback (*Gasterosteus aculeatus*). *Can. J. Zool.* 73:898–906.

Kurenkov, S.I. (1978) Two reproductively isolated groups of kokanee salmon (*Oncorhynchus nerka kennerlyi*), from Lake Kronotskiy. *J. Ichthyol.* 17:526–534.

Kynard, B.E. (1972) *Male breeding behavior and lateral plate phenotypes in the threespine stickleback* (Gasterosteus aculeatus *L.*) Ph. D. Dissertation, University of Washington, Seattle.

Kynard, B.E. (1978) Breeding behavior of a lacustrine population of threespine sticklebacks (*Gasterosteus aculeatus* L.). *Behaviours* 67:178–207.

Kynard, B.E. (1979) Nest habitat preference of low plate number morphs in threespine sticklebacks (*Gasterosteus aculeatus*). *Copeia* 1979:525–528.

Lavin, P.A. and McPhail, J.D. (1985) The evolution of freshwater diversity of threespine stickleback (*Gasterosteus aculeatus*): site specific differences of trophic morphology. *Can. J. Zool.* 63:2632–2638.

Lavin, P.A. and McPhail, J.D. (1986) Adaptive divergence of trophic phenotype among freshwater populations of threespine stickleback (*Gasterosteus aculeatus*). *Can. Fish. Aquat. Sci.* 43:2455–2463.

Lavin, P.A. and McPhail, J.D. (1987) Morphological divergence and the organization of trophic characters among lacustrine populations of the threespine stickleback (*Gasterosteus aculeatus*). *Can. J. Fish. Aquat. Sci.* 44:1820–1829.

Leary, R.F., Allendorf, F.W., Phelps, S.R. and Knudsen, K.L. (1989) Genetic divergence and identification of seven cutthroat trout subspecies and rainbow trout. *Trans. Amer. Fish. Soc.* 116:580–587.

Liem, K.F. (1973) Evolutionary strategies and morphological innovations: cichlid pharyngeal jaws. *Syst. Zool.* 22:425–441.

Lindsey, C.C. (1962) Experimental study of meristic variation in a population of threespine stickleback, *Gasterosteus aculeatus*. *Can. J. Zool.* 40:271–312.

Lindsey, C.C. (1981) Stocks are chameleons: plasticity in gill rakers of coregonid fishes. *Can. J. Fish. Aquat. Sci.* 38:1497–1506.

Lindsey, C.C. and McPhail, J.D. (1986) Zoogeography of fishes of the Yukon and Mackenzie basins. *In:* C. Hocutt and E.O. Wiley (eds): *The Zoogeography of North American Freshwater Fishes.* Wiley, New York, pp 639–674.

Loudenslager, E.J. and Gall, G.A.E. (1980) Geographic patterns of protein variation and subspeciation in cutthroat trout, *Salmo clarki*. *Syst. Zool.* 29:27–42.

MacDonald, J.F., Bekkers, J., MacIsaac, S.M. and Blouw, D.M. (1995) Intertidal breeding and aerial development of embryos of a stickleback fish (*Gasterosteus*). *Behaviour* 132:1183–1206.

Martins, E.P. and Garland, Jr., T. (1991) Phylogenetic analyses of the correlated evolution of continuous characters: a simulation study. *Evolution* 45:534–557.

Mayr, E. (1942) *Systematics and the Origin of Species*. Columbia University Press, NY.

McCune, A.R. (1996) Biogeographic and stratigraphic evidence for rapid speciation in semio-notid fishes. *Paleobiolog* 22:34–48.

McDonald, C.G., Reimchen, T.E. and Hawryshyn, C.W. (1995) Nuptial colour loss and signal masking in *Gasterosteus*: an analysis using video imaging. *Behaviour* 132:963–977.

McDowall, R.M. (1970) The galaxiid fishes of New Zealand. *Bull. Mus. Comp. Zool.* 139:341–432.

McDowall, R.M. (1988) *Diadromy in Fishes: Migrations Between Freshwater and Marine Environments*. Croom Helm, London.

McDowall, R.M. (1990a) *New Zealand Freshwater Fishes: a Natural History and Guide*. Hene-mann Reed MAF Publishing Group. Auckland/Wellington.

McDowall, R.M. (1990b) When galaxiid and salmonid fishes meet – a family reunion in New Zealand. *J. Fish. Biol.* 37 (Suppl A):35–43.

McPhail, J.D. (1963) Geographic variation in North American ninespine sticklebacks, *Pungitius pungitius*. *J. Fish. Res. Bd. Can.* 20:27–44.

McPhail, J.D. (1984) Ecology and evolution of sympatric sticklebacks (*Gasterosteus*): morphological and genetic evidence for a species pair in Enos Lake, British Columbia. *Can. J. Zool.* 62:1402–1408.

McPhail, J.D. (1992) Ecology and evolution of sympatric sticklebacks (*Gasterosteus*): evidence for a species pair in Paxton Lake, Texada Island, British Columbia. *Can. J. Zool.* 70:361–369.

McPhail, J.D. (1993) Ecology and evolution of sympatric sticklebacks (*Gasterosteus*): origin of the species pairs. *Can. J. Zool.* 71:515–523.

McPhail, J.D. (1994) Speciation and the evolution of reproductive isolation in the sticklebacks (*Gasterosteus*) of south-western British Columbia. *In:* M.A. Bell and S.A. Foster (eds): *The Evolutionary Biology of the Threespine Stickleback*. Oxford University Press, Oxford, pp 399–437.

McPhail, J.D. and Lindsey, C.C. (1970) Freshwater fishes of northwestern Canada and Alaska. *Bull. Fish. Res. Bd. Can.* 173:1–381.

Mina, M.V. (1991) *Microevolution of Fishes: Evolutionary Aspects of Phenetic Diversity* (Transl. by Indira Kohli) A.A. Balkema, Rotterdam.

Moodie, G.E.E. (1972) Predation, natural selection and adaptation in an unusual threespine stickleback. *Heredity* 28:155–167.

Moodie, G.E.E. and Reimchen, T.E. (1976) Phenetic variation and habitat differences in *Gasterosteus* populations of the Queen Charlotte Islands. *Syst. Zool.* 25:49–61.

Moodie, G.E.E., McPhail, J.D. and Hagen, D.W. (1973) Experimental demonstration of selective predation in *Gasterosteus aculeatus*. *Behaviour* 47:95–105.

Narver, D.W. (1969) Phenotypic variation in threespine sticklebacks (*Gasterosteus aculeatus*) of the Chignik River system, Alaska. *J. Fish. Res. Bd. Can.* 26:405–412.

Nelson, J.S. (1968) Distribution and nomenclature of North American kokanee (*Oncorhynchus nerka*). *J. Fish. Res. Bd. Can.* 25:409–414.

Nelson, J.S. (1971) Absence of the pelvic complex in ninespine sticklebacks, *Pungitius pungitius*, collected in Ireland and Wood Buffalo National Park Region, Canada, with notes on meristic variation. *Copeia* 1971:707–717.

Nelson, J.S. (1994) *Fishes of the World,* 3rd edition. John Wiley and Sons, New York.

Nielsen, J.L. (ed.) (1995) *Evolution and the Aquatic Ecosystem: Defining Unique Units in Population Conservations*. American Fisheries Society Symposium 17, Bethesda.

Nordeng, H. (1983) Solution to the "char problem" based on Arctic char (*Salvelinus alpinus*) in Norway. *Can. J. Fish. Aquat. Sci.* 40:1372–1387.

Northcote, T.G. and Ward, F.J. (1985) Lake resident and migratory smelt, *Retropinna retropinna* (Richardson), of the Lower Waikato River system. *N. Z. J. Fish. Biol.* 27:113–129.

Olsen, P.E. (1986) A 40-million-year record of early Mesozoic orbital climatic forcing. *Science* 235:842–848.

Ortí, G., Bell, M.A., Reimchen, T.E. and Meyer, A. (1994) Global survey of mitochondrial DNA sequences in the threespine stickleback: evidence for recent migrations. *Evolution* 48:608–622.

Ovenden, J. and White, R.G. (1990) Mitochondrial and allozyme genetics of incipient specia-
    tion in a landlocked population of *Galaxias truttaceus* (Pisces: Galaxiidae). *Genetics* 124:
    701–716.
Paepke, H.-J. (1983) *Die Stichlinge*. Ziemsen Verlag, Wittenberg.
Parker, B.J. (1988) Status of the deep water sculpin, *Myoxocephalus thompsoni*, in Canada.
    Can. Field-Natur. 102:126–131.
Parker, H.H. and Johnson, L. (1991) Population structure, ecological segregation and repro-
    duction in non-anadromous Arctic charr, *Salvelinus alpinus* (L.), in four unexploited lakes in
    the Canadian high Arctic. *J. Fish. Biol.* 38:123–147.
Partington, J.D. and Mills, C.A. (1988) An electrophoretic and biometric study of Arctic charr,
    Salvelinus alpinus (L.), from ten British lakes. *J. Fish. Biol.* 33:791–814.
Pease, C.M., Lande, R. and Bull, J.J. (1989) A model of population growth, dispersal and
    evolution in a changing environment. *Ecology* 70:1657–1664.
Pianka, E.R. (1966) Latitudinal gradients in species diversity: a review of concepts. *Amer. Nat.*
    100:33–46.
Plafker, G. and W.O. Addicott (1976) Glacioimoraine deposits of Miocene through Holocene
    age in the Yagataga Formation along the Gulf of Alaska margin. *In:* T.P. Miller (ed.): *Recent
    and Ancient Sedimentary Environments in Alaska*. Alaska Geological Society, Anchorage,
    pp Q1–Q23.
Potter, I.C. (1980) The Petromyzoniformes with particular reference to paired species. *Can.
    J. Fish. Aquat. Sci.* 37:1595–1615.
Reimchen, T.E. (1980) Spine deficiency and polymorphism in a population of *Gasterosteus
    aculeatus*: an adaptation to predators? *Can. J. Zool.* 58:1232–1244.
Reimchen, T.E. (1983) Structural relationships between spines and lateral plates in threespine
    stickleback (*Gasterosteus aculeatus*). *Evolution* 37:931–946.
Reimchen, T.E. (1989) Loss of nuptial color in threespine sticklebacks (*Gasterosteus aculeatus*).
    Evolution 43:450–460.
Reimchen, T.E. (1991a) Trout foraging failures and the evolution of body size in stickleback.
    Copeia 1991:1098–1104.
Reimchen, T.E. (1992) Injuries and survival of *Gasterosteus* from attacks by a toothed predator
    (*Oncorhynchus*) and some implications for the evolution of lateral plates. *Evolution* 46:
    1224–1230.
Reimchen, T.E. (1994) Predators and morphological evolution in threespine stickleback. *In:*
    M.A. Bell and S.A. Foster (eds): *The Evolutionary Biology of the Threespine Stickleback*.
    Oxford University Press, Oxford, pp 240–276.
Reimchen, T.E. (1995) Predator-induced cyclical changes in lateral plate frequencies of
    Gasterosteus. *Behaviour* 132:1079–1094.
Reimchen, T.E. and Nelson, J.S. (1987) Habitat and morphological correlates to vertebral num-
    ber as shown in a teleost, *Gasterosteus aculeatus*. *Copeia* 1987:868–874.
Reimchen, T.E., Stinson, E.M. and Nelson, J.S. (1985) Multivariate differentiation of parapatric
    and allopatric populations of threespine stickleback in the Sangan River watershed, Queen
    Charlotte Islands. *Can. J. Zool.* 63:2944–2951.
Reist, J.D. (1980) Predation upon pelvic phenotypes of brook stickleback, *Culaea inconstans*,
    by selected invertebrates. *Can. J. Zool.* 58:1253–1258.
Ricker, W.E. (1940) On the origin of kokanee, a fresh-water type of sockeye salmon. *Trans. Roy.
    Soc. Can.* 34:121–135.
Ricklefs, R.E. (1996) Phylogeny and ecology. *Trends Ecol. Evol.* 11:229–230.
Ridgway, M.S. and McPhail, J.D. (1984) Ecology and evolution of sympatric sticklebacks
    (*Gasterosteus*): mate choice and reproductive isolation in the Enos Lake species pair. *Can.
    J. Zool.* 62:1813–1818.
Ridgway, M.S. and McPhail, J.D. (1988) Raiding shoal size and a distraction display in male
    sticklebacks (Gasterosteus). *Can. J. Zool.* 66:201–205.
Riget, F.F., Nygaard, K.H. and Christensen, B. (1986) Population structure, ecological segrega-
    tion, and reproduction in a population of Arctic char (*Salvelinus alpinus*) from Lake
    Tasersuaq, Greenland. *Can. J. Fish. Aquat. Sci.* 43:985–992.
Robinson, B.W. and Wilson, D.S. (1994) Character release and displacement in fishes: a neglec-
    ted literature. *Amer. Nat.* 144:596–627.
Rosenzweig, M.L. (1995) *Species Diversity in Space and Time*. Cambridge University Press,
    Cambridge.

Rowland, W.J. (1994) Proximate determinants of stickleback behaviour: an evolutionary perspective. *In:* M.A. Bell and S.A. Foster (eds): *The Evolutionary Biology of the Threespine Stickleback.* Oxford University Press, Oxford, pp 297–344.

Roy, K., Valentine, J.W., Jablonski, D. and Kidwell, S.M. (1996) Scales of climatic variability and time averaging in Pleistocene biotas: Implications for ecology and evolution. *Trends Ecol. Evol.* 11: 458–463.

Ryman, N., Allendorf, F.W. and Ståhl, G. (1979) Reproductive isolation with little genetic divergence in sympatric populations of brown trout *(Salmo trutta). Genetics* 92: 247–262.

Schluter, D. (1993) Adaptive radiation in sticklebacks: size, shape and habitat use efficiency. *Ecology* 74: 699–709.

Schluter, D. (1994) Experimental evidence that competition promotes divergence in adaptive radiation. *Science* 266: 798–801.

Schluter, D. (1995) Adaptive radiation in sticklebacks: trade-offs in feeding performance and growth. *Ecology* 76: 82–90.

Schluter, D. (1996 a) Ecological causes of adaptive radiation. *Amer. Nat.* 148 (Suppl.): 540–564.

Schluter, D. (1996 b) Ecological speciation in postglacial fishes. *Phil. Trans. R. Soc. Lond.,* Ser. B, 531: 807–814.

Schluter, D. and McPhail, J.D. (1992) Ecological character displacement and speciation in sticklebacks. *Amer. Nat.* 140: 85–108.

Schluter, D. and McPhail, J.D. (1993) Character displacement and replicate adaptive radiation. *Trends Ecol. Evol.* 8: 197–200.

Schluter, D. and Nagel, L.M. (1995) Parallel speciation by natural selection. *Amer. Nat.* 146: 292–301.

Simpson, G.G. (1944) *Tempo and Mode in Evolution.* Columbia University Press, New York.

Simpson, G.G. (1980) *Splendid Isolation, the Curious History of Mammals in South America.* Yale University Press, New Haven.

Skúlason, S. and Smith, T.B. (1995) Resource polymorphisms in vertebrates. *Trends. Ecol. Evol.* 10: 366–370.

Skúlason, S., Noakes, D.L.G. and Snorrason, S.S. (1989) Ontogeny of trophic morphology in four sympatric morphs of arctic charr *Salvelinus alpinus* in Thingvallavatn, Iceland. *Biol. J. Linn. Soc.* 38: 281–301.

Smith, G.R. (1978) Biogeography of intermontain fishes. *Great Basin Natur. Mem.* 2: 17–42.

Smith, G.R. and Todd, T.N. (1984) Evolution of species flocks of fishes in north temperate lakes. *In*: A.A. Echelle and I. Kornfield (eds): *Evolution of Fish Species Flocks.* University of Maine, Orono, pp 45–68.

Snorrason, S.S., Skúlason, S., Jonnson, B., Malmquist, H.J., Jónasson, P.M., Sandlund, O.T. and Linden, T. (1994) Trophic specialization in Arctic charr *Salvelinus alpinus* (Pisces: Salmonidae): morphological divergence and ontogenetic niche shifts. *Biol. J. Linn. Soc.* 52: 1–18.

Sparholt, H. (1985) The population, survival, growth, reproduction and food of arctic charr, *Salvelinus alpinus* (L.), in four unexploited lakes in Greenland. *J. Fish. Biol.* 26: 313–330.

Ståhl, G. (1987) Genetic population structure of Atlantic salmon. *In:* N. Ryman and F. Utter (eds): *Population Genetics and Fishery Management.* University of Washington Press, Seattle, pp 121–140.

Stanley, S.M. (1975) A theory of evolution above the species level. *Proc. Natl. Acad. Sci. USA* 72: 646–650.

Stern, W.L. (1971) *Adaptive Aspects of Insular Evolution.* Washington State University Press, Seattle.

Svärdson, G. (1961) Young sibling fish species in northwestern Europe. *In:* W.F. Blair (ed.): *Vertebrate Speciation.* University of Texas Press, Austin, pp 498–513.

Svärdson, G. (1979) Speciation of Scandinavian *Coregonus. Rep. Inst. Freshwater Res. Drottninholm* 57: 1–95.

Svedäng, H. (1990) Genetic basis of life-history variation of dwarf and normal Arctic charr, *Salvelinus alpinus* (L.), in Stora Rösjön, central Sweden. *J. Fish. Biol.* 36: 917–932.

Swain, D.P. (1992 a) The functional basis of natural selection for vertebral traits of larvae in the stickleback *Gasterosteus aculeatus. Evolution* 46: 987–997.

Swain, D.P. (1992 b) Selective predation for vertebral phenotype in *Gasterosteus aculeatus:* reversal in the direction of selection at different larval sizes. *Evolution* 46: 998–1013.

Swain, D.P. and Lindsey, C.C. (1984) Selective predation for vertebral number of young stickle-backs, *Gasterosteus aculeatus. Can. J. Fish. Aquat. Sci.* 41:1231–1233.

Taniguchi, N., Honma, Y. and Kawamata, K. (1990) Genetic differentiation of freshwater and anadromous threespine sticklebacks (*Gasterosteus aculeatus*) form northern Japan. *Japanese J. Ichthyol.* 37:230–238.

Taylor, E.B. and Bentzen, P. (1993a) Evidence for multiple origins and sympatric divergence of trophic ecotypes of smelt (*Osmerus*) in northeastern North America. *Evolution* 47:813–832.

Taylor, E.B. and Bentzen, P. (1993b) Molecular genetic evidence for reproductive isolation between sympatric populations of smelt *Osmerus* in Lake Utopia, south-western New Brunswick, Canada. *Mol. Ecol.* 2:345–357.

Taylor, E.B. and Foote, C.J. (1991) Critical swimming velocities of juvenile sockeye salmon and kokanee, the anadromous and non-anadromous forms of *Oncorhynchus nerka* (Walbaum). *J. Fish. Biol.* 38:407–419.

Taylor, E.B., Foote, C.J. and Wood, C.C. (1996) Molecular genetic evidence for parallel life-history evolution within a Pacific salmon (sockeye salmon and kokanee, *Oncorhynchus nerka*). *Evolution* 50:401–416.

Taylor, E.B., McPhail, J.D. and Schluter, D. (1997) History of ecological selection in stickle-backs: uniting experimental and phylogenetic approaches. *In:* T.J. Givnish and K.J. Sytsma (eds): *Molecular evolution and adaptive radiation.* Cambridge University Press, Cambridge, pp 511–534.

Terborg, J. (1973) On the notion of favorableness in plant ecology. *Amer. Nat.* 107:481–501.

van den Assem, J. and Sevenster, P. (eds) (1985) Fifty years of behavior study in sticklebacks. Papers read at the First International Conference on Stickleback Behaviour. *Behaviour* 93, viii + 277 p.

Verheyen, E., Rüber, L., Snoeks, J. and Meyer, A. (1996) Mitochondrial phylogeny of rock-dwelling cichlid fish reveals evolutionary influence of historical lake fluctuations of Lake Tanganyika, Africa. *Phil. Trans. R. Soc. Lond.,* Ser. B, 531:797–805.

Vladykov, V.D. (1985) Does neoteny occur in Holarctic lampreys (Petromyzontidae)? *Syllogeus* 57:1–3.

Vladykov, V.D. and Kott, E. (1979) Satellite species among the Holarctic lampreys (Petromyzo-nidae). *Can. J. Zool.* 57:860–867.

Vuorinen, J.A., Bodaly, R.A., Reist, J.D., Bernatchez, L. and Dodson, J.J. (1993) Genetic and morphological differentiation between dwarf and normal size forms of lake whitefish (*Coregonus clupeaformis*) in Como Lake, Ontario. *Can. J. Fish. Aquat. Sci.* 50:210–216.

Walker, A.F., Greer, R.B. and Gardner, A.S. (1988) Two ecologically distinct forms of Arctic charr *Salvelinus alpinus* (L.), in Loch Rannoch, Scotland. *Biol. Conserv.* 43:43–61.

Walker, J.A. (1995) *Morphometric and functional analysis of body shape evolution in three-spine stickleback.* Ph. D. Dissertation, State University of New York, Stony Brook.

Walker, J.A. (1996) Principal components of shape variation within an endemic radiation of threespine stickleback. *In:* L.F. Macus, M. Corti, D. Slice and G. Naylor (eds): *Advances in Morphometrics.* Plenum, New York, pp 321–334.

Walker, J.A. (1997) Ecological morphology of lacustrine threespine stickleback *Gasterosteus aculeatus* L. (Gasterosteidae) body shape. *Biol. J. Linn. Soc.* 61:3–50.

Webb, P.W. (1994) The biology of fish swimming. *In:* L. Maddock, Q. Bone and J.M.V. Rayner (eds): *Mechanics and Physiology of Animal Swimming.* Cambridge University Press, Cambridge, pp 42–62.

Weihs, D. (1989) Design features and mechanics of axial locomotion in fish. *Amer. Zool.* 29:151–160.

Whoriskey, F.G. and FitzGerald, G.J. (1985) Sex, cannibalism and sticklebacks. *Behav. Ecol. Sociobiol.* 18:15–18.

Williams, G.C. (1992) *Natural Selection: Domains, Levels, and Applications.* Oxford University Press, Oxford.

Wilson, E.O. (1961) The nature of the taxon cycle in the Melanesian ant fauna. *Amer. Nat.* 95:169–193.

Wood, C.C. (1995) Life history variation and population structure in sockeye salmon. *Amer. Fish. Soc. Symp.* 17:195–216.

Wood, C.C. and Foote, C.J. (1990) Genetic differences in the early development and growth of sympatric sockeye salmon and kokanee *(Oncorhynchus nerka)* and their hybrids. *Can. J. Fish. Aquat. Sci.* 47:2250–2260.

Wood, C.C., Riddell, B.E., Rutherford, D.T. and Withler, R.E. (1994) Biochemical genetic survey of sockeye salmon (*Oncorhynchus nerka*) in Canada. *Can. J. Fish. Aquat. Sci.* 51 (Suppl.1):114–131.

Wootton, R.J. (1976) *The Biology of the Sticklebacks*. Academic Press, London.

Wootton, R.J. (1984) *A Functional Biology of the Sticklebacks*. Croom Helm, London, and University of California Press, Berkeley.

Zanandrea, G. (1959) Speciation among lampreys. *Nature* 184:380.

Ziuganov, V.V. (1983) Genetics of osteal plate polymorphism and microevolution of threespine stickleback (*Gasterosteus aculeatus*). *Theor. Appl. Genet.* 65:239–246.

Ziuganov, V.V. (1991) The family Gasterosteidae of world fish fauna. *Fauna of the USSR, Fishes,* New Ser. 137, 5(1):1–254.

Ziuganov, V.V. (1995) Reproductive isolation among lateral plate phenotypes (low, partial, complete) of the threespine stickleback, *Gasterosteus aculeatus*, from the White Sea basin and the Kamchatka Peninsula, Russia. *Behaviour* 132:1173–1181.

Ziuganov, V.V. and Bugayev, V.F. (1988) Isolating mechanisms between spawning populations of the threespine stickleback, *Gasterosteus aculeatus*, of Lake Azabachije, Kamchatka. *Voprosy Ikhthiologii* 2:322–325. [English translation].

Ziuganov, V.V. and A.A. Zotin (1995) Pelvic girdle polymorphism and reproductive barriers in the ninespine stickleback *Pungitius pungitius* (L.) from northwest Russia. *Behaviour* 132:1095–1104.

Subject Index